NATURAL SCIENCE AND INDIGENOUS KNOWLEDGE

How do we combine the areas of intersection between science and Indigenous knowledge, but without losing the totality of both? This book's objective is to consider how Indigenous populations have lived and managed the landscape, specifically how their footprint was a result of the combination of their empirical knowledge and their culture. The chapters are divided into four groups: The first deals with reintegrating cultures and natural landscapes and the role of kinship and oral tradition. The second group approaches the landscape as a living university of learning and managing, discussing the ethnobotany of how to grow more responsibly and assess and project the harvest. The third group deals with the managing of fire in an anthropogenic plant community and how to integrate Indigenous agriculture in hydrology and dry regions. The fourth group consists of studies of how science and Indigenous knowledge can be taught in schools using land-based studies.

EDWARD A. JOHNSON is Faculty Professor of Ecology in the Department of Biological Sciences at the University of Calgary. He was Director of the Biogeoscience Institute at the University of Calgary from 1991 to 2017, and from 2003 to 2007 the G8 Legacy Chair in Ecology. His research interests relate to the coupling of physical processes to ecological processes, specifically natural disturbances. He has completed fieldwork in Australia, Canada, Belize, Sri Lanka, Sweden, and the USA. He has published extensively in refereed scientific journals and has authored four books: *Fire and Vegetation Dynamics* (1992, Cambridge University Press), *Forest Fires: Behavior and Ecological Effects* (2001, Academic Press), *Ecological Education and Environmental Advocacy* (2005, Cambridge University Press), *A Biogeoscience Approach to Ecosystems* (2016, Cambridge University Press), and *Plant Disturbance Ecology: The Process and the Response*, 2nd ed. (2023, Elsevier Press). He has been part of the Natural Sciences and Engineering Research Council of Canada's Network of Centres of Excellence (NSERC-NCEs) in Sustainable Forest Management and NSERC-NCEs on GEOIDE and Complex Data Systems, and two National Science Foundation networks: the Community Surface Dynamic Modeling System (CSDMS) and National Ecological Observatory Network (NEON). He received the William S. Cooper Award of the Ecological Society of America for research on wildfire effects on boreal forest dynamics. He is a Fellow of the Royal Society of Biology and the Ecological Society of America. He was the Editor-in-Chief of the *Bulletin of the Ecological Society of America* from 2008 to 2019 and has been a subject editor of *Botany, Canadian Journal of Forest Research, Vegetation Science,* and *Ecology/Ecological Monographs.*

SUSAN M. ARLIDGE has been a certified Alberta teacher, environmental educator, and outdoor guide for over thirty years. She graduated with a BSc (Physical Geography) and a BEd (Secondary) from the University of Calgary and is a trained hiking and snowshoeing guide. Susan was School Program Consultant at the Biogeoscience Institute at the University of Calgary from 2000 to 2022. Her experience includes heightening student engagement in field science and using novel environments to engage learners and place-based learning in schools. As well as teaching field trips and an extensive list of workshops, she has contributed papers and teacher manuals, including *Guiding Students Toward Open Inquiry in a Novel Outdoor Setting* (2018), *Using the Understanding Science Flowchart to Illustrate and Bring Students' Science Stories to Life* (2017), and *Five-Minute Field Trips: Teaching About Nature in Your Schoolyard* (2000). She has taught and has volunteered extensively with her neighbors, the Stoney Nakoda Nation.

NATURAL SCIENCE AND INDIGENOUS KNOWLEDGE

The Americas Experience

Edited by

EDWARD A. JOHNSON
University of Calgary

SUSAN M. ARLIDGE
University of Calgary

CAMBRIDGE
UNIVERSITY PRESS

Shaftesbury Road, Cambridge CB2 8EA, United Kingdom

One Liberty Plaza, 20th Floor, New York, NY 10006, USA

477 Williamstown Road, Port Melbourne, VIC 3207, Australia

314–321, 3rd Floor, Plot 3, Splendor Forum, Jasola District Centre,
New Delhi – 110025, India

103 Penang Road, #05–06/07, Visioncrest Commercial, Singapore 238467

Cambridge University Press is part of Cambridge University Press & Assessment,
a department of the University of Cambridge.

We share the University's mission to contribute to society through the pursuit of
education, learning and research at the highest international levels of excellence.

www.cambridge.org
Information on this title: www.cambridge.org/9781009416672

DOI: 10.1017/9781009416665

First published 2024

A catalogue record for this publication is available from the British Library

Library of Congress Cataloging-in-Publication Data
Names: Johnson, Edward A. (Ecology Professor), editor. | Arlidge, Susan M., editor.
Title: Natural science and indigenous knowledge : the Americas experience / edited by Edward Johnson,
University of Calgary, Susan Arlidge, University of Calgary.
Description: 1. | New York : Cambridge University Press, 2024. | Includes bibliographical references and
index.
Identifiers: LCCN 2023047350 (print) | LCCN 2023047351 (ebook) | ISBN 9781009416672 (hardback) |
ISBN 9781009416658 (paperback) | ISBN 9781009416665 (epub)
Subjects: LCSH: Ethnoscience – Research. | Traditional ecological knowledge.
Classification: LCC GN476 .N38 2024 (print) | LCC GN476 (ebook) | DDC 500.897–dc23/eng/20231229
LC record available at https://lccn.loc.gov/2023047350
LC ebook record available at https://lccn.loc.gov/2023047351

ISBN 978-1-009-41667-2 Hardback

Contents

Contributors

Bert Adams, Sr.
Dry Bay-Alsek River

Bertrand (Bert) Adams, Sr., is an elder of the Tlingit L'uknax̱.ádi clan, which hails from the Dry Bay-Alsek River basin. Bert carries the Tlingit name Kaadashaan and spent much of his life in Dry Bay, commercial fishing and subsistence food gathering. He has authored several books and penned numerous essays as a columnist for the *Juneau Empire*. As a tribal leader, including as president of the Yakutat Tlingit Tribe for many years, Bert worked tirelessly with the National Park Service and other entities to conserve Tlingit cultural ties to Dry Bay, especially by bringing young people back to this spectacular and foundational ancestral landscape in the history of Yaakwdáat and G̱unaax̱oo people.

Susan M. Arlidge
Exshaw, Alberta, Canada

Susan (Sue) M. Arlidge is a teacher with the Christ The Redeemer School Board in Canmore, Alberta. For twenty years, she was School Consultant at the University of Calgary's Biogeoscience Institute, where she taught on grades 6–12 ecology field trips, using student inquiry as a portal to understand the natural world. Sue's career has focused on inspiring wonder in and scientific questioning of the world outdoors. She has collaborated with numerous educational and recreational initiatives with the young people in the Stoney Nakoda Nations.

Richard Arnold
Pahrump Paiute Tribe, Pahrump, Nevada, USA

Richard Arnold is Southern Paiute and serves as Chairman of the Pahrump Paiute Tribe and Spokesperson for the Consolidated Group of Tribes and Organizations (CGTO), which comprises sixteen tribes from Nevada, California, Utah, and

Arizona. He focuses on expanding government-to-government interactions between tribes and federal and state agencies. Arnold participates in several technical working groups, where he facilitates discussions and evaluates research, development, and implementation methods that integrate tribal perspectives and collective wisdom of tribal people.

Rochelle Bloom
Department of Anthropology, University of Victoria, Victoria,
British Columbia, Canada

Rochelle Bloom is a doctoral student in the Department of Anthropology at the University of Victoria. She has served as a research associate at Portland State University and spent several years based in Yosemite National Park conducting ethnographic research. Her research focuses on ethnographic mapping, cultural landscapes, and Indigenous–state relations within a national park context.

Douglas Deur
Anthropology Department, Portland State University, Portland, Oregon, USA
Adjunct Professor, School of Environmental Studies, University of Victoria,
Victoria, British Columbia, Canada

Douglas Deur serves as research faculty in the Portland State University Anthropology Department and as adjunct faculty in the University of Victoria School of Environmental Studies, and formerly served as senior research staff in the University of Washington School of Environmental and Forest Sciences. He works closely with, and often for, Indigenous communities of western North America to document their traditional ecological knowledge, environmental values, and resource management practices. He is also the primary academic researcher to assist the US National Park Service in documenting Indigenous peoples' enduring connections to national parks in the west coast states of North America, from the desert Southwest to the Alaskan arctic.

John W. Ives
Department of Anthropology, University of Alberta, Edmonton,
Alberta, Canada

John W. (Jack) Ives (PhD, University of Michigan, 1985) is Emeritus Professor in the Department of Anthropology, University of Alberta, having been part of the department since 2007, and is Adjunct Professor at Simon Fraser University and the University of Saskatchewan. He was the founding director of the Institute of Prairie Archaeology (2008) and was Landrex Distinguished Professor from 2012 to 2017. From 1979 to 2007, Ives served with the Archaeological Survey of Alberta, the Royal Alberta Museum, and the Historic Resources Management Branch, with senior management responsibilities as Alberta's Provincial Archaeologist for twenty-one years. He won the University of Michigan's Distinguished Dissertation Award, published as *A Theory of Northern Athapaskan Prehistory* (1990, Westview/University of Calgary Press) and is coeditor of *Holes in Our Moccasins, Holes in Our Stories* (2022, University of Utah Press), an edited volume on Apachean origins.

Edward A. Johnson

Department of Biological Sciences, University of Calgary, Alberta, Canada

Edward A. Johnson is Faculty Professor of Ecology in the Department of Biological Sciences at the University of Calgary. He was also Director of the Biogeoscience Institute at the University of Calgary from 1991 to 2017 and was the G8 Chair in Ecology from 2003 to 2007. His research interests are the coupling of physical processes to ecological processes. In 1986, he received the William S. Cooper Award of the Ecological Society of America. He is a Fellow of the Royal Society of Biology and a Fellow of the Ecological Society of America.

Taal Levi

Department of Fisheries, Wildlife, and Conservation Sciences, Oregon State University, Corvallis, Oregon, USA

Taal Levi is Associate Professor in the Department of Fisheries, Wildlife, and Conservation Sciences at Oregon State University. He was trained in physics and biology at UC Berkeley and went on to receive a PhD in Environmental Studies from UC Santa Cruz. He has diverse research interests in wildlife ecology, conservation biology, and disease ecology in tropical, temperate, and boreal ecosystems. A consistent theme of his research is the implementation of quantitative and molecular methods to applied ecology and conservation issues.

Carlos A. Peres

Centre for Ecology, Evolution, and Conservation (CEEC), Schools of Environmental and Biological Sciences, University of East Anglia, Norwich, UK

Carlos A. Peres is Professor of Conservation Ecology in the School of Environmental Sciences of the University of East Anglia. His research interests include wildlife community ecology in Amazonian forests, the population ecology of key tropical forest resource populations, and the biological criteria for designing large nature reserves. He has published extensively on neotropical forest ecology and conservation at scales ranging from populations to landscapes and to continents. In 1995, he received a Biodiversity Conservation Leadership Award from the Bay and Paul Foundation (USA) and, in 2000, he was elected an Environmentalist Leader for the

New Millennium by *Time* magazine and CNN network. In 2023, he won the inaugural Frontiers Planet Prize.

Glenn H. Shepard
Department of Anthropology, Museu Paraense Emílio Goeldi,
Belém, Pará, Brazil

Glenn H. Shepard is an ethnobotanist, medical anthropologist, and filmmaker with more than thirty years of field experience in the Amazon. He is currently a staff researcher in the Human Sciences Division at the Museu Paraense Emílio Goeldi in Belém, Brazil. He has a blog name "Notes from the Ethnoground" (ethnoground.blogspot.com).

Richard W. Stoffle
School of Anthropology, University of Arizona, Tucson, Arizona, USA

Richard W. Stoffle is Professor of Anthropology in the School of Anthropology at the University of Arizona. He has worked with more than 100 Native American tribes and pueblos and has lived and worked with the Caribbean coastal peoples in the Dominican Republic, Antigua, Saint Lucia, and Barbados. He argues that respect for traditional ecological knowledge of these native and traditional peoples is key for both understanding how sustainable coadaptations with the environment are developed and how environmental and social restoration is possible.

Thomas F. Thornton
University of Alaska-Southeast, Alaska Coastal Rainforest Centre, Juneau,
Alaska, USA

Board on Environmental Change and Society, National Academies of Sciences,
Washington, DC, USA

Thomas F. Thornton (PhD, University of Washington) is an environmental anthropologist with over thirty years of experience working with communities around the North Pacific, Arctic, Subarctic, and Eurasia. He is currently Director for the US National Academies of Sciences Board on Environmental Change and Society and Professor of Environment and Society at the University of Alaska, where he has also served as Dean, Vice-Provost, and Director of the Alaska Coastal Rainforest Center. From 2008 to 2018, he directed the Environmental Change and Management graduate program at Oxford University, where he is currently Honorary Research Fellow. He has published over 100 peer-reviewed articles and seven books, most recently *Herring and People of the North Pacific: Sustaining a Keystone Species* (coauthored with Madonna Moss, 2000, University of Washington Press) and the *Routledge Handbook of Indigenous Environmental Knowledge* (coedited with Shonil Bhagwat, 2021, Routledge). Tom is an adopted member of the Tlingit Kaagwaantaan clan.

Nancy J. Turner
Environmental Studies, University of Victoria, Victoria,
British Columbia, Canada

Nancy J. Turner is an ethnobotanist who has worked with Indigenous Elders and cultural specialists in western Canada for over fifty years, learning about plants and environments. As Distinguished Professor Emerita in Environmental Studies at the University of Victoria, she has written/edited thirty books and over 150 papers, and is a member of the Order of Canada, the Order of British Columbia, and the Royal Society of Canada. She also has honorary degrees from four British Columbia universities.

Kathleen Van Vlack
Living Heritage Research Council, Cortez, Colorado, USA

Kathleen Van Vlack is Assistant Professor at Northern Arizona University. She has participated in a range of ethnographic studies such as cultural landscape studies, ethnographic overview and assessments, ethno-ecological studies, and cultural heritage preservation studies for the Bureau of Land Management, the National Park Service, the US Forest Service, the Department of Defense, and the Department of Energy. She has worked with over fifty tribes from across the United States on over thirty federally funded projects. She also cofounded the non-profit Living Heritage Research Council in an effort to continue working with Indigenous communities in the United States. Her personal research has focused on cultural heritage management, cultural landscapes, and pilgrimage trails.

Preface

This book is for people in the sciences or interested in the sciences and serves as an introduction to the overlap between the sciences and certain parts of Indigenous knowledge. We want to stress that the overlap does not account for all of science or Indigenous knowledge. Each author will define the overlap as it suits their subject.

Increasingly, countries around the world are realizing that they owe to Indigenous peoples more recognition for the management of land on which these people have treaty rights and other historical, cultural, and spiritual connections. There is also belated recognition both of Indigenous people's knowledge of the land's natural processes and of the land's cultural and spiritual importance to them. As yet, we have no clear understanding of how Indigenous knowledge can contribute to science and of how science can contribute to Indigenous knowledge. This book is a modest attempt to lay out different understandings of the overlap between these areas by different individuals. Although much work has been done on our different understandings and knowledge of the land, much work is still needed in this area. We stress that the book is about the shared areas of knowledge between science and parts of Indigenous knowledge.

The arrival of humans in the Americas came in two stages. The first people entered the Americas from Asia-Siberia during the Pleistocene, arriving in at least three waves, and these people entered a continent empty of humans. The second stage happened in the last six centuries and was primarily European colonizers and slaves from Africa, who entered a continent occupied by Indigenous peoples. Both of these arrivals had major impacts on the landscape, but in different ways. The first arrivals resulted in changes in plants, animals, and terrain, leaving cultural footprints. These cultural footprints were often taken by European colonizers as the natural landscape. The second colonization had significant effects on Indigenous populations: major reductions in populations caused by disease and genocide, a loss of culture, and reduced access to the land. Indigenous peoples had a functionally organized property system based on multiple (open) access to resources, while Europeans had a spatially organized property system. This led to serious misunderstandings as to the meaning of treaties, and the two sides differed in their economics, language, culture, and spirituality.

Indigenous knowledge is an understanding of local landscapes. This knowledge can be understood in different ways as a result of language, spiritual beliefs, kinship, economic systems, and past history. Consequently, there has never been and there still is not homogeneous Indigenous knowledge. However, Indigenous knowledge holders did face similar problems with respect to acquiring food and managing areas to conserve and sustain their food supplies. Indigenous knowledge also provides spiritual values, kinship systems, and the means of preserving social values.

We now know that, in the beginning, all hominin came out of Africa so, in this sense, we all have been both colonized and colonizers. Our origin stories are, however, varied. In the case of the Americas, we have continents that were unoccupied by humans until late in hominin history some 30,000 to 12,000 years ago. Therefore, as a group, these early arrivers are Indigenous, and it is important to recognize this group's cultural and kinship diversity.

In this book, our interest is in the area where science knowledge and Indigenous knowledge overlap. It is necessary for scientists to listen to people who have lived on the land and who have depended on its products to sustain themselves and their kin. This knowledge of the land's close observers provides both information on and insight into scientific and Indigenous knowledge. However, Indigenous and local knowledge can be overwhelmed by science, which is a worldwide undertaking. Yet, there is not just a power difference but also a difference in approach between science and Indigenous knowledge. The question is how to make the exchange between the science and Indigenous communities a learning and mutually beneficial experience for both sides. This book will mostly be about the overlap in knowledge in the Americas. The collaboration is ongoing.

1

What Do Indigenous People Have to Tell Us about the Cultural Landscapes They Have Created?

EDWARD A. JOHNSON AND SUSAN M. ARLIDGE

1.1 Introduction

Cultural landscapes are the imprint of human land-uses on physical and biological processes at both the small scale of meters and the large scale of kilometers and at both a seasonal scale and the scale of millennia.

Indigenous populations of the Americas have experimented for over 30,000 years with adaptations to different environments, but all of these populations experienced similar constraints. These populations were usually small, with a few exceptions in which agriculture played an important role. They also had oral traditions, no written language, no metal tools, few domesticated animals or beasts of burden, only the native plants to cultivate and domesticate into agricultural production, and no previous human land-use on the areas that they first occupied. These boundaries are fairly constraining and should be remembered in any discussion of the limits in creating cultural impacts. However, within these constraints, Indigenous populations have developed multiple cultural landscapes after generations of experimenting with different technologies, cultures, and spiritual rules on how to produce a sustainable environment.

Indigenous knowledge evolved because of this people's continuous, attentive exposure to local landscapes. This experiential knowledge has been accumulated through hunting, fishing, food gathering, agriculture, other forest and grassland crafts, and outdoor living. Through these ways of living and cultures, there have been many generations of exposure to local and regional knowledge and there has been an oral tradition for passing on this knowledge. Knowledge has also been passed on about cultural and spiritual matters related to the relationship with the local landscape. This knowledge about the landscape allowed the development of landscape-scale practices to sustain the Indigenous lifestyle. Kinship and gender play roles in this knowledge. Knowledge is accumulated by observation and by experimental manipulation of the environment. Special knowledge is often carried by valued individuals who are known for their depth of understanding of the land, cultural and spiritual endowment, communication, and leadership skills. All of this is passed on by oral tradition.

The environments of the Americas changed radically during the 30,000 years over which Indigenous people established themselves. Consequently, the cultural landscape's footprint changed as they discovered more sustainable methods or they migrated, assimilated into other groups, or went extinct. It appears that, through these evolutionary adaptations, Indigenous people produced an interactive system of empirical information about how to adapt and survive in any particular location and a set of cultural and spiritual norms to remain within that adaptive boundary while still being willing to adjust to the often rapidly changing environments. In modern terminology, Indigenous populations must have experimented with modifications of their cultural landscape so as to maintain a sustainable number of ecosystem services.

Notice that this is considerably different from the traditional view that European colonizers had of the Americas when they arrived. European colonizers saw a land that they thought was largely empty of people. This was because the introduction of contagious European diseases prior to their arrival or at the same time had killed more than 50 percent of the Indigenous population. The reaction of European colonizers, like the Pilgrims, was to say, "and God made the land empty for us." They commented on how open the forests were and how bountiful the nut-bearing trees were and sometimes they noticed that there were areas that looked remarkably like crop fields. This was because European colonizers believed that Indigenous people were Stone Age peoples with no written language and therefore with no sophisticated culture and certainly not with European religious values. Some of this misunderstanding was not helped by the fact that the decimation of populations of Indigenous people by the epidemic diseases allowed the loss of many Indigenous secular and sacred practices because of the loss of elders and knowledge keepers, the main carriers of oral tradition.

The situation was further exacerbated by colonial efforts to destroy Indigenous oral culture by not allowing them to speak their language, carry out ceremonies, or practice old methods of land management (e.g., controlled burns to manage vegetation cover of the landscape). Europeans did not seem to recognize that Indigenous people had created several very important domestic crops from native plants in the Americas. There are a large number of these domesticated plants, not just the most famous three sisters of corn, squash, and beans. The myopia of Europeans to these cultural landscapes seems today unbelievable considering some of these would certainly have been familiar in Europe, as many studies today confirm (Rackham, 1980; Birks *et al.*, 1988; Williams, 2003).

Today, natural and social sciences are companions to Indigenous knowledge. Science is not about Truth with a capital T and is not the only form of knowledge. In fact, the name "science" is a relatively new term from the mid-1800s. Before this, science was divided into natural history and natural philosophy. The role of natural history was mainly to observe, describe, and find some natural order (classification) of objects in terms of their essential characteristics. The role of natural philosophy was to study and try to understand the processes in the natural world. Today, natural philosophy has largely been replaced by the term science. However, there are some science disciplines that still have viable natural history components.

Science today is a systematic methodology of inquiry used to obtain an understanding of how the natural universe works. It has a long history in Europe and an even longer history in China, India, and Arabia. Today, science is a worldwide way of understanding the natural world through empirical knowledge, contributed by all peoples of the world. Science seeks to consider relationships and processes by observing, measuring, manipulating, experimenting, and forming models. Models play an important role in science and are generally of two types: representations of actual things designed to study processes, for example laboratory flumes, and sets of assumptions about the processes. Models of this latter type are often mathematical equations. Mathematics is used because it allows one to check the logic in the model of the processes and to display the relationships more clearly in the processes.

However, to be successful, models cannot just be logical-mathematical equations but must have empirical content and have the possibility of being falsified by comparison with the phenomena. Giving equations empirical content is usually done by what are called correspondence rules, that is, the use of either previous science to define the empirical data or new and empirical ways of measuring. This research program looks for logical and empirical consistency with the universe. Notice that natural science does not allow or deal with objects that, by their very definition, can never be observed or falsified.

1.2 Overlap between Indigenous Knowledge and Science

In brief, this section discusses the overlap and some differences between science and Indigenous knowledge. This discussion is not definitive, and other comparisons are available (e.g., Aikenhead and Mitchell, 2011).

1. In many ways, science and Indigenous knowledge do things in very similar ways. The difference is primarily that Indigenous knowledge is transmitted through an oral, not a written, tradition and not only includes empirical observations but is also united with metaphysical worlds.

Science has a written procedure for passing on science knowledge and often uses mathematics to represent these ideas, has methods for collecting data, and empirically tests if the data conform to the ideas being tested. Finally, the results are evaluated by independent scientists who may try to reproduce the results. These ideas, methods, and tests are then organized into a paper that is sent to be considered for publication in a scientific journal. These journals send the paper out to be reviewed by knowledgeable scientists from anywhere in the world on the particular subject. These referees explain their concerns and make suggestions. Some of these papers are found deficient in either their ideas, methods, or testing. About a third of the papers submitted to journals are found to not meet the standard of rigor and are not published. Science tries to synthesize the current scientific knowledge through this review process in the refereed, scientific journals and in further conversations that occur in the literature about previous papers and in gatherings, workshops, and conferences.

Indigenous knowledge uses an oral tradition and material objects (e.g., sacred bundles) to pass on ideas, methods, and how to collect and retain knowledge. The ideas are preserved by elders who have demonstrated their understanding and ability to pass on information carefully and from wide experience. Information is often passed on in songs as is and has been done in many parts of the world, for example the Iliad, the Odyssey, and the Kalevala. In Indigenous knowledge, information is passed on usually to younger individuals who have shown understanding, leadership, wise judgment, discipline, and an ability to learn the information and the traditions. Often, a circle of elders will tell a story from each of their understandings and backgrounds.

2. Indigenous knowledge is a place-based understanding of the world with both empirical and spiritual components. Science, on the other hand, has only an empirical understanding of the world. Other ways of knowing are left to others.

Science has been organized internationally in recent centuries. Scientists meet to present and discuss information and also do so through scientific journals. Science, from its beginning, has always been a group process in which everyone may contribute. It works toward knowledge that is generalized so that it can apply to other parts of the empirical world.

Indigenous knowledge is organized locally, with a place-based understanding. Some Indigenous knowledge is not totally place-based, as many Indigenous groups have, in their long history, moved to different places and have traded goods and ideas, often over long distances. They still carry some of this knowledge from other places and other groups. However, Indigenous knowledge has an empirical and rational approach to understanding the environment combined with a spiritual understanding.

3. Science within its frame of knowledge defines the natural world as knowable. It therefore does not explore anything spiritual and unmeasurable. Indigenous knowledge is fundamental to Indigenous peoples' understanding of the physical and biological worlds and is a holistic view with no separation between empirical knowledge, arts, religion, animistic, etc., that is, other ways of knowing.

4. Science knowledge is collected in a qualitative and quantitative manner so that it can be transferred to others in a form that can be evaluated. Indigenous knowledge, because it encompasses empirical and spiritual components, generally does not see the world as reducible to empirical understanding only. It sees the world as intertwining both the empirical world and the spiritual world. The Indigenous way of knowing does not prevent science from cooperating with Indigenous knowledge, in the same way that science does not prevent an appreciation of other spiritual and artistic traditions.

5. Both science and Indigenous knowledge see the world as dynamic and changing. The difference is primarily the role that spiritual knowledge plays in Indigenous knowledge. Furthermore, Indigenous knowledge believes in a relationship between all of the natural world, whether it is human, animal, plant, rock, mountain, or even a particular view. All of the world has spiritual content and should be respected, and the world is seen as animistic. Science defines the natural world only through its knowable parts.

6. Indigenous knowledge sees time as flowing into the past and into the future in a cyclic manner in terms of both days and seasons.

In conclusion, science and Indigenous knowledge share an empirical understanding of nature, but Indigenous peoples also have an integrated spiritual view. This spiritual view should not interfere with cooperation between and respectfully sharing the different kinds of knowledge, and such cooperation can lead to working together productively.

1.3 Objectives and Chapters

This book's overarching objective is to consider how Indigenous populations have lived and managed the landscape, specifically how the landscape's footprint is a result of the combination of Indigenous peoples' empirical knowledge and spiritual culture.

Science has carefully tried to remove unmeasurable things/interactions. Indigenous knowledge deals with observable and experiential attributes but also the spiritual realm. This discrepancy is apparent in the fact that, although science does not say there are not other ways of knowing, the scientific approach explores only within certain limits. Science has been very effective at knowing within these limits, but its practitioners are not without ethical and moral values. However, we all have a certain world view that has hidden biases and, through the existence of a diversity of knowledge, we can be made aware of this.

How do we combine the areas of intersection between science and Indigenous knowledge without losing the totality of both? Each author in this book has been asked to provide background on both Indigenous knowledge and science, so that the overlap, similarities, and differences between them, as they understand them, are explained.

The chapters are divided into four groups. The first group deals with reintegrating cultures and the natural landscapes and the role of kinship and oral tradition in more distant pasts. The second group of chapters approaches the landscape as a living university of learning and managing, the ethnobotany of how to grow and hunting and trapping more responsibly, and assessing and projecting the harvest. The third group deals directly with managing fire in an anthropogenic plant community and how to integrate Indigenous agriculture in hydrology and dry regions. Finally, the fourth group consists of studies of how science and Indigenous knowledge can be taught in schools using land-based studies.

References

Aikenhead, G. and Mitchell, H. (2011). *Bridging Cultures: Indigenous and Scientific Ways of Knowing Nature*. Don Mills, Ontario: Pearson Education.

Birks, H. H., Birks, H. J. B., Kaland, P. E., and Moe, D. (eds.) (1988). *The Cultural Landscape: Past, Present and Future*. Cambridge, UK: Cambridge University Press.

Rackham, O. (1980). *Ancient Woodlands*. London: Arnold.

Williams, M. (2003). *Deforesting the Earth: From Prehistory to Global Crisis*. Chicago, IL: University of Chicago Press.

2

Reintegrating Cultural and Natural Landscapes
Indigenous Homelands of the Alsek-Dry Bay Region, Alaska

THOMAS F. THORNTON, DOUGLAS DEUR, AND BERT ADAMS, SR. [*]

2.1 Introduction: Rethinking "Natural" Landscapes

As this volume helps to illuminate, Indigenous environmental knowledge is rooted in local landscapes. Landscapes themselves are a coevolutionary product of human and other-than-human (often so-called natural) systems. In this chapter, we turn to how Indigenous peoples' cultural landscapes are the products of local communities' systems of knowledge, perception, and practices on the land over time. This chapter explores how Indigenous knowledge and oral tradition help us to understand the nature and culture of landscapes through a detailed examination of the historical, ethno-, and political ecologies of the Alsek-Dry Bay region of southeast Alaska and western Canada. This region lies at the intersection of the northern Tlingit and Athabaskan cultural worlds, and the dynamic convergence of subarctic and northern Northwest Coast biocultural regions. Home to the G̲unaax̲oo Ḵwáan (among the Interior Peoples) Tlingit, with their roots in both Tlingit and Athabaskan tradition, this has long been a place of cultural encounter, contestation, and synthesis like so many places where vast Indigenous biocultural regions meet. This is a place of international significance for its cultural heritage – being, among other things, the locus of some of the largest concentrations of sites associated with epic "Raven" trickster and worldmaker story cycles in Native North America.

Simultaneously, the Alsek-Dry Bay region is home to towering peaks that descend abruptly to the Pacific Ocean shore – many above 4,000 m in height, including the 5,489 m (18,009 ft) Mount Saint Elias – and some of North America's

[*] The authors would like to thank the Yakutat Tlingit Tribe, especially J. P. Buller, Bert Adams, and Fred White, as well as other tribal members who participated in the project to document cultural landscapes in the Alsek-Dry Bay region. We are also indebted to the National Park Service, particularly Mary Beth Moss and the staff at Glacier Bay National Park and Preserve, for making the project possible, including visits to Yakutat and Dry Bay. Professor Thornton thanks the Hokkaido University Global Station for Indigenous Studies and Cultural Diversity for the opportunity to present this research at its January 2023 International Symposium on Indigenous Heritage and Research Ethics. Finally, the authors wish to thank Jamie Hebert for her assistance in preparing the final manuscript. This research was funded by the National Partk Service under CESU Master Cooperative Agreement P16AC00003.

largest glaciers, many now in active retreat. Isostatic rebound and rapid coastal sedimentation reroute entire river courses (Loso *et al.*, 2021), while seismic activity regularly reshapes the land and contributes to occasional tsunamis. The largest tsunami ever recorded was in 1958, just south of the study area at Lituya Bay, and it swept trees from mountainsides up to approximately 524 m above normal sea level. Simply put, this is one of the most stunningly rugged, complex, and dynamic landscapes on Earth. The regional landscape of Alsek-Dry Bay has come to be officially viewed, in both Canada and the United States, as a set of "natural" landscapes, preserved and managed as national parks and protected areas and captured collectively as a World Heritage Site of "outstanding universal value." Management and public discourse about the Alsek-Dry Bay region has therefore focused on geological and other strictly "natural" phenomena, rather than on the region's diverse and dynamic set of cultural landscapes forged with and by Indigenous peoples and other-than-human inhabitants of the cosmos.

The Kluane/Wrangell–St. Elias/Glacier Bay/Tatshenshini-Alsek World Heritage Site encompasses four vast protected areas, namely Canada's Kluane National Park and Reserve and Tatshenshini-Alsek Provincial Park, as well as Wrangell–St. Elias National Park and Preserve and Glacier Bay National Park and Preserve in the United States. Together these areas represent some 98,391 square kilometers (37,989 square miles) of contiguous park lands straddling the international divide. For perspective, this is an area considerably larger than Scotland or some 4.7 times the size of Wales. The United Nations Educational, Scientific and Cultural Organization (UNESCO) is the entity that grants World Heritage Site status and manages the World Heritage list – shaping both the management and the public perception of places possessing such unique global significance. The UNESCO description of the Kluane/Wrangell–St. Elias/Glacier Bay/Tatshenshini-Alsek World Heritage Site gives flesh to the specific qualifying criteria by which a World Heritage Site of outstanding universal value is conceptualized and assessed (Table 2.1). As summarized by UNESCO (2022a):

The Kluane / Wrangell–St. Elias / Glacier Bay / Tatshenshini-Alsek national parks and protected areas along the boundary of Canada and the United States of America contain the largest non-polar icefield in the world as well as examples of some of the world's longest and most spectacular glaciers. Characterized by high mountains, icefields and glaciers, the property transitions from northern interior to coastal bio-geoclimatic zones, resulting in high biodiversity with plant and animal communities ranging from marine, coastal forest, montane, sub-alpine and alpine tundra, all in various successional stages. The Tatshenshini and Alsek river valleys are pivotal because they allow ice-free linkages from coast to interior for plant and animal migration. The parks demonstrate some of the best examples of glaciation and modification of landscape by glacial action in a region still tectonically active, spectacularly beautiful, and where natural processes prevail.

Table 2.1 *World Heritage Site criteria and qualifying properties of the parks (UNESCO, 2022a)*

World Heritage Site criteria	Qualifying properties of the parks
Criterion (vii): to contain superlative natural phenomena or areas of exceptional natural beauty and aesthetic importance	The joint properties encompass the breadth of active tectonic, volcanic, glacial, and fluvial natural processes from the ocean to some of the highest peaks in North America. Coastal and marine environments, snow-capped mountains, calving glaciers, deep river canyons, fjord-like inlets, and abundant wildlife abound. It is an area of exceptional natural beauty.
Criterion (viii): to be outstanding examples representing major stages of Earth's history, including the record of life, significant ongoing geological processes in the development of landforms, or significant geomorphic or physiographic features	These tectonically active joint properties feature continuous mountain building and contain outstanding examples of major ongoing geologic and glacial processes. Over 200 glaciers in the ice-covered central plateau combine to form some of the world's largest and longest glaciers, several of which stretch to the sea. The site displays a broad range of glacial processes, including world-class depositional features and classic examples of moraines, hanging valleys, and other geomorphological features.
Criterion (ix): to be outstanding examples representing significant ongoing ecological and biological processes in the evolution and development of terrestrial, fresh water, coastal, and marine ecosystems and communities of plants and animals	The influence of glaciation at a landscape level has led to a similarly broad range of stages in ecological succession related to the dynamic movements of glaciers. Subtly different glacial environments and landforms have been concentrated within the property by the sharp temperature and precipitation variation between the coast and interior basins. There is a rich variety of terrestrial and coastal/marine environments with complex and intricate mosaics of life at various successional stages from 500 m below sea level to 5,000 m above.
Criterion (x): to contain the most important and significant natural habitats for in situ conservation of biological diversity, including those containing threatened species of outstanding universal value from the point of view of science or conservation	Wildlife species common to Alaska and northwestern Canada are well represented, some in numbers exceeded nowhere else. The marine components support a great variety of fauna including marine mammals and anadromous fish, the spawning of which is a key ecological component linking the sea to the land through the large river systems. Populations of bears, wolves, caribou, salmon, Dall sheep, and mountain goats that are endangered elsewhere are self-regulating here. This is one of the few places remaining in the world where ecological processes are governed by natural stresses and the evolutionary changes in a glacial and ecological continuum.

Although this is an inhabited landscape of pronounced cultural significance to the G̲unaax̲oo K̲wáan Tlingit and other Indigenous communities, universal value here equates with "natural processes," nonhuman species, and landscapes other than cultural ones. We could deconstruct what "universal" really means in the UNESCO context then, with World Heritage status being contingent on the culture-bound concepts emanating from the industrialized West, such as "superlative" and "outstanding," or "important" and "significant," "natural" and "beauty," or even "conservation" and "biodiversity" (UNESCO, 2022a, b).

However, our purpose here is to inform and bridge these culture-bound notions with reference to Indigenous knowledge and perspectives relating to the same landscape. The juxtaposition of Indigenous and Western scientific knowledge, the informing of each narrative by the other, offers us the potential to see the landscape in its fuller context, to perceive the intersection of human and not-human actors as coproduction – allowing us at once to decolonize our views of landscapes and to consider more appropriate and meaningful ways to engage landscapes such as the Alsek-Dry Bay region. Indeed, Indigenous North American perspectives of the region tend to view the "universal" perspective expressed by UNESCO's assessment as a rather odd, scientistic mis-valuation of a geography that their cultural heroes, including Raven (or sometimes "Crow"), helped cocreate as a unique homeland and *kwáan* (a dwelling place for Tlingits, as well as Southern Yukon Athapascans). From the G̲unaax̲oo K̲wáan perspective, this distinctly settler framing of the "natural" character of this landscape continues to distort and undermine Indigenous relations to it as a World Heritage Site and National Park and Preserve today. The tensions between these two views of the geography bring the greater significance of the landscape into sharper relief. In bringing these contrasting perspectives together in search of reconciliation and points of commonality, we seek to demonstrate how one might achieve a more balanced reframing of the nature of landscape itself – a necessary step in bridging Indigenous and scientific knowledge systems in service of conservation and other collaborative aims.

Building on geographer Carl Sauer's famous redefinition of landscape as "an area made up of a distinct association of forms, both physical and cultural," subject to evolution and change, and constituting an integrated whole rather than merely the sum of features (Sauer, 1963: 321), we suggest three ecological frames for viewing landscapes *in* and *as* culture, using the Alsek-Dry Bay region as a case study. First, we acknowledge a historical ecological frame that explicitly posits landscapes in the inhabited world as the coevolutionary outcome of natural and cultural processes (Baleé, 2013). Second, we deploy an ethno-ecological frame, which considers Indigenous environmental knowledge and practice relating to the "natural" world, either implicitly or explicitly applying universalist scientific or other non-Indigenous understandings of the Indigenous experience of place and landscape

(Thornton, 2008). Finally, we explore a third ecological frame addressing the political ecology of landscapes (Neumann, 2011), which regards landscapes as sources of power, identity, and destiny, or *historia* (Wallace, 2005; Thornton, 2014). These perspectives are analyzed in contentious juxtaposition with colonial or other competing, largely non-Native, forces that seek to stake claims on the perspective, identity, and destiny of particular landscapes. Combining these perspectives, this moves us beyond Sauer's view to a more profound ecological view of landscapes like Alsek-Dry Bay as loci of both heritage and destiny. Such a view is anticipated by the works of such scholars as Tim Ingold (2000: 54), who noted:

First, [landscape] is not a given substrate awaiting the imprint of activities upon it but is itself the congelation of past activity . . .. Second it is not so much a continuous surface as a typologically ordered network of places, each marked by some physical feature, and the paths connecting them. Thirdly, the landscape furnishes its human inhabitants with all the lineaments of personal and social identity, providing each with a specific origin and a specific destiny. And therefore, fourthly, the movement of social life is itself a movement in (not on) a landscape. In short, the landscape . . . is . . . life's enduring monument.

In fact, the Tlingit of Alsek-Dry Bay and beyond have a term for the combined notion of heritage and destiny: *shagóon* (Dauenhauer and Dauenhauer, 1987, 1994; Thornton, 2008). Looking both backward and forward in time, the term literally means "our ancestry" or "our ancestors," but its broader meaning extends well beyond this to suggest, among other things, that we each inherit the outcomes of innumerable "natural" and human events and relationships from the past, live within landscapes that manifest those events and relationships today, and care for this inheritance in ways that anticipate a future of similar proportions and complexity. As such, the term meshes nicely with Ingold's corrective, while also problematizing the Western notion of a purely natural landscape.

Considering these frames, the term "cultural landscape" appears redundant, as every landscape is a cultural landscape at some level; every part of the Earth is affected materially by human action, especially but not solely in the context of global climate change. Yet each landscape is also understood and valued by human communities through their own culturally bound lenses. As Sauer noted nearly a century ago, "Landscape [is] landshape, in which the process of shaping is by no means . . . simply physical" (Sauer, [1925] 1963: 321). No landscape is completely "natural," then, despite what UNESCO's version of World Heritage might suggest. "Natural" and "cultural" retain their utility only as terms of relative preponderance. Of course, UNESCO does not deny the existence of places of combined natural and cultural significance, but it treats them as a discrete and relatively minor category of the landscapes commemorated and protected by the organization. As designated in Article 1 of the 1992 World Heritage Convention, such cultural landscapes represent the "combined works of nature and of man." Of well over 1,100 World Heritage Sites, only 121 World Heritage cultural landscapes exist on the planet, and one in the United States: Papahānaumokuākea.[†] The category is growing, however – both reflecting and bolstering a growing international appreciation of the untenable nature/culture dichotomy as a conceptual binary. Indigenous knowledge of places such as the Alsek-Dry Bay region and the domains of historical, ethno-, and political ecologies help us understand why.

2.2 Historical Ecology

The historical ecological perspective on landscape has its roots in both environmental history and anthropology. In terms of North American Indigenous perspectives, William Cronon's (1978) seminal *Changes in the Land: Indians, Colonists and the Ecology of New England* is a good starting point. Despite the challenges of reconstructing a deep history of New England's landscapes prior to colonization, Cronon demonstrates the intertwining cultural and natural processes and material relationships that shaped the New England landscape in the precontact and colonial periods. He contrasts different modes of landscaping as "landshaping" – drawing a contrast between the mobile, rotating, and overlapping usufruct tenure of the Algonkian Indians, which created a mosaic of varied landscapes with high biodiversity, and the more fixed, settled, and commodified tenure systems of the colonists, which led to expansive clearing and degradation of the forest ecosystem. A key insight from Cronon is that the "natural" forests of the region were always, in fact, cultural landscapes.

By working directly with Indigenous knowledge holders, in combination with archeological, historical, and other ethnographic sources, anthropologists have expanded on this historical ecology perspective. William Balée's (2013) *Cultural Forests of the Amazon: An Historical Ecology of People and Their Landscapes* is a fine example of this approach. By demonstrating the human role in shaping even virgin forests of the Amazon, Balée thoroughly deconstructs the myth of the "pristine" (Denevan, 1992) or "natural" forest. On the Northwest Coast, our own work (cf. Deur and Turner, 2005; Thornton and Deur, 2015; Thornton *et al.*, 2015; Thornton and Moss, 2021; Chapter 5) and the work of others (cf. Lepofsky and Lertzman, 2008; Lepofsky, 2009; Turner *et al.*, 2013, 2021; Lepofsky *et al.*, 2015; Langdon, 2020) has taken a similar approach, yielding insights into the human coproduction of land-, river-, and seascapes over time among these so-called complex hunter-gatherers. These works demonstrate the coproduction of such environments through a wide range of cultivation techniques, to the extent that

[†] https://whc.unesco.org/en/list/1326.

the term "hunter-gatherer" strains credulity in describing livelihoods in Tlingit country and other parts of northwestern North America.

Indigenous knowledge of the Alsek-Dry Bay region posits this place as a dynamic, shifting, convergent, and thoroughly inhabited and coproduced landscape. The identity of this place rests in its position at the coastal margins between peoples and landscapes. The Tlingit term for the Alsek-Dry Bay region – Gunaax̱oo or "Among the Athabascan (interior) People" – marks the nature of its inhabitation at an intersection of Native nations. It's a region defined by the coastal Tlingits dwelling and interacting among people from the upstream interior, who migrated to the coast and converged with them at Dry Bay. Similarly, the term Alsek (Aalseix̱, literally meaning "Where it [Alsek River's water] comes to Rest") is not merely a reference to the Alsek River. Instead, the term is a reference to the uniquely expansive, shallow, and coursing estuary where this major transboundary river and many smaller rivers converge and literally "come to rest" before entering the Pacific Ocean. The people too came to rest there during migrations long ago, at first perhaps temporarily, before heading out into the Gulf of Alaska or upriver to the interior of Alaska and the Yukon, and then more permanently in settlements whose names came to dot the Indigenous landscape (see Thornton, 2012) and to define the human dwelling spaces of the Gunaax̱oo Kwáan.

Landscape features of Gunaax̱oo-Alseix̱ are tied not only to "natural" physical processes, then, but also to cultural and even more widespread supra-cultural processes. In this place, the supra-cultural features are most notably associated with the trickster-worldmaker figure of Raven (also known as Crow in the interior), who dominates the worldmaking narratives in the Northwest Coast culture area from northern British Columbia to Prince William Sound. It was Raven that opened the Box of Daylight at Dry Bay, releasing the sun, moon, stars, and other elements necessary for survival, transforming not just the landscape but the cosmos into the basic form we recognize today. Preparing the world for humans, his epic deeds shaped not only physical landscapes, but also the moral landscapes navigated by the Tlingit and other tribes (Swanton, 1909). While his deeds are worldmaking and his ego and appetites world sized, most of Raven's doings are exercised at the landscape scale. Accounts of his stories are told across the region, and in derivative and popular accounts across the wider world.

Raven was born before the Great Flood, an event triggered by his own (maternal) uncle in Tlingit versions of the story, who, in jealously, tries to do away with his nephew as the heir to his wealth. In one version of the Raven cycle (Swanton, 1909), Raven is said to have endured the Great Flood by turning into rock and, in other versions, by clinging to the atmospheric layer of the sky, before dropping to Earth again in Yéil T'ooch' (literally "Black Raven"), that is, the Pacific Ocean (or Gulf of Alaska), where he set about transforming the dark flooded world into the contemporary land- and seascape. According to Yakutat elders Bert Adams and Fred White and other descendants from Alsek-Dry Bay (see de Laguna, 1972), it was at the Dry Bay estuary at the mouth of the Alsek River where Raven opened the Box of Daylight that he stole from his grandfather at Nass River (de Laguna, 1972: 84). In doing so, not only did he release the sun, moon, and stars into the cosmos, but he frightened everything else – even the rocks and the mountains – off the face of the land from Dry Bay up to Ocean Cape, leaving the unique sandy (glacial) forelands along this part of the otherwise steep and rocky coast. In another portion of the Raven story cycle, Raven is described as pulling ashore a vast canoe full of all manner of animals; in doing so he leaves enduring marks on the landforms of the forelands and fills the land so cleared with all the species later used by humans for sustenance (de Laguna, 1972: 84).

Among the interior Tlingit and the Tagish and Southern Tutchone Athapaskans, the stories are similar. Elder Annie Ned described to anthropologist Julie Cruikshank (2005: 15) how:

Raven, also known as Crow, originally configured the drainages from the interior to coast at the beginning of time, tipping his wings to orient them in the opposite directions; some lakes and rivers now flow north to the Yukon River and hence to the Bering Sea and others pour south to the Gulf of Alaska through the Alsek-Dry Bay drainage.

Thus, Raven is not only the instigator of the Earth-changing flood, but also the shaper of post-disaster watersheds, mountains, coastlines, and seascapes, and the ultimate source of the Pacific tides.

Place names linked to Raven's works constitute a distinct subset of the Indigenous geographic nomenclatures of the Tlingit, and nowhere are Raven toponyms denser than the Alsek River-Dry Bay basin. In Dry Bay, even those toponyms that don't reference Raven directly seem to be the result of his interventions. For example, Bear Island, the major island feature in Dry Bay, is known as Yáay (Yáay X'áat'i) (see No 17 in Figure 2.1) – (Whale Island). But this is no simple metaphoric reference to the island's physical resemblance to a humpback whale or metonymic association to the presence of whales around the island. Rather, it is *the* whale that Raven is said to have harpooned first at Kayak Island, near Cordova, with his line and float attached (see Nos 1, 2, and 3 in Figure 2.1). Tlingit oral tradition reports that Raven then entered the whale through its blowhole, feasted all winter on its blubbery innards, and finally beached his host at Dry Bay, where he "wished the Whale to strand on a fine sandy beach" – and thus the Alsek delta is saidto be sandy as a result (de Laguna, 1972: 84). The people that lived on the east side of Dry Bay, known as Whale's Fat (Yáay Taayí), heard Raven calling and, after initially being scared by this sound, dragged the whale further ashore and flensed it, eventually opening a hole big enough for Raven to fly away. As he emerged, he

Figure 2.1 Reference map of place names related to the trickster-worldmaker Raven in the Alsek-Dry Bay region. (Reproduced from Thornton *et al.*, 2019.)

exclaimed his customary "<u>K</u>aa!" cry of trickster triumph, having cheated the people out of some of the whale meat and fat. The landmark of Raven's work endured at this place ever since, even playing a role in later cycles of the Tlingit oral narratives relating to Alsek-Dry Bay. Significantly, these narratives tell us that, when the people at the Dry Bay village of Gus'ei<u>x</u> faced a later tsunami-like flood – caused not by Raven but a local clan's mistreatment of a seagull – the whale's "fin," a marked protrusion on the island, served as a mooring for people to tie their canoes to as the flood waters washed the settlement away (de Laguna, 1972: 76). Such oral traditions have transmitted across generations many forms of place-based knowledge: knowledge of environmental phenomena such as tsunamis that sweep over the forelands every few generations, for example, or of environmental ethics such as appropriate relations with other species, all linked with enduring mnemonic bonds to specific landforms.

Raven's quest is also evident in the tracks he left on the west side of Dry Bay, which were produced when he dragged the famous food canoe ashore then built a repository or jumping fish house (Kudatankahídi; see No 13 in Figure 2.1) at Dry Bay. The food canoe, or "ark" as some modern Tlingits call it, which also supported a kelp-bed edifice to house marine creatures, was exceedingly heavy. Raven's efforts to drag it ashore were arduous, so his deep tracks remain as depressions in the Earth at Yéil Áx Daak Uwanugu Yé (see No 12 in Figure 2.1) – literally meaning Where Raven Scooted Back [Kicking Up the Sand] – and Yéil Áa Yoo Akaawajiyi Yé (see No 16 in Figure 2.1), also known as Yéil Áx Daak Akawujiyi Yé – Where Raven's Feet Worked into the Mud Dragging. According to Yakutat elder Fred White (2001), who spent significant time in Dry Bay as a child, you can see Raven's tracks at the sand dunes between the Ustay and Akwe rivers, the dunes themselves having resulted from when Raven dragged the canoe to shore, working his feet into the mud and kicking up sand. As de Laguna (1972: 84) recorded:

Raven with a cane shaped like a devilfish [octopus] tentacle drew ashore an "ark" filled with all kinds of food animals. Canoe Prow House of the [L'uknax̱.ádi clan] refers to the enclosed prow of this canoe, and the [L'uknax̱.ádi] use a dance paddle shaped like the cane. The point, "Atuqka [probably Áa Tú<u>k</u> X'aa – Point at Lower End of Lagoon (Island point between Cannery and Muddy Creeks)] . . . was "a place just like the prow of a canoe . . . ."

Bert Adams (1998), a L'uknax̱.ádi clan Tlingit elder, adds:

When he [Raven] pulled the ark ashore, the birds were released and that is why we have birds in the world today. As he pulled the ark from the ocean he left his foot prints in the sand hills along the Akwe River and that is where Gus'ei<u>x</u> [village] is. This goes along with the story that the Lukaa<u>x</u>.ádi [another clan] people were the earliest people in the Dry Bay area and initially built their house in Gus'ei<u>x</u>.

Raven also lured ashore the first king salmon (*Oncorhynchus tshawytscha*) in Dry Bay. Working his way up the Alsek River, Raven comes to be associated with other unique or anomalous features of the landscape as he pursues his quest for food, such as Gateway Knob – a prominent island in Alsek Lake.

These landmarks represent just a sampling of the dozens of places associated with Raven in the Alsek-Dry Bay area, and they paint a very different portrait of the historical ecology of the region from the prevailing characterizations provided by UNESCO for this World Heritage Site. As a central domain of Raven's landshaping, Alsek-Dry Bay is not only the source of all creatures in the world, but the source (or one of the sources) of the order of the world itself, a place where the planetary system was transformed to support people's life on Earth. It is from the primordial time of Raven's work, and because of his work, that people have been able to live well and sustainably – a fact of central cosmological importance within the Tlingit worldview. Equally important is that Raven not only is the source of foundational transformations of the land and sea, but also manifests the region's dramatic disruptions, such as tectonic upheavals, and its incredible dynamism and fluidity. To the extent that Raven "[showed] all the Tlingit what to do for a living" (Swanton, 1909: 83), his human counterparts have also continued to shape this land with their presence, with nature and culture coevolving across time – defining the historical ecology of the Alsek-Dry Bay region.

2.3 Ethno-ecology

Ethno-ecology posits that peoples' ways of knowing and relating to nonhuman species and other elements of the Earth may vary, leading not only to different knowledge, perspectives, beliefs, and practices, but also to different forms of communication about and with nature (Bateson, 1973). The nature–culture dichotomy critiqued in the previous sections is itself a culturally bounded ethno-ecological perspective of the world shaped by Western science and scientism, which seeks to separate nature from culture and to dissect the natural world into component parts as a way to make sense of its assumed essential order. In this mechanistic view, little space remained for Raven's disruptive doings and his transformations of nature. Instead, the separation and dissection led to a "stale dichotomy of nature and culture" without, as Ingold (2000: 9) terms it, "the dynamic synergy of organism and environment" necessary to develop "a genuine ecology of life" in the spirit of Bateson's (1973) *Steps to an Ecology of Mind*. Indeed, this synergy lies in the very interactions and relationships between beings in the processes of inhabiting the world. Among Earth's constituent beings, including humans, natural processes are never completely divorced from interactive processes. Glaciers advance destructively because humans violate protocols for interaction and disrespect them (cf. Cruikshank, 2005; Thornton, 2008); salmon and herring come because, like Raven, humans cultivate and lure them in to be taken, as well as showing due respect (Thornton *et al.*, 2015). This perspective is widespread in Indigenous societies, especially as they relate to so-called cultural keystone species (Garibaldi and Turner, 2004), which play an especially significant role in people's livelihoods.

Thus, historical ecology makes sense of the interactive, coevolutionary processes of inhabitation; ethno-ecology, by contrast, sheds light on the balance and weight of interactions among humans and other beings in shaping diverse landscapes and livelihoods. We cannot focus on every line of evidence in the rich corpus of ethno-ecology relating to the Alsek-Dry Bay region; therefore, it can be useful to start with the unique and outstanding species in Indigenous ways of knowing, as sources of ecological understanding. At Alsek-Dry Bay, that means focusing on the five major species of Pacific salmon, most of which are staples of central cultural importance and therefore are considered to be what we might term cultural keystone species among Indigenous inhabitants of Dry Bay.

Among these, king salmon (*O. tshawytscha*; *t'á* in Tlingit) is the prize and is the reason Raven spent considerable time luring king salmon to the shores of Dry Bay where this fish could be taken.

As de Laguna (1972: 51) noted in her mid-twentieth century research:

King salmon average 15–23 pounds, although many reach 50 or 60 pounds, and a few giants of 100 pounds have been noted. They live from 3–9 years [and] usually breed only in the larger rivers, such as the Alsek or Copper River [though Yakutat and Dry Bay are exceptional in having many smaller rivers, including the bountiful Situk River, with king salmon, perhaps a unique result of the rapid deglaciation and braiding of rivers there]. Spawning runs begin about the last week of April and may continue until the fall (when the king salmon are particularly fat), which was when the natives formerly caught them. While still in salt water the king salmon usually stays close to shore and may be taken by trolling, but this method was not employed until modern times [but see Langdon, 2006, for evidence of aboriginal trolling techniques]. Sometimes king salmon appear in Yakutat Bay as early as February.

Missing in this detailed description is the ethno-ecological connection that local Tlingits make between Dry Bay and Yakutat (the latter some fifty miles northwest of Dry Bay and the mouth of the Alsek River) in producing highly prized early-season king salmon, known locally as "Dry Bay kings." Yakutat Tlingits prize these late-winter and early-spring kings because they are fatter and better tasting than other king salmon. They are fatter, it is said, because they need greater fat stores to ascend the long and mighty Alsek River far into the interior, where they spawn. They fatten themselves for the journey by feeding on prodigious late-winter herring and smelt in Yakutat Bay – often very close to the village – before heading southeast to Dry Bay, where they ascend the Alsek River in May. In fact, their reputation for tastiness is so unique and widespread that it is a high tradition among Yakutat people to give away the first Dry Bay king they catch to someone they wish to please or honor. Similarly, each year, residents may share the first Dry Bay king salmon meal they prepare with wide circles of relatives and friends to mark the arrival of fresh salmon after the long, lean winter. This tradition, originating in feasts at Dry Bay, continues in Yakutat, and Dry Bay king salmon are also shipped to cherished friends and relatives in places throughout Alaska and beyond. As Yakutat elder Elaine Abraham remarked, these salmon are special: "That's your first salmon that starts coming into the village, is your Dry Bay kings" (in Ramos and Mason, 2004: 18).

It was Raven who taught the Tlingit how to lure king salmon to shore, as Yakutat elders related to de Laguna in the 1950s. Indeed, the place where Raven tricked king salmon into coming to shore at Dry Bay is associated with an important contemporary fishing site. Elder Harry K. Bremner (in de Laguna, 1972: 867), referencing both a place-name and an iconic heraldic carved screen that reference the Raven and King Salmon story, stated:

In Tlingit the name they call it – *t'a*, king salmon. T'a yAna kU-li-t' Ul Yel [T'ayaana Kulit'ool Yéil] – "Raven Makes the King Salmon Come Ashore." That's the name of it [the screen]. The story of it: When Old Raven created the world, he see the king salmon fin [moving], and he put up a green stone [*neixinté*] on top of the rock, and he tell the king salmon: "That green stone tell you this and that, and name or everything." That's what he tell the king salmon. . . . [Pretending he was the stone, Raven said] Mean things. So the king salmon could get mad and come in to shore. King salmon swim in again and Raven ask [king salmon], "You not going to do anything to that stone?" And Old Raven said, "I'm going to help you fight that stone if you coming ashore." He got his club in his hand. Old Raven. When the king salmon coming ashore to fight that stone. Old Raven hit him, the king salmon, on the head. He kill it. That's the xin [*x'éen*: screen].

Later, people learned to lure king salmon not with insults but by trolling with hooked lures along their pathways into the Alsek River and its tributaries. From the late nineteenth century onwards, Tlingit fishermen also used gillnets and utilized other new fishing sites favorable to these methods. Weirs were also employed for salmon fishing, with the oldest dating back some 4,000 years (cf. Langdon, 2006; Thornton and Moss, 2021), although less commonly for king salmon, which were typically taken with hooks, spears, or gaffs. A site favorable for landing king salmon was a precious place, typically named and claimed as property by matrilineal clans or their subdivisions, known as houses (*hít*). Numerous examples exist at Alsek-Dry Bay. For example:

On the lower part of the Alsek River in Dry Bay is a small place called KunagX'a [Kunaga.áa?, perhaps "Going into a Lagoon," see Thornton, 2012: 24, #270.1], where people living on the east side of the bay used to go to put up king salmon. "We always go up to the [Alsek] glacier here and . . . then go to KunagA'a[?] in Dry Bay over here," . . . Emmons (MS.) reports "Ku-nar Ka-ha" as a sand flat at the mouth of the Alsek where people caught king salmon in the early summer. Possibly it was on the north shore of the bay, opposite the entrance to Tebenkov's "Kunakagi" or what we now call East River.

(de Laguna, 1972: 84)

This site was used for subsistence and commercial fishing into the contemporary period because king salmon swim close to shore and can be easily caught. In the swift, murky, debris-filled waters emptying from the Alsek, skill is required to capture this cultural keystone species, whether by harpoon, lure, or net. Yet, the challenges of the Alsek-Dry Bay waters are more than offset by the bountiful and extended salmon season, for May is just the beginning of the salmon runs. Following the king/Chinook salmon (*O. tshawytscha*; *t'á*) come sockeye/red salmon (*O. nerka*; *gaat*), followed by humpback/pink (*O. gorbushka*; *chaas'*), chum/dog (*O. keta*; *teel*), and the coho/silver (*O. kisutch*; *l'ook*) salmon in late summer and fall. Thus, salmon are present during all four seasons of the year, preceded in late winter/early spring by copious runs of the eulachon (*Thaleichthys pacificus*), prized and traded for its rich oil, affording dwellers rich sources of fish nearly all year round.

Alsek-Dry Bay king salmon could be dried, stored, and traded in a way that was uniquely linked to the microclimatic conditions of the Alsek watershed. de Laguna (1972: 87) reports that, as the king salmon moved upstream, Dry Bay people (particularly the Lukaax.ádi Raven clan), like Raven:

. . . used to go up the Alsek, and take their canoes over a point called Gel'k'w, through a V-shaped notch. They would go up every fall or winter to YewAltcA hin [Yei Wal'ji Héen,

"Stream that Breaks Downward"], a river where there is a glacier and the ice breaks in spring. [There, or somewhere else up the Alsek, is a place called "King Salmon Bone," T'aketci (possibly t'ax'ici, "dried king salmon").] When they went up in the fall they would hang fish to dry there [just covering it with cottonwood boughs] and it would take care of itself. Nothing happens to it. They would spend the winter in the interior [trading] and come back in the spring. [He also mentioned Yel tsunayi (Yéil Dzoonáyi, Raven's Bola)] "a place up the Alsek where pebbles keep falling all the time, but they won't hit the canoe unless someone is going to die. Yel [Yéil, Raven] told his wife the pebbles would fall outside the canoe – that's why . . . When they came down [under Alsek Glacier], they would put on their best clothes, and after they had passed the glacier they would yell, and it would break behind them, because they were so happy." He also reported that they sang.

Arid conditions for air-drying salmon were rare in the coastal rainforests of Southeast Alaska. This highlights another critical feature of the Alsek-Dry Bay landscape: its diversity. Alsek-Dry Bay is diverse not merely in terms of its topography and resulting microclimates, but also in terms of its biodiversity, which, in turn, fostered an exceptional diversity of cultural practices, even within the domain of salmon production.

In the wetter downstream rainforest environs of Alsek-Dry Bay, salmon could only be dried in smokehouses. Here, we find that the processing of salmon was attuned to river conditions through other cultural protocols in order to guarantee their return upstream each year:

A number of supernatural regulations formerly governed the cutting and drying of salmon. Thus, the head of the fish was turned upstream when the most important cuts were made, and the women apparently faced downstream. "Don't set towards the water. Among the fishes, it's like you set naked. You have to set sidewise to the river . . .. When you have finished splitting the fish out, you hang it with its head upstream." This was obviously to insure the return of the reincarnated salmon in another run. The smokehouse is built with the door facing the river, and the sticks on which the fish are hung run across the house, according to this informant's sketch. The poles on any outdoor fishrack also run parallel to the river bank, and therefore the fish can be hung or saddled on the pole with the head upstream.

(de Laguna, 1972: 400; see also fig. 19, p. 304)

Most places where king salmon and other salmon massed and rested before ascending streams were also known and named. Generically, deep holes at the mouths of streams where salmon collected to adjust their metabolism from salt water to fresh water were known as *ish*, often translated as "fish [or salmon] hole [or pool]." *Ísh* were recognized not only as unique microclimates, but also as places where salmon prepared for the osmotic challenges of moving from salt water to fresh; there they "took a drink of fresh water" (*héen awé a x'eit awdinuk*) (Sealaska Heritage Institute, 2006) and began to sacrifice their fat to muster their strength for the arduous upstream journey to their natal spawning grounds. Elder George Ramos of Yakutat (Sealaska Heritage Institute, 2006) notes: "*Gwaats'ílaa yoo áwé duwasakw wé ish. Ldakaat a saayi kutseeti wé Dry Bay aayi tsu ya Kóokjéinik sé heench duwasá wé heen."* (The *ísh* behind our village [Yakutat] is known as Gwaats'ílaa. There are names of these waters, even the one at Dry Bay, the waters are named Kooxjéinik, also.) The "drinking of fresh water" that occurred at the *ísh* is also said to have improved the quality of the fish for drying, presumably in part by moderating the oil and fat content of the fish (cf. Langdon, 2006).

From this brief ethno-ecological survey focusing on the ethno-ecology of salmon, especially king salmon, we see just how intimate and coevolutionary human and environmental systems were in the development of the Alsek-Dry Bay region as a landscape and nexus for human–salmon relations. But the Tlingit ethno-ecological perspective goes even deeper in its attention to relations, not only between humans and salmon but also among salmon species. The status and relations between salmon "tribes" (species) as nonhuman persons is well developed among the Tlingit, as illustrated in the Salmon Boy story (also known as *Aak'wtaatseen*) and other narratives (c.f. Swanton, 1909: 889; Thornton, 2012; Thornton *et al.*, 2015).

At Dry Bay, we also find the concept of salmon – dog salmon especially – as ecological engineers. Already, we have alluded to the advantageous extended temporal portfolio of all five species of Pacific salmon at Dry Bay. Important among these were late runs of dog salmon, which came to Dry Bay's East River (and Dohn River) in great numbers. De Laguna (1972: 51) notes:

[Dog or chum salmon] is relatively unimportant to the Gulf Coast Indians, although the Tlingit of southeastern Alaska regard it as the best to smoke for the winter, and recognize the Dog Salmon as the crest of a Raven sib. I was told that there were few dog salmon in the Yakutat area, but that they could be caught east of Dry Bay, where the Dohn River or a tributary is called "Dog Salmon Stream" [Tilhéeni].

Late autumn runs of dog salmon can be especially important for extending fresh fish supplies and boosting winter dry fish stores (Thornton *et al.*, 2015). Moreover, according to Tlingit elders we interviewed concerning salmon fishing in the Alsek-Dry Bay region, dog salmon also create a habitat for other fish, including other species of salmon. Yakutat Tlingit elder Sam Demmert speaks about the important role of dog salmon in maintaining the quality of the streambed of East River at Dry Bay. This river became commercially important for sockeye salmon when an earthquake rerouted the Dohn River to enlarge the East River system, but it is now largely overgrown due, non-Native scientists had assumed, to isostatic rebound, sedimentation, and forest encroachment across the glacial landscape. However, Sam Demmert suggests that dog salmon are also partly responsible, based on his Indigenous knowledge of dog salmon behavior:

I've always maintained that chums . . . are an aggressive spawner and they'll go a lot of other places where other fish won't go. And it was after they [commercial fisheries] wiped out the chums [at East River] that some people said it was glacial upheaval [that caused the stream to fill in and become less viable for spawning salmon], but I never did buy it – that idea of glacial [retreat and isostatic rebound]. I thought it was because they got rid of chums – aggressive spawners that cleaned out the beds you know, for other fish also [to spawn] – that we lost that fishery. . . . . Once they got rid of the chums, then the weeds and everything took over the East River and basically killed off our [salmon] fishery.

In short, the viability of the river for multispecies salmon spawning may have been predominately a function of the behavior of the prodigious chum salmon, which by their aggressive *en masse* movements upstream literally cleared river channels for salmon migration and created beds suitable for spawning. However, when the market selected against dog salmon in favor of sockeye salmon, state and federal fisheries' managers tipped the balance against chums, effectively managing them out of their ecological engineering niche. As a result, all salmon in the East River suffered. This is not to rule out the influence of isostatic rebound or warmer stream temperatures due to global warming, but rather also to place salmon themselves in a causal role in maintaining the river system, as evidenced in the Native perspective of close observations of the impact of dog salmon on stream conditions over generations. From an Indigenous perspective, misreading the role of the dog salmon in the East River led to management priorities that undermined this productive social-ecological system.

This brings us to the lesser-known political ecology of commercialized salmon fishing in the Alsek-Dry Bay area of the World Heritage Site, which became a major development in the late twentieth century.

2.4 Political Ecology

Political ecological analysis seeks to understand how and why competition over landscape perspectives, resources, and other prerogatives plays out the way it does – especially, but not exclusively, between colonizing Western institutions and the societies of Indigenous peoples. We have already seen the competition among cultural models for how to characterize Alsek-Dry Bay i.e., the UNESCO vision of a World Heritage Site of natural forces and wilderness versus the Tlingit view of a cultural landscape cocreated by Raven and Indigenous peoples working within nature. The former is the official national (park) and international (UN) perspective on the meaning of the landscape, while the latter represents a subordinate, subaltern Indigenous perspective. These competing perspectives extend not only to the fundamental nature and origin of the landscape, but also to the domain of specific resources, most notably salmon.

In part because of its rich, diverse salmon fisheries, Dry Bay was originally settled by Tlingit and Athabascan peoples and evolved with time to become a vibrant human community for the G̲unaax̲oo Tlingit. By the early period of American settlement in Southeast Alaska, many Tlingits found themselves drawn into commercial salmon fisheries as a point of entry into the industrial cash economy, although subsistence salmon fishing persisted alongside these commercial pursuits. By the early twentieth century, the Indigenous inhabitants of the Alsek-Dry Bay area were pressured to relocate, especially because of the absence of schools – with assimilationist federal and state policies requiring either local schooling or the forced relocation of Native young people to distant residential schools. In this context, many families moved away. By the early twentieth century, schools and services were consolidated at Yakutat, some fifty miles up the Gulf of Alaska coast, and most Dry Bay families settled there with members of other Tlingit communities (Deur *et al.*, 2015). Yet, for these families, seasonal commercial salmon fishing provided a way to sustain their connections to Dry Bay. Seasonal commercial fishing supplemented the rich subsistence opportunities that Alsek-Dry Bay had afforded them. People continued to live seasonally in proximity to their fishing and subsistence camps (typically in cabins or tents) through the mid-twentieth century. In 1958, the earthquake and subsequent tsunami that so transformed Lituya Bay also disrupted the Alsek-Dry Bay geography such that rivers like the Doame and East Rivers, and salmon runs, literally changed course, significantly affecting the fisheries.

At the same time, from the early twentieth century on, non-Native fishers increasingly moved in to take advantage of commercial fishing opportunities in the region. This non-Native population grew after World War II, when increased access and statehood brought more people to Alaska. Many of these people built cabins and other structures in Dry Bay to serve as bases for their commercial fishing operations, even as many Tlingit families lacked the resources to build or maintain such structures from their new homes in Yakutat. In 1980, the US Congress enlarged Glacier National Park under the terms of a new US law called the Alaska National Interest Lands Conservation Act to encompass a preserve encompassing the southeastern portion of Alsek-Dry Bay. Allowing for fishing in the preserve by Native and non-Native fishermen alike, this allowed commercial fishing to flourish once again. The surge in non-Native commercial fishing occurred not only on the Alsek River delta but on the East River, which had become a focal point for non-Native fishing after the transformative 1958 earthquake, when its waters became clearer and the waterway became a viable home for all five species of Pacific salmon. Initially, the US National Park Service was not enthusiastic about taking over management of this area (Gmelch and Gmelch, 2018). Commercial fishing was not consistent with the park's notion of Glacier Bay National Park and Preserve as a "wilderness" (i.e., as a pristine landscape needing to be shielded from commercial extraction of its native species in order to retain its wild character).

However, as long as the fisheries were sustainable and fishers obtained permits, state law governing navigable waters was favorable toward commercial fishing on Dry Bay. Commercial fishing persisted in Alsek-Dry Bay and East River, but on unequal terms. Permanent structures within the preserve, built largely by non-Natives, were grandfathered into the new Glacier Bay Park and Preserve, while new cabin construction was severely restricted. As Sam Demmert (interview), a commercial fisherman from Yakutat, relates:

Those people that had permanent cabins were somehow grandfathered in but they had to meet restrictions on height. Well, you know very few people build a cabin they can raise and lower. And some people did, but not the locals [i.e., Tlingits]. And then it even came to the point where you couldn't even put up a tent frame anymore. So that was the beginning of the end on Yakutat['s participation] . . . in the East River fishery. We refer to them [non-Native fishers] as "non-shareholders" [; they] are able to get the permits to build their cabins and continue fishing down there whereas locals can't.

In the 1980s, competition between Indigenous and non-Native fishers at East River intensified and diversified. In addition to the increasingly lucrative salmon fisheries, another catalyst for contention was the development of Alaska's limited-entry permit system in the mid 1970s. Under this system, the state of Alaska limited permits in various regions to conserve fisheries and the sustainable livelihoods of fishers, with the permits being marketable and transferable. This had the effect of alienating permits from fishers in rural villages who were typically more cash poor and vulnerable to having to sell their permits in the face of rising operational expenses or debt. Often, rural Indigenous fishers had to sell their permits to remain solvent. As a result, most of the commercial setnet permits became concentrated in non-Native hands.

In Yakutat, however, Native fishers continued to fish streams throughout the region, including the Alsek River and East River, using unique methods of surf or "breaker" fishing – from small boats in the tumbling ocean surf – which they had pioneered and evolved over many years of negotiating rough outer-coastal waters. In doing so, they were guided by their intimate knowledge of the currents and other environmental patterns and processes of their Alsek-Dry Bay homeland, and in turn this granted Tlingit fishermen options for adaptation and resilience in their fisheries in the face of growing non-Native competition. According to elder Walter Johnson, "breaker fishing is a very special type of fishing that only in Yakutat they do this." Yakutat fishermen mastered the art of negotiating large surf waves with their canoes and, later, rowboats. In the early- to mid-twentieth century, with the arrival of small outboard motors, fishermen further refined their techniques, learning to net salmon directly in the rolling waves along the outer-coast beaches. They would thrust, idle, and maneuver their motorized skiffs strategically to remain stable amid the surf, while deftly deploying their fishing nets in the breaking waves (Deur *et al.*, 2015).

On this point, too, there is emergent Tlingit oral history. Again, quoting Walter Johnson:

We set where the waves are coming in, we set our nets right in the waves and the breakers hit the net and then wash right over and then we're picking it with the skiff. And, quite exciting. After you get through it you know you're alive, let's put it that way – you know very definitely that you're alive, your heart is beating, you're just shaking, you're scared, you're vibrating all over but then you know you're alive after you get done. And these are not little waves like this, this is twenty, thirty-foot seas, waves that we take all the time. We've done it many times.

He explains that the surf-fishing technique traditionally used by Yakutat developed in the latter half of the twentieth century:

The breaker fishing that is taking place when I first came to Yakutat in 1957, after I graduated out of high school I came here. ... We were rowing all over the river. We rowed from the camp all the way around the point, all the way down, set our nets, come back and row back up with the tide. We'd rowed until we hit the tide then we'd started walking it up, fish and all. Within three years they started introducing motors onto the back of the skiffs. And they would tow the skiffs down, the breaker skiffs down and anchor the outboards out and row them in. After a while they started using the outboards to go set their nets and stuff and brought it around to what it is today.

(In Deur et al., 2015)

Unlike the non-Native fishers, who remained at East River, Tlingit fishers carried out breaker fishing between the Alsek River and East River, depending on the strength of the runs. Some sought to develop cabin sites along the East River, as non-Natives had done, but with few exceptions they were prohibited from doing so, thus further committing them to surf fishing between the Alsek and East Rivers.

When the sockeye fisheries in East River began to decline, the Alaska Department of Fish and Game began to tighten regulations on surf fishing, even though it could not be demonstrated that this fishing was contributing to the decline in sockeye runs. Yet, as is common, when salmon runs dwindle, user groups will point fingers at perceived competitors. Because surf fishing provided access to salmon before they entered the rivers, non-Native instream fishers in East River lobbied for the practice to be prohibited. Feeling the pressure, the Alaska Department of Fish and Game sought to restrict the practice of breaker fishing, including an outright ban in 1981 covering the Alsek-Dry Bay's East River sockeye fishery (Ramos and Mason, 2004: 56). Tlingit surf fishing, already an adaptive response to competition and an arena of unique tribal competence, was threatened with instant extinction. In response, Native fishermen filed a lawsuit: "[We] filed suit against the state of Alaska for when they were going to take away the breaker fishing" (Walter Johnson in Deur *et al.*, 2015).

In the end, the Native fishermen prevailed, but it was a limited victory, as the salmon runs in East River have not returned to levels to make fishing economically viable for most Natives. In order to surf fish the area, Yakutat Natives have to expend considerably more fuel than instream fishers and haul their own camping supplies from Yakutat to Dry Bay each season. Consequently, most Native fishers remain sidelined from the East River and Dry Bay sockeye fisheries. From a political, ecological, and economic standpoint, Indigenous fishers from Alsek-Dry Bay have lost ground in maintaining subsistence and commercial fishing livelihoods, and consequently have experienced an erosion of their enduring connections to their homeland, despite the legal victory for surf fishing. As such events accumulate, they become part of a continuously emerging oral history and a continuously evolving relationship of human and nonhuman actors, in which the presence and disruptions of non-Native peoples on Indigenous livelihoods and landscapes are an increasingly central theme. This theme is very much subordinated and elided, however, in the nationalist and internationalist conservation good news preservation narratives that bolster the efficacy of US National Parks and World Heritage Sites.

2.5 Conclusions

The history of G̲unaax̲oo Tlingit people, the narratives of Raven, and the epic accounts of landscape coproduction, have much to teach us. The natural landscapes of the Alsek-Dry Bay region are surely dramatic, dynamic, and exist at monumental scales. And, to be sure, UNESCO is correct in identifying this as a land of sprawling glaciers, towering mountains, rapidly shifting river courses, emergent habitats of high diversity, and tremendous salmon abundance – as well as recurring monumental earthquakes, tsunamis, and landslides that, while surely transformative, do not figure quite so prominently in their pronouncements on the region's natural splendor.

Yet, the Alsek-Dry Bay region is so much more, for this landscape is the meeting point of Tlingits and Athabascans, the homeland of G̲unaax̲oo K̲wáan. It is the place of Raven's work, where oral traditions shared along the entire coastline converge – where Raven brought forth the world and cosmos as we know it, and evidence of his work remains visible on the land. These are not strictly natural landscapes or cultural landscapes, but are both and more. Across generations, humans have valued and shaped the nonhuman world, and the nonhuman world has shaped the human experience in countless ways. These coevolutionary exchanges have left indelible imprints on human and nonhuman events, like enduring Raven tracks on the beach. These relationships remain encoded in Tlingit oral history, and this history continues to be written into present times, as the relationship between the human and nonhuman worlds is complicated and contested by the peoples of the outside world.

The juxtaposition of Native and non-Native understandings of the Alsek-Dry Bay region illuminates the cultural and ideological underpinnings of the UNESCO vision of this landscape and helps to destabilize the presumed neutrality of Western scientific ways of knowing. In his essay on political ecology and theorizing landscape, Neumann (2011: 846) points out how some of the more pernicious misreadings of landscape (cf. Fairhead and Leach, 1996) by (often colonizing) scientists have been corrected only when Indigenous and local relations with these landscapes have been carefully researched and juxtaposed to mainstream understandings of (via science) and responses to (via management) people, lands, and resources. While Neumann made his comments with reference to so-called New Cultural Geography studies, in anthropology we find a similar pattern. Anthropological studies of landscape excel particularly in their attention to ethnohistorical and ethno-ecological details (once part of the "new ethnography"), many of which prove exceedingly relevant to political ecological understandings of contemporary contestations over landscapes and resources, as our case study of Alsek-Dry Bay illustrates.

What then are the lessons to be drawn for scientists (natural and social) who aim to engage with Indigenous environmental knowledge in their work? First, we suggest that just as important as the "integrating," "bridging," or "braiding" (cf. Kimmerer, 2013) frames that have come to characterize approaches to relating Indigenous knowledge and science more respectfully and equitably (as opposed to extractively) is the landscape frame, which stresses the coevolution of beings and places. A landscape frame encourages integrative ecological thinking in terms of the evolution of landscapes as (ecological) systems cocreated by natural, human, and other-than-human processes of being and relating over time. Our Alsek-Dry Bay case shows that a landscape is not just its physical features – uplands, rivers, bays – but rather the whole complex of fluid watersheds, rivers, tributaries, estuaries, geologies, and beings and their actions that contribute to the dynamics of Alsek-Dry Bay as a living system. What is more, these relations also shape peoples' interactions with other (interior and coastal) systems used by other peoples (e.g., interior Athabaskan groups) and other species (e.g., salmon). One cannot fully understand Gunaaxoo-Dry Bay culture outside this relational, ecological, and communicative (cf. Bateson, 1973) context, just as one cannot fully apprehend this landscape without understanding how it has been so valued and shaped by Gunaaxoo people and their constellation of other-than-human beings, especially Raven. To miss all this is to risk misreading the landscape and becoming insensitive to the generative wisdom with which the coevolutionary landscape perspective speaks at a level beyond the discrete scientific "facts" and "data."

A second lesson is to consider the baselines and temporal frames of any landscape-scale study in terms of a robust historical ecology. In Alaska, the temporal frames and baselines used in scientific studies and management of salmon and other animal populations are often shallow, dating only from the first application of Western scientific protocols in the last century. Yet, these baselines may have already shifted due to human impacts such as overfishing and extreme climatological or seismic events prior to formal data collection. This can produce "shifted baselines" (Pauly, 1995; Thornton and Moss, 2021) far below aboriginal conditions, thus giving a distorted sense of the history and productivity of landscape or seascape. Again, we see this at Alsek-Dry Bay. To counter this bias requires a historical ecological lens to see changes in the system over time and how they correlate with relations between agents, including Indigenous peoples, who carefully observed these landscapes long before the advent of scientific data gathering. Most shifted baselines in nature are from abundance to relative scarcity due to human impacts. By contrast, in the case of Dry Bay sockeye salmon, we see that the population actually boomed after mid-twentieth century geological events reconfigured the East River, aided by large runs of aggressive, stream-shaping chum salmon, which endowed the river with the capacity to support large runs of sockeye and other salmon. Shifting baselines are inherently slippery things, and the Alsek-Dry Bay landscape is a potent reminder of this fact.

Finally, a third lesson is that it is important to track Indigenous perspectives of scientific species of interest in a truly ethno-ecological way. This means, in part, that we should not simply seek information about the species in question, but also seek to understand its relations to other species and dimensions of the landscape or ecosystem, with reference to long-term observations of complexly linked environmental phenomena. It may be attractive to focus just on sockeye salmon at East River because they are of high commercial value, but all five salmon species live there; and, as Tlingit residents observe, these sockeye salmon play a role in the evolution of the landscape and salmon's place within it. The Tlingit ethno-ecological perspective on the relationships between salmon species in producing habitat within the dynamic Alsek-Dry Bay landscape is an important example of how Native knowledge and scientific knowledge intersect. If chum salmon have indeed sustained certain channels in a way that enhances the survival and output of other salmon species, this is a novel finding and one not recoverable from Western scientific sources; rather, this requires recourse to Tlingit oral tradition that encodes observations made by people living in intimate multigenerational proximity to the land and its nonhuman denizens. Their observations provide access to a world beyond the limited observations of formal scientists who only rarely visit Dry Bay. These observations also raise many testable hypotheses that hold the capacity to enrich scientific knowledge and, by extension, enhance the efficacy of the management of natural resources in such settings. By Native people sharing their expertise with Western scientists and those scientists sharing their findings with Native people in a spirit of collaborative and coequal exchange, both might gain a deeper appreciation of the full richness of the natural world. All ways of knowing are culturally bound. Tlingit and Western scientific traditions might yet learn to communicate across cultural divides; each has lessons to teach the other. Each provides a pathway toward certain shared truths, helping us have more meaningful conversations across cultural divides and to more successfully care for the landscapes in our mutual care.

We see, then, that the Alsek-Dry Bay region is indeed a place of "superlative natural phenomena" and "exceptional natural beauty," of significant "ongoing geological processes" and significant "natural habitats," as UNESCO affirmed in its formal designation of the Kluane/Wrangell–St. Elias/Glacier Bay/Tatshenshini-Alsek World Heritage Site. Yet, our findings make it clear that a view of the Alsek-Dry Bay region that focuses on these attributes alone is an incomplete vision – an impoverished vision – because it misses the full richness and meaning of this cultural landscape. For this is the home of Raven, whose actions transformed the Earth, whose stories are celebrated even today across the world. This is also the home of the G̲unaax̲oo Tlingit, whose villages once dotted the shoreline (and remain in archaeological form today), whose rich oral tradition and intimate familiarity with this dynamic landscape still has the capacity to transform our shared knowledge of the land and its constituent biota and other features. To ignore or subordinate this cultural inheritance is to misread the superlative significance of this astonishing corner of the North American continent. Contrary to the terms of its designation, this World Heritage Site is at once an exceptional cultural and natural landscape. Through ongoing conversations between Western scientists and Native environmental knowledge holders; between Native and non-Native resource managers; between UNESCO assessments and oral histories of Raven, king salmon, and the people, we can yet locate a truer, deeper vision of the meaning of the landscapes of Alsek-Dry Bay, and beyond.

References

Adams, B. (1998). Interview and Stories of Dry Bay, Alaska, Project Jukebox. Fairbanks, AK: University of Alaska. https://jukebox.uaf.edu/p/3510 (accessed October 23, 2023).

Baleé, W. (2013). *Cultural Forests of the Amazon: An Historical Ecology of People and Their Landscapes*. Tuscaloosa, AL: University of Alabama Press.

Bateson, G. (1973). *Steps to an Ecology of Mind: Collected Essays in Anthropology, Psychiatry, Evolution, and Epistemology*. Chicago, IL: University of Chicago Press.

Cronon, W. (1978). *Changes in the Land: Indians, Colonists and the Ecology of New England*. New York: Hill and Wang.

Cruikshank, J. (2005). *Do Glaciers Listen? Local Knowledge, Colonial Encounters and Social Imagination*. Vancouver, BC: UBC Press and Seattle, WA: University of Washington Press.

Dauenhauer, N. M. and Dauenhauer, R. (1987). *Haa Shuká, Our Ancestors, Tlingit Oral Narratives*. Classics of Tlingit Oral Literature. Seattle, WA: University of Washington Press.

Dauenhauer, N. M. and Dauenhauer, R. (1994). *Haa Kusteeyí, Our Culture: Tlingit Life Stories*. Classics of Tlingit Oral Literature. Seattle, WA: University of Washington Press.

De Laguna, F. (1972). *Under Mount Saint Elias: The History and Culture of the Yakutat Tlingit*. Smithsonian Contributions to Anthropology 7. 3 vols. Washington, DC: Smithsonian Institution Press.

Denevan, W. (1992). The pristine myth: the landscape of the Americas in 1492. *Annals of the Association of American Geographers*, **82**(3), 369–85.

Deur, D., Thornton, T., Lahoff, R. and Hebert, J. (2015). *Yakutat Tlingit and Wrangell–St. Elias National Park and Preserve: An Ethnographic Overview and Assessment*. Anchorage, AK: USDI National Park Service, Alaska Region.

Deur, D. and Turner, N. J. (2005). *Keeping it Living: Traditions of Plant Use and Cultivation on the Northwest Coast of North America*. Seattle, WA: University of Washington Press.

Fairhead, J. and Leach, M. (1996). *Misreading the African Landscape: Society and Ecology in a Forest-Savanna Mosaic*. Cambridge, UK: Cambridge University Press.

Garibaldi, A. and Turner, N. J. (2004). Cultural keystone species: implications for ecological conservation and restoration. *Ecology and Society*, **9**(3), 1.

Gmelch, G. and Gmelch, S. B. (2018). *In the Field: Life and Work in Cultural Anthropology*. Berkeley, CA: University of California Press.

Kimmerer, R. W. (2013). *Braiding Sweetgrass: Indigenous Wisdom, Scientific Knowledge and the Teachings of Plants*. Minneapolis, MN: Milkweed Editions.

Ingold, T. (2000). *The Perception of the Environment: Essays on Dwelling, Livelihood, and Skill*. London: Routledge.

Langdon, S. J. (2006). *Traditional Knowledge and Harvesting of Salmon by Huna and Hinyaa Tlingit*. Study Number: FIS 02–104. Juneau, AK: US Fish and Wildlife Service, Office of Subsistence Management and Central Council of Tlingit and Haida Indian Tribes.

Langdon, S. J. (2020). Tlingit engagement with salmon: the philosophy and practice of relational stability. In T. Thornton and S. Bagwhat, eds., *The Routledge Handbook of Indigenous Environmental Knowledge*. London: Routledge, pp. 169–85.

Lepofsky, D. (2009). Traditional resource management: past, present and future. *Journal of Ethnobiology* (Special issue: Indigenous Resource Management: Past, Present and Future, D. Lepofsky, ed.), **29**(2), 184–212.

Lepofsky, D. and Lertzman, K. P. (2008). Documenting ancient plant management in the northwest of North America. *Botany*, **86**, 129–45.

Lepofsky, D., Smith, N. F., Cardinal, N., *et al.* (2015). Ancient shellfish mariculture on the northwest coast of North America. *American Antiquity*, **80**(2), 236–59.

Loso, M. G., Larsen, C. F., Tober, B. S., *et al.* (2021). Quo vadis, Alsek? Climate-driven glacier retreat may change the course of a major river outlet in southern Alaska. *Geomorphology*, **384**, 107701.

Neumann, R. P. (2011). Political ecology III: theorizing landscape. *Progress in Human Geography*, **35**(6), 843–50.

Pauley. D. (1995). Anecdotes and the shifting baseline syndrome of fisheries. *Trends in Ecology and Evolution*, **10**, 430.

Ramos, J. and Mason, R. (2004). *Traditional Ecological Knowledge of Tlingit People Concerning the Sockeye Salmon Fishery of the Dry Bay Area*. Anchorage and Yakutat, AK: USDI National Park Service Alaska Region and the Yakutat Tlingit Tribe.

Sauer, C. O. (1963). *Land and Life: A Selection from the Writings of Carl Ortwin Sauer*. Berkeley, CA: University of California Press.

Sealaska Heritage Institute (SHI) (2006). Transcript of discussion of Ish conducted by the Council of Traditional Scholars, November 2005. Copy in William Paul Archives. Juneau, AK: Sealaska Heritage Institute.

Swanton, J. R. (1909). Tlingit myths and texts. Smithsonian Institution. *Bureau of American Ethnology Bulletin*, **39**, 1–451.

Thornton, T. F. (2008). *Being and Place among the Tlingit*. Seattle, WA: University of Washington Press.

Thornton, T. F., ed. (2012). *Haa Léelk'w Has Aaní Saax'ú: Our Grandparents' Names on the Land*. Seattle, WA: University of Washington Press and Juneau, AK: Sealaska Heritage Institute.

Thornton, T. F. (2014). A tale of three parks: Tlingit conservation, representation, and repatriation in southeastern Alaska's National Parks. In S. Stevens, ed., *Indigenous Peoples, National Parks, and Protected Areas: A New Paradigm Linking Conservation, Culture, and Rights*. Tucson, AZ: University of Arizona Press, pp. 108–29.

Thornton, T. F. and Deur, D. (2015). Introduction to the special section on marine cultivation among Indigenous peoples of the Northwest Coast. *Human Ecology*, **43**(2), 187.

Thornton, T. F., Deur, D., and Adams, B. (2019). Raven's work in Tlingit ethno-geography. In G. Holton and T. F. Thornton, eds., *Language and Toponymy in Alaska and Beyond: Papers in Honor of James Kari*, Language Documentation & Conservation Special Publication No. 17. Honolulu, HI: University of Hawaii Press, pp. 39–55.

Thornton, T. F., Deur, D., and Kitka, H., Sr. (2015). Cultivation of salmon and other marine resources on the northwest coast of North America. *Human Ecology*, **43**(2), 189–99.

Thornton, T. F. and Moss, M. (2021). *Herring and People of the North Pacific: Sustaining a Keystone Species*. Seattle, WA: University of Washington Press.

Turner, N. J., Armstrong, C. G., and Lepofsky, D. (2021). Adopting a root: documenting ecological and cultural signatures of plant translocations in northwestern North America. *American Anthropologist*, **123**(4), 879–97.

Turner, N. J., Deur, D., and Lepofsky, D. (2013). Plant management systems of British Columbia's First Peoples. *BC Studies* (Special issue: Ethnobotany in British Columbia: Plants and People in a Changing World, N. J. Turner and D. Lepofsky, eds.), **179**, 107–33.

UNESCO (2022a). Kluane / Wrangell–St. Elias / Glacier Bay / Tatshenshini-Alsek. https://whc.unesco.org/en/list/72/ (last accessed April 2022).

UNESCO (2022b). The criteria for selection. https://whc.unesco.org/en/criteria/ (last accessed April 2022).

Wallace, A. F. C. (2005). The consciousness of time. *Anthropology of Consciousness*, **16**(2), 1–15.

White, F. (2001). Interview on Dry Bay, Alaska, Project Jukebox. Fairbanks, AK: University of Alaska. https://jukebox.uaf.edu/interviews/3520 (accessed October 23, 2023).

3

"My Uncle Was Resting His Country": Dene Kinship and Insights into the More Distant Past

JOHN W. IVES

3.1 Introduction

*Intellectual reasons or reasons of curiosity ... are ignored, as though
these people had no inner life. There's all this language that paints people
on the move and migrating as if they weren't these fully self-actualized
human beings who also had curiosity, who laughed, who had interesting
kinship dynamics, who had joy in their lives.*
*(Kim Tallbear, speaking of scientific approaches to understanding the
peopling of the Americas, in Gannon, 2019)*

Although many might situate the most important traditional Indigenous knowledge in the realm of perceptive ecological awareness, there is a compelling argument to be made that there is a more critical, underlying realm of knowledge. As important as traditional ecological knowledge and effective, naturally sustained technologies are, sophisticated knowledge involving kinship can legitimately be regarded as *the* critical factor that made possible the rich and complex history of at least 14,000 years of Indigenous life in the Americas. In settings like the Great Basin or boreal forests, extensive webs of kin relationships were essential in ensuring human presence throughout the millennia. Here, I sketch a profile of what can be learned from Dene (Athapaskan) kinship, which is pivotal in managing not just Subarctic but a wide range of environments in western North America. A "thought model" derived from Dene kin principles is helpful in grasping the unusual circumstances that founding Indigenous populations encountered when they departed Pleistocene Beringia for the Western hemisphere, which was entirely without human inhabitants: an epic human journey. Such thought models can, with judicious use, intersect with rapidly unfolding genetic and archaeological findings and thereby can be applied to further our understanding of the dawn of the Indigenous presence in the Americas.

The title for this chapter came from a conversation with the late Terry Remy Sawyer. Terry was from Tsiigehtchic (Arctic Red River), but she had spent considerable time in Edmonton by the 1990s, at a time when I was Assistant Director at what is now the Royal Alberta Museum. We were developing an aboriginal culture gallery and Terry, highly skilled at sewing traditional clothing, was making beautiful Dene leather garments for a northern fish camp diorama. Terry was a fluent speaker of Gwichyah Gwich'in, a "high" version of which she had learned directly from her grandparents. From earlier research, I had a deep interest in Dene principles of kinship; at one point in the course of things, Terry provided a clear directive: that we were to sit down and make a detailed recording of her knowledge of Gwichyah Gwich'in kin terms.

The anthropological side of this involved asking questions like "What would you call your mother's brother's daughter or your father's sister's daughter's daughter?" and so on, until all the logical positions on a genealogical grid (256 instances in this case) were filled out. Terry would often then think of a relative at that genealogical position and indicate the kin term she would use. Terry had many interesting stories, and from time to time we would digress into one of them. It was in this way that Terry made the remark in the title: that her uncle was visiting her family and that he was "resting his country." By this, Terry meant that he was purposefully absenting himself from the use of game, fish, and firewood resources there, allowing them to come back into balance. It is difficult to imagine a more cogent, evocative contrast between this knowledgeable yet light touch upon the land, informed by a web of kin relations, and our highly extractive, unsustainable Western ways.

Securing a better understanding of kin systems applied in Subarctic contexts is of course valuable in its own right. Implausible though it might sound, it can also be argued that a thorough understanding of kin systems as they are applied in more recent, small-scale, highly mobile societies can provide fruitful "thought models" for the constructive exploration of archaeological records connected with the first peopling of the Americas, at the end of the Pleistocene. In a world of Paleo-Indigenous[1] studies that has primarily been concerned with technological, subsistence, and environmental matters, there are signs of a shift toward a broader cultural perspective. Pitblado (2021) noted that we have "largely failed to tap into Paleo-Indigenous intellectual, emotional, and social lives," a situation that can change only by more consistently "mobilizing our own human capacity to creatively interrogate the deep past" (cf. Amick, 2017). Bradley and Collins (2013) pondered whether the 13,000-year-old Clovis phenomenon might reflect a cultural revitalization movement, focused upon ritualized hunting of Pleistocene megafauna and linking disparate though related groups. Cannon and Meltzer (2022) recently showed that optimizing models involving intensive pursuit of large game

---

[1] The term "Paleo-Indian" is deeply ingrained in the early period literature for the Americas. In agreement with Pitblado (2021) that an alternative is definitely needed, here I will use her term "Paleo-Indigenous" instead.

(the late Pleistocene megafauna), followed by population movements after local declines in prey abundance, cannot explain the rapid expansion of Indigenous ancestors in the Americas after roughly 14,000 years ago. They concluded that yet other social and demographic factors must have been at play.

Programmatic statements urging that we make a more thorough consideration of social factors in this deep time frame are a welcome change. Yet, it is one thing to urge such considerations and quite another to chart a constructive way forward. Of the many potential social factors that could be involved at the dawn of the human history of the Americas, there is one human constant that most assuredly must have been pivotal: kinship. A volume dedicated to critical intersections of natural science and traditional knowledge is a most suitable place to engage with those interesting kin dynamics and their consequences for Paleo-Indigenous socioeconomic organization, as Tallbear quite rightly encourages. One path forward would be to take seriously the topic of kinship and its patterned variability, once so central to anthropological thought, in these deeper time contexts.

There is much to be learned from the kin dimension of the visit that Terry Sawyer's uncle paid to her family. The Gwichyah Gwich'in kin terms that Terry recounted render everyone in one's generation as blood relatives (Ives *et al.*, 2010). As we will see, those kin semantics ensure that one must marry outside one's coresident group. Those exogamous marriage practices have created a vast web of kinship ties among Dene speakers of the Mackenzie Basin. I have had the privilege of listening as Dene women from distant communities, upon first meeting and as strangers to each other, explore that web of relationships, finding connections over an immense geography. I would assert that this particular form of traditional knowledge was *the* singular ingredient that made for hundreds of generations of successful life in Subarctic environments that could at times be decidedly challenging. Above all else, we can learn from examples like this that communities use kin principles in making conscious human choices, choices that Paleo-Indigenous peoples would have exercised in embarking upon one of the most extraordinary journeys in human history.

3.2 Elementary Structures of Kinship in the Americas

Kin systems reflect an intersection of two worlds: One world might be termed the biology of human reproduction and the other would be the domain of culture and society (Figure 3.1). While kinship might commonly be thought to reflect biological relatedness, kin systems in every human society are cultural constructions. Kin systems are not simply nomenclatures, but are logically interrelated sets of relationship terms in particular languages and societies. In various fashions, kin systems trace biological relatedness among consanguine or blood relatives; they are

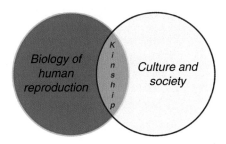

Figure 3.1 Human kin systems reflect biological reproduction, but are in every instance cultural constructions, where human biology and human social life intersect.

nevertheless also always concerned with affines, those with whom we are related by marriage. As is the case elsewhere in the world, kin systems in the Americas manifested significant, yet predictable, variability, for human minds have fixed upon a finite degree of variability in applying kin terms such that there are a more limited number of kin system alternatives.

In his massive 1871 treatise founding kinship studies, *Systems of Consanguinity and Affinity of the Human Family*, Lewis Henry Morgan (1966 [orig. 1871]) noted two polarities in kin systems – those that were descriptive in character and those that were classificatory in character. Descriptive kin systems tend to apply individual terms to each specific location on the genealogical grid of kin positions. So, for example, there might be distinct terms for "father," "father's brother," "mother's sister's husband," and "mother's brother," which could be expressed as F≠FB≠MZH≠MB (Table 3.1 shows a simple kin notation). Other kin systems are categorical, grouping different locations on the genealogical grid together (paralleling our first example, a single term might be applied, so that FB=MB=FZH=MZH). The kin system with which we are familiar in English has both qualities, lumping together FB, MB, FZH, and MZH under the term "uncle," for instance. Relatedness, in terms of descent, however, tends to radiate away from an individual ego: One has a brother and a sister and, at one remove, cousins and, at another remove, second cousins, and so forth. In English, affinal relationships apart from "husband" and "wife" are captured by "in-law" usages, such as mother- or father-in-law or brother- or sister-in-law.

The next highly influential treatment of kinship came with Lévi-Strauss' *The Elementary Structures of Kinship* (*Les structures élementaire de parenté*) first published in 1949, and subsequently available in English in 1969. Lévi-Strauss (1969 [orig. 1949]) engaged with fundamental issues surrounding logical ways to construct the origins of society. His enterprise involved examining a global range of societies in which there were positive marriage rules accompanied by unilineal descent reckoning (particularly matrilineal or patrilineal forms), societies that did

Table 3.1 *A simple kinship notation*

B	Brother
D	Daughter
F	Father
H	Husband
M	Mother
S	Son
Sp	Spouse
W	Wife
Z	Sister
=	Marries
≠	Is not the same as or must *not* marry
e	Elder
y	Younger

not have positive marriage rules but that did feature unilineal descent, and societies that had neither positive marriage rules nor unilineal descent. *The Elementary Structures of Kinship* treated binary patterns of marriage, such as the Dravidian kin systems we will consider in greater detail, but also Iroquoian and Crow-Omaha kin systems with more elaborate means of dispersing marriage alliances (see Whitely and McConvell [2021] for a current treatment of intergenerational skewing of kin terms in Crow-Omaha kin systems, with preferential matriliny or patriliny).[2]

In a most astute comment, Eggan (1980: 188) observed that "There is a whole world of social structures 'underneath' Claude Lévi-Strauss' *Elementary Structures of Kinship.*" In *Kinship and the Drum Dance*, concerning Dehcho Dene or Slavey kinship in the Mackenzie Basin, Asch (1988) stressed that Lévi-Strauss' analytical framework was actually missing a fourth logical category of considerable importance in the world of foragers, hunter-gatherers, or band societies. That other logical category would be societies that do *not* have unilineal descent but that *do* have positive marriage preferences – this we will find is true in a number of Subarctic

---

[2] Iroquoian and Crow-Omaha kin terminologies appeared when there were demographic stresses involving high population densities in North America and Australia (and elsewhere), as Whitely and McConvell (2021) and others have shown convincingly. To survey all of the variability in kin systems for the Americas would be far beyond the scope of a single chapter, particularly as more complex kin systems arose later in time in connection with higher degrees of sedentism, an increasing role of horticultural practices, and the formation of complex social systems, as exemplified by societies along the Northwest Coast, large polities like Cahokia and Chaco Canyon, and of course the civilizations of Central and South America. From the outset, I wish to be clear that the semantic frameworks provided by systems of kin terminologies can be featured in a wide range of contexts. The reconstructed Mayan kin terminology, for example, has specific similarities to the Dene and other band or forager society kin systems that we will consider, but the Mayan kin system was applied in a social setting involving one of the more spectacular civilizations of the Americas, very different from a small-scale human society (Hage, 2003). Here we will be specifically concerned with the impacts that kin semantics have on small-scale human societies.

Dene cases (and quite a number of others in North and South America and elsewhere). These societies will provide an early focus for this chapter.

3.3 Modalities in Hunter-Gatherer Group Sizes

Before moving to these Dene case studies, however, it is important to recognize that many of the kin distinctions noted earlier arose from ethnographic studies of societies with comparatively settled communities and, often, high population densities (in places such as South India or the Amazon Basin). The 1960s saw the emergence of an empirical framework for smaller scale human groups, variously referred to as hunter-gatherers, foragers, or band societies. In contrast with more populous and settled communities, these were highly mobile communities in regions with low human population densities, where they dealt with arid, cold, or other environments that presented significant seasonal challenges to their inhabitants. This "man the hunter" generation of research articulated an empirical basis for "magic numbers" of roughly 25 and 500 persons in band societies globally (Lee and DeVore, 1966; Damas, 1969). The smaller number reflected the typical size of coresident groups (local groups or microbands) using the landscape over most of the year, whereas the larger number reflected the population of all the smaller groups in a larger regional entity – a regional group, macroband, or regional marriage isolate, to use various terms from the literature. Regional groups could gather only seasonally, when food resources permitted.

There have since been a number of more sophisticated elaborations of this basic distinction (e.g., Binford, 2001; Hamilton *et al.*, 2007; Hill *et al.*, 2011). Binford's (2001: 213) scheme, for example, featured three group sizes: Group 1 he defined as "the social unit camping together during the most dispersed phase of the settlement-subsistence system," Group 2 comprised "the camp-sharing groups during the most aggregated phase of the subsistence settlement system," and Group 3 involved "social aggregations occurring annually or every several years that assemble for other reasons than strictly subsistence-related activities." While there can be these finer distinctions, a "microband–macroband" dichotomy (or variants thereof) has great value in helping us to understand hunter-gatherer socioeconomic organization in a wide variety of circumstances for foraging societies. There is a direct relationship between small-scale populations and prospects for founder events. Tournebize *et al.* (2022) found that, in over half of the human populations they analyzed, there was evidence for recent founder events, associated with geographic isolation, modes of sustenance, or cultural practices such as endogamy.

In essence, the microband–macroband dichotomy can be construed as a formula for both socioeconomic *and* demographic success. In many of the biomes that foragers inhabit, it is simply not possible for large aggregations to persist over long periods of time. For a Subarctic example, think of how widely dispersed moose populations are in boreal forest settings. Outside cow–calf pairs, rutting, or the occasional winter phenomenon of "yarding," success in securing one moose will mean that any other moose is some distance away. A moose kill is a significant resource for a group of roughly twenty-five persons, but it cannot feed a group as large as several hundred persons for long. Smaller groups or microbands are well suited to securing and sharing boreal forest food resources, circulating widely in hunting and fishing activities in the dispersed subsistence-settlement system phase to which Binford referred.

Still thinking on this empirical plane, it is important to note that the ethnographic literature shows time again that these smaller local groups or microbands have fluid membership, but are not simply random assortments of individuals. Whether we speak of Arctic, Subarctic, or Great Basin examples in North America, or even globally, we invariably find that there is a nexus of kin relationships at the heart of these smaller groups. There is a strong Western inclination to think in terms of individuals, conjugal pairs, and family units. Seen from Indigenous perspectives, however, there is usually a "sibling core" to such groups, with various central arrangements of brothers and sisters (or, as we shall see, classificatory relatives who are in sibling relationships) about which microbands tend to form (see, e.g., Ives, 1998; Hill *et al.*, 2011).

Several simulation studies using actual hunter-gatherer demographic data have concluded that the small population sizes typical of microbands are highly suscep- tible to stochastic variability in birth events: Model results indicate that group sizes in the range of twenty-five to seventy-five persons do not persist beyond two to four centuries at best, and usually less (e.g., Weiss, 1973; Wobst, 1974; Ammerman, 1975; Anderson and Gillam, 2000; Moore, 2001).[3] This is a matter of chance alone, never mind the other kinds of challenges that small groups might face. Consequently, microbands require access to an entity the size of the macroband – the more aggregated phase of a subsistence-settlement system. The several hundred individuals in the macroband are far less susceptible to these purely stochastic effects. The microband is the hunter-gatherer entity that manages economic needs throughout the majority of a year; the macroband exists in a demographic range that more readily allows for the persistence of a regional population. Macrobands gather when resources permit – when major fish spawns or migrating caribou pass key locales in Subarctic settings, for example. Then, friendships can be renewed,

---

[3] Moore (2001) showed how variable gender can be in strings of births in band societies, where the occasional predominance of sisters within a family provides considerable capacity for having children, while, correspondingly, there can be strings where brothers predominate in a family, with little or no capacity for having children.

ceremonies can be conducted, information about game resources can be shared, and spouses may be found.

This dichotomy in group sizes is not a recent development in human history. It is very likely that microband–macroband dynamics were a key element in the emergence of anatomically and culturally modern human populations in high-latitude settings during late or Upper Paleolithic time frames. Along with authors such as Whallon (1989, 2006) and Soffer (1994), I would argue that it was this basic, highly adaptable demographic framework that allowed human populations to expand into all of the world's environments, including those that were most challenging. Viewed from the perspective of the peopling of the Americas, this is vital: Late Pleistocene conditions in Northeast Asia, particularly during the Late Glacial Maximum (LGM, roughly 24,000 to 18,000 years ago), required that Indigenous ancestors pass through an arctic "filter" requiring high-quality tailored clothing, immaculate fire-making skills, and sophisticated lithic and organic technologies. Critically, they would have had that adaptable band society demographic framework that allowed them to thrive in difficult, cold, open, high-latitude environments that archaic hominins like Neanderthals and Denisovans never fully solved.

These last factors appropriately continue to influence archaeological and genetic thinking, but the semantic dimensions and group-forming principles noted in the previous section have received far less attention. Those semantic dimensions are in part anthropological constructs, but they can be much better informed by taking into account the traditional precepts that Indigenous peoples themselves have applied in ensuring the intergenerational reproduction of their societies. Whiteley and McConvell (2021) have suggested that kin systems can be regarded as flexible "social technologies" capable of creating advantageous outcomes. While there is a danger of shifting the analysis in an overtly functionalist direction, there could be no more instrumental role for a kin system than that of determining the social and geographic locus of the incest taboo in any given society. All of these perspectives can be woven together, so that we can now move to some informative, Subarctic Dene cases. Marriage patterns are very much under the conscious control of knowledgeable persons in those societies in ways that would have been equally apparent to the ancestors of the founding Indigenous populations of the Americas many thousands of years before.

3.4 Two Subarctic Dene Cases

Two Subarctic Dene cases provide important insights into how kin semantics and principles of group formation are intimately related to socioeconomic strategies with critical ecological implications. The Dehcho Dene (South Slavey) of the Northwest Territories and the Dunne-za (Beaver Dene) of northeastern British

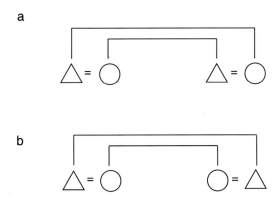

Figure 3.2 (a) An unlike or opposite-sex sibling core in which a brother and sister have married another sister and brother. (b) A like or same-sex sibling core in which two brothers have married two sisters. Apart from that salient difference, the two sibling cores have the same empirical attributes.

Columbia, as reflected in excellent ethnographic studies by Michael Asch (1980, 1988, 1998) and Robin Ridington (1968a, b, 1969), explored in different ways an underlying theme that can be traced back to the mid-Holocene origins of the Athapaskan (Dene)-Eyak-Tlingit language family of northwestern North America (Ives, 1990, 1998, 2022; Ives *et al.*, 2010).[4] As mentioned previously, Western-oriented frames of reference have typically involved analyses of conjugal pairs or nuclear families (e.g., MacNeish, 1960; Helm, 1961, 1965). Yet, as noted earlier, there are frequently sibling sets at the core of coresident local groups in foraging societies and, as we shall see, arrangements of sibling sets are integral to Dene conceptions of those groups.

In order to understand how this can be so, consider the two sibling sets in Figure 3.2; in (a), a brother and a sister have married a sister and a brother and, in (b), two brothers have married two sisters. If we take a strictly empirical approach, there would be little to choose between the two situations. There are two men and two women, each person has one sibling and one in-law, and so on. How could there be any meaningful difference between the two group compositions with regard to Subarctic life? It is here, however, that the cultural construction of kin systems asserts itself. Unlike previous scholars conducting Mackenzie Basin Dene ethnographies, Asch (1980) recognized that the semantic structure of the Slavey kin

---

[4] During the last two millennia, Dene ancestors have spread widely in western North America from an Alaska-Yukon homeland first to Pacific Coast settings extending from Washington state to northern California and, later, to the American Southwest and southern Plains. In doing so, they would undertake significant variations on this kinship theme, variously adopting clans and Crow-like kin conventions in some circumstances and alternative strategies in other cases, but adapting kin principles creatively in the different sociopolitical circumstances they encountered (Ives, 2022).

terminology closely paralleled that of what would be called Dravidian kin systems, so named because their formal properties were first recognized in South India (Dumont, 1953; Trautmann, 1981).

The underlying logic revealed in Dravidian kin terminologies arises directly from a framework for bilateral or symmetrical cross-cousin marriage. This form of marital exchange is illustrated in Figure 3.3(a), in which its classificatory properties are evidenced in terminological equations (like MB=FZH=WF, MBD=FZD=BW=W or WZ, MZD=FBD=WBW=Z, or ZD=SW) that would flow from intergenerational patterns of cross-cousin marriage. In this example, a male ego has married a person who is at once his MBD *and* FZD. In rudimentary terms, cross-relatives are those to whom you are linked by intervening relatives of the opposite sex. Parallel relatives are those to whom you are linked by intervening relatives of the same sex.

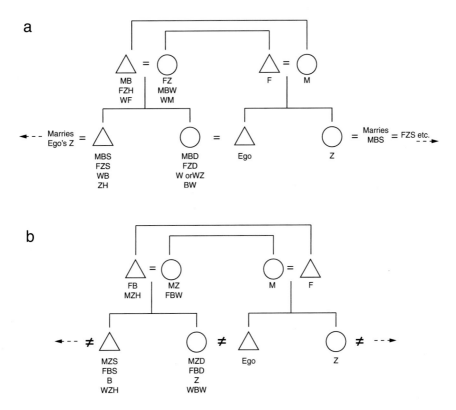

Figure 3.3 The same two sibling cores as in Figure 3.2, but with the next generations of children and a kin terminology applying the principles of bilateral or symmetrical cross-cousin marriage, with characteristic terminological equations. (a) The cross-cousin children of an unlike-sex sibling core can marry. (b) The parallel cousin children of a like-sex sibling core must not marry, as they are regarded as siblings.

Viewed from this more culturally appropriate perspective, we can see that the two sibling core compositions expanded in Figure 3.3, and derived from the same underlying semantic framework, send powerfully different messages: When a brother and sister have married another sister and brother, the children of those marriages will be cross-cousins – permitted or in many cases desirable marriage partners. This opposite- or unlike-sex sibling core composition signals that endogamous marriages *can* take place or may actually be the intended outcome, even at this tiny group size. When two brothers have married two sisters, the children of those marriages will be termed (and will treat each other) as siblings who *must not* marry. This same-sex sibling core is a way of asserting the need for a local group or microband to be exogamous in its marriage practices – and literally to reach out into the regional group domain in creating affinal alliances.

We can now turn to the two case studies to see how these different strategies were fully realized in Subarctic environments. Asch's Dehcho Dene (South Slavey) work was informed by the detailed knowledge of Jessica Hardisty, who provided a clear account of the kin terminology used in Pehdzéh Kı̨ (Wrigley, Northwest Territories). Asch's (1980, 1988) rendering of the kin terminology she provided is given in Figure 3.4, which uses a format developed by Trautmann (1981) for laying out a Dravidian kin system. These kin systems make categorical distinctions of relatives in a classificatory mode, allowing "generation" and "elder versus younger" relationships to be captured in rows, while distinguishing male and female kin persons as well as cross- and parallel categories in columns. Many of the diagnostic "equations" we expect of a Dravidian kin terminology are present in Figure 3.4.

Jessica Hardisty also explained that the Slavey local group of which she was a part conceived of itself as a group of brothers who had married a group of sisters. This conceptualization involving a sibling core is much more culturally appropriate than sociological constructs such as conjugal pairs. Here, it is important to recognize, as Trautmann (1981) did, that Dravidianate kin distinctions are "tyrannical." Once a person assumes a cross- or parallel status with respect to one individual, that logic extends to that individual's entire social setting, ordering everyone as cross or parallel, even for more distant kin persons. For example, in a Dravidian framework (or Type A crossness, as Trautmann also called this pattern when he sought a more neutral term), second cousins such as MMBDD or FFZSD will receive either cross-cousin or sibling terms, respectively. This includes *fictive* kinship, whereby kin statuses can be ascribed to those who are not initially either consanguine or affinal relatives. Conceptualizing the entire coresident group as having a like-sex sibling core then leaves the terminological structure intact, but continues to enforce strict local group exogamy.

In the real world, ideal circumstances do not always obtain. There are Slavey situations in which there are unlike-sex sibling cores, where brothers and sisters and

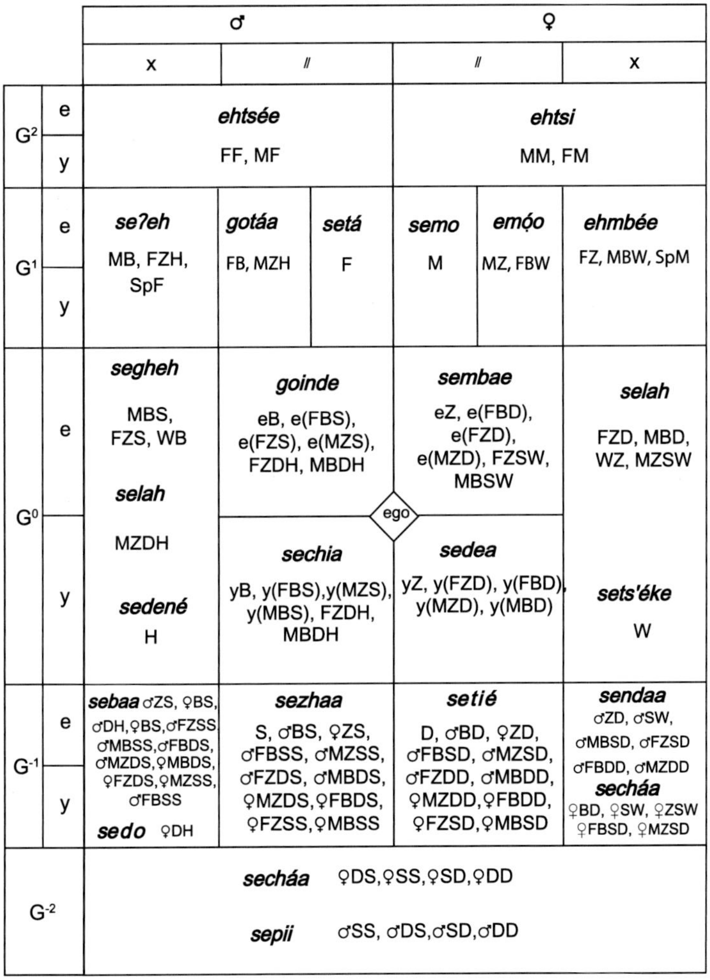

		♂		♀	
		x	ǁ	ǁ	x
G²	e y	*ehtsée* FF, MF		*ehtsi* MM, FM	
G¹	e	*se?eh* MB, FZH, SpF	*gotáa* FB, MZH *setá* F	*semo* M *emǫó* MZ, FBW	*ehmbée* FZ, MBW, SpM
	y				
G⁰	e	*segheh* MBS, FZS, WB *selah* MZDH	*goinde* eB, e(FBS), e(FZS), e(MZS), FZDH, MBDH	*sembae* eZ, e(FBD), e(FZD), e(MZD), FZSW, MBSW	*selah* FZD, MBD, WZ, MZSW
	y	*sedené* H	*sechia* yB, y(FBS), y(MZS), y(MBS), FZDH, MBDH	*sedea* yZ, y(FZD), y(FBD), y(MZD), y(MBD)	*sets'éke* W
G⁻¹	e	*sebaa* ♂ZS, ♀BS, ♂DH,♀BS,♂FZSS ♂MBSS,♂FBDS, ♂MZDS,♀MBDS, ♀FZDS,♀MZSS, ♂FBSS	*sezhaa* S, ♂BS, ♀ZS, ♂FBSS, ♂MZSS, ♂FZDS, ♂MBDS, ♀MZDS,♀FBDS, ♀FZSS,♀MBSS	*setié* D, ♂BD, ♀ZD, ♂FBSD, ♂MZSD, ♂FZDD, ♂MBDD, ♀MZDD,♀FBDD, ♀FZSD,♀MBSD	*sendaa* ♂ZD, ♂SW, ♂MBSD, ♂FZSD ♂FBDD, ♂MZDD
	y	*sedo* ♀DH			*secháa* ♀BD, ♀SW, ♀ZSW ♀FBSD, ♀MZSD
G⁻²		*secháa* ♀DS,♀SS,♀SD,♀DD			
		sepii ♂SS, ♂DS,♂SD,♂DD			

Figure 3.4 The Dehcho Dene or Wrigley Slavey kin terminology that Jessica Hardisty explained to Michael Asch. It features characteristic Dravidian terminological equations throughout the medial generations.

other brothers and sisters (real, classificatory, or even fictive) are present. In these cases, the overriding ideology remains one of local group exogamy. That imperative is so strong that the form of the kin terminology itself is shifted. "Mackenzie Basin" kin terminologies, as Spier (1925) styled them, feature the hallmark equations of a cross–parallel semantic framework in the first ascending (e.g., MB=FZH, FB=MZH) and first descending generations from ego. In ego's own generation, however, *all* siblings and cousins (whether cross or parallel) are termed siblings, again assuring exogamy. Such a framework continues to value the uniting of sibling sets in forming local groups, but equally reveals a strong insistence on subsequent generation exogamy. This is actually quite a common pattern in the Mackenzie Dene world and elsewhere in North America. It is not, as has sometimes been suggested, the consequence of epidemic disease or the interference of missionary figures. In a remarkably complete, assiduously collected kin terminology that Robert Kennicott gathered as part of Morgan's global outreach campaign, we see the same pattern of extinguishing crossness in ego's generation. Kennicott insisted upon consulting with older, well-informed individuals, so that his mid-nineteenth schedule would reflect Fort Liard Slavey kin usages typical of the early nineteenth century and perhaps even the later eighteenth century (Ives, submitted).

Dunne-za or Beaver Dene kinship as explored by Ridington (1968a, b, 1969) uses the same Dravidian semantic framework, but takes the opposite tack. There, opposite- or unlike-sex sibling cores are favored and marriages *do* take place within the local group, making them agamous (neutral with respect to endogamy or exogamy) or distinctly endogamous. In this case, prospective marriage partners can be systematically retained with a local group. Here we see an important contrast that is very much a deliberate choice, drawing the boundary for incest closer in one case, within the coresident local group, or pushing that boundary outward, creating a web of external alliances, in the other case.

Before leaving this passage, it is important to note both the intricacy and significance of kin-related knowledge. While it is relatively straightforward to reconnoiter the consequences of Dravidian-type kin distinctions in a person's proximate range of relatives, cross–parallel reckoning intricacies ramify quickly. For instance, for a male ego, the second cousin MMBDD is a cross-relative and a prospective marriage partner. MMZDD, on the other hand, is a parallel relative who will be referred to by a sibling term and would not be marriageable. Tjon Sie Fat (1998) applied associative matrices to capture that complexity in algebraic equations such as the following:

$$\male M(BD) = \male C_{\neq}^{+1}\left(C_{\neq}{}^{\circ}C_{\neq}{}^{-1}\right) = {}_{\male}C_{=}{}^{+1}A_{\neq}{}^{-1} = \male A_{\neq}{}^{\circ}$$

The elements of this equation can be read as follows: for a man speaking, the intergenerational associative bracketings (in which "C" represents a parallel relative and "A" a cross-relative) yield the result that MBD is a zero-generation affine.

For many anthropologists and for virtually all archaeologists, when other mathematically sophisticated models have been applied (such as for optimal foraging), this essential traditional knowledge has gone unnoticed or at best is regarded as esoteric. Such knowledge is not evenly distributed across a society. Asch (1998: 145) emphasized that it was senior Dene women in particular who used this information and its theoretical framework to construct the social universe, providing a significant source of their political power – a power conventionally withheld from men.

3.5 Principles of Group Formation and Developmental Processes in Subarctic Settings

Back then up there, Native women gave birth to not just one, two or three,
but twelve, fourteen even eighteen children . . . For us northerners of
a certain epoch, it is par for the course, tradition, part of the culture.
(Highway, 2022: 14)

In order to see how kin semantics are enmeshed in matters of economic and ecological significance, we will need to consider both the practicalities of living in the boreal forest ecosystems and a broader range of group-forming principles. While capable of providing plenty, boreal forest environments can nevertheless be challenging. Those challenges come in three principal forms: (1) There are serious constraints over human mobility, (2) food resources are widely dispersed, and (3) food resources fluctuate both regularly and unpredictably over the course of the year and at longer intervals (for a more comprehensive treatment of these factors, see Ives, 1990).

Subarctic Dene and Algonquian ancestors met these challenges in a variety of ways. Deeper, softer boreal forest snows, extensive muskeg, and vast lake and waterway systems created genuine mobility issues that were met with effective, naturally sustained, traditional technologies including snowshoes, toboggans, and birch bark canoes. Subarctic societies were indeed highly mobile, in terms of both frequent moves of residential encampments and wide-ranging subsistence practices (see Ives, 1990, for a broader treatment). These high degrees of mobility over vast ranges were matched with groups that had extremely low population densities.

Apart from seasonally available berries, plant food sources in the boreal forest are severely limited, so successful human subsistence relied heavily upon large and small game and fish resources, but these were not uniformly available for a variety of reasons. Game animal resources fluctuated critically through time, perhaps most famously illustrated by the snowshoe hare–lynx population cycle, in which hare populations, tracked by their principal predator (the lynx), regularly rise and fall dramatically at an interval of roughly a decade. The hares themselves were an important small game food source, but, even more than that, the hare–lynx cycle had resonances throughout the entire food web of the boreal forest, triggering population cycles in other animals (such as ruffed grouse and muskrats) through effects ranging from overbrowsing to predator–prey shifting.

Large game animals such as caribou and moose could be the subject of population shifts over decades and, even more critically, these animals did not have equivalent food value throughout the entire year. Late winter and early spring months typically found large game animals of both genders in very poor condition, having variously passed through rut, carrying calves, and beginning to nurse – all high physiological demands leading to a time of the year when food sources scarcely meet the basic metabolic needs of game animals. Known in the bush as "rabbit starvation," extremely lean meat costs the human body energy to digest, so that no matter how much fat-poor protein might be consumed, one remains unnourished (e.g., Speth and Spielman, 1983; Speth, 2010). Subarctic populations had to deal with these difficult late-winter and early-spring conditions on an annual basis, further hampered by limits to mobility from ice break-up and melting snows.

As noted previously, one key ungulate species, the moose, is typically widely dispersed. Other key resources, such as woodland caribou and wood bison, occur in more widely dispersed and much smaller groups than their barren-ground and plains relatives. A key food source in offsetting the "rabbit starvation" problem would be the beaver, whose seasonal energy regime is quite different, resulting in beavers remaining fat during the winter (spring break-up being their energetically demanding time of year as they emerge from lodges and repair dams). Yet, beavers are also spaced periodically along water courses. To deal with these challenges, Subarctic Dene and Algonquian peoples matched high degrees of mobility with vast traditional ranges for which there were extremely low population densities. Most vitally of all, these groups were linked by a web of kin ties that provided people with economic options in times of both abundance and scarcity. It is here that we begin to see how the underlying kin semantics that we have been discussing can be pressed in one direction or another, in ways that were within the conscious control of different societies.

Many studies of foraging societies have tried to show how group sizes have stayed roughly the same in essentially a synchronous perspective. I find it more productive to take a longitudinal perspective, considering how local groups within a regional group form, persist for various periods of time, and, in many cases, eventually dissolve. Subarctic local groups truly can begin with groups as tiny as those ideal models in Figure 3.2, but they trigger differing developmental processes (see Ives, 1990, 1998, for a broader treatment of these factors). Ridington's (1968a, b, 1969) ethnography showed that some small groups with unlike-sex sibling cores can set out to make a living. The developmental trajectory of any small group will be determined by the balance in those stochastic forces involved in natality and mortality, as well as recruitment or loss of other personnel. Imagine that a small group of brothers and sisters married to other sisters and brothers enjoys real hunting success. Others may seek to join this successful group, in which a Dravidian kin terminology will quickly order everyone's cross or parallel status. Subarctic ethnographies do note instances of many children being born to a couple, as Tomson Highway (2022) reflected in his autobiography.

We can further imagine circumstances, then, in which a small local group begins to grow. What next happens as their children mature when there has been an unlike-sex sibling core? Many of those children will be in a cross-cousin relationship and would therefore be legitimate potential spouses. Those children and their partners can thus be retained within the local group, causing further growth in numbers. Ridington (1968a) used genealogical and historical data to show how, in some circumstances, an originally small coresident group could grow so large as to constitute the entire regional group. Of course, Subarctic environments and other cultural factors do not allow limitless growth. Difficult seasons or even entire years of challenging conditions, as well as the death of founding members of a group, act to diminish numbers. There is also the idea of scalar stress, suggesting that, when the size of coresident groups increases, tensions and frictions among its members can emerge (e.g., Johnston, 1982). In the Beaver Dene world, Ridington (1968b) spoke of "medicine fights," in which spiritually powerful individuals could come into conflict with each other. These and related factors could lead to group fissioning – in the case of medicine fights, propelling groups apart.

When principles of group formation favor exogamy, different developmental pathways would be followed out. A small group of brothers married to sisters – or at least a group in which coresident children are regarded as siblings who cannot marry – can also start up, and it too could enjoy success in the numbers of children born, perhaps encouraging recruitment. That group could grow, but in contrast with our endogamous alternative, something different will take place. Those maturing children cannot marry each other and are enjoined to find spouses elsewhere. For the Dehcho Dene, Asch (1980) found a pattern he termed "unilocality." If a brother married outside and left the group, then the other brothers would do so too. The sisters within the group would tend then to have male spouses from outside marry into their group. The pattern would be reversed if it was sisters who had married out. As founding members of the group aged, daughters tended to care for them; therefore, those founding members too might have left if the pattern of unilocality was one in which sisters were the ones marrying outside. Triggered by a significant proportion of younger group members marrying outside the group, these several factors were self-limiting for local groups. These groups therefore cycled in a smaller size range or in fact tended to disappear in a generation or two. By this, I don't mean that an entire populace would disappear, but rather that a particular group might cycle into and out of existence while yet other groups formed and began the first parts of the developmental process again.

It might be argued that perhaps these kin-structured, differing principles of group formation were somehow the consequence of latter-day colonial processes that did indeed disrupt traditional lifeways (cf. Asch, 1988). This does not appear to be the case. For entirely different, business-related purposes, fur trade personnel kept detailed account records, noted the size of trading parties coming into the post, and actually undertook region-wide censuses in the 1820s, enumerating the men in local groups trading into a post, as well as their spouses and dependents (Ives, 1990). Using post records for Beaver and Slavey Dene populations between 1786 and 1845, it is possible to show in statistically significant terms that Beaver local groups were larger on average and had a greater range of sizes than were contemporary Slavey groups, which were smaller on average and occupied a narrower size range (Ives, 1990).

As with any human choices, there were advantages and disadvantages to each of these alternatives. A very small endogamous group, such as our ideal model in Figure 3.2(a) would be constrained to be boreal forest foragers – highly mobile, seeking a broader range of large and small game as well as fish resources. Should that group grow into several dozen or even more than 100 people, other economic alternatives would emerge. One of these would be to engage in communal hunting, as Beaver Dene ancestors did with respect to Peace River Basin wood bison (Goddard, 1916). In this size range, there would be sufficient personnel to execute a communal hunting strategy and, when successful, to undertake the intensive butchering, meat preparation, and hide preparation that must quickly follow a large kill. On the other hand, boreal forest settings, while often amply providing for people, could at times impose severe constraints. Factors such as the many ecological resonances of the hare–lynx cycle, the deleterious impacts of rain-crusted snow for ungulates when wolves can have great impacts, or situations such as failing to get winter fish nets set because early, intense cold caused lake ice to freeze to great depths all could result in hardship and starvation. When marriage ties were focused internally, fewer external ties limited avenues where help could be sought. One can also see that, in settings west of the Rockies, where Pacific salmon are an important resource, the ability of a group to "fix" a valuable location for weirs or other forms of fishing by perpetuating internal marriage alliances could provide other compelling reasons for group-forming principles that favored endogamous marriages.

In the exogamous alternative, external ties were actively promoted, at times creating vast webs of kinship. Should hardship strike, these ties could be activated to seek assistance elsewhere. For many Dene people, for example, intercepting barren-ground caribou as they made their seasonal movements across the treeline provided an effective lifestyle. At times, however, the particular narrows, pennisu-lae, or river crossings where caribou herds had normally crossed could suddenly shift. That broad web of kin ties provided a socioeconomic "safety net." This was actually the more common approach for many Mackenzie Basin Dene peoples – a kin-structured exogamous strategy that made for ten millennia of successful life in the Subarctic world. One can also begin to see the ecological implications of such a strategy, with small local groups forming, persisting for one or a few generations, and then disappearing, freeing up a significant portion of the boreal forest landscape that could later again become the traditional use area for another group that would form. And this would also explain on a deeper level what Terry Sawyer's uncle was doing when he was "resting his country." He was purposefully lightening his impact in his area of traditional use by visiting his relatives.

This is a simplified treatment of complex topics with many subtle distinctions. That being said, similar themes were explored not just in Dene societies but in High Arctic Inuit societies, in eastern Subarctic Algonquian societies, and in the Numic societies of the Great Basin (cf. Stevenson, 1997; Ives, 1998). Apart from the great value in understanding these particular instances, the foregoing passages show the extent to which small-scale societies invariably make conscious choices about socioeconomic strategies and demographic arrangements – choices that are kin-structured. This can be an area in which scientific and Indigenous knowledge flows together to illuminate our understanding of how Indigenous peoples entered the Americas so many millennia ago in one of the most extraordinary chapters of human history. Let us now turn for a moment to the current archaeological and genetic evidence.

3.6 Setting the Stage for Founding Indigenous Populations of the Americas

As Binford (1962) observed six decades ago, we will never excavate a kin system. Yet, in developments that were difficult to foresee six decades ago, ancient DNA (aDNA) studies now allow empirical tracking of the genetic proximity of hominin mating. Such findings document the degree of exogamy or endogamy taking place. Current evidence indicates that the modern human incursion into southern Europe began some 54,000 years ago, with France's Mandrin cave sequence showing alternating Neanderthal and *Homo sapiens* occupations from earlier and for a longer period of time (until ~41,000 years ago) than had previously been known (Slimak *et al.*, 2022). Ultimately, it would be anatomically and culturally modern humans who would persist in Eurasia, with a trace of Neanderthal heritage. After 50,000 years ago, modern human populations began to probe high-latitude Arctic environments in Siberia – environments characterized not simply by severe cold, but equally by vast open environments. Until the advent of Holocene conditions, Siberian *Homo sapiens* populations (with shifting genetic affinities) responded to the ebb and flow of glacial conditions, tending to be pushed southward at the height of glacial periods and responding with northward movements when glacial conditions abated (e.g., Pitulko *et al.*, 2016).

It is worth considering for a moment why it would be that earlier hominins (Neanderthals and Denisovans) apparently failed to enter the Americas.

Lalueza-Fox *et al.* (2011) have shown through mitochondrial (mtDNA) and Y chromosome studies that the El Sidron (Asturias, Spain) Neanderthal males carried the same mtDNA (maternal) lineages, while El Sidron Neanderthal females did not, implying that this small population was a patrilocal one in which closely related males remained together whereas women circulated more widely.

Prüfer *et al.* (2014) provided another extraordinary insight into archaic hominin life through genomic sequencing of a Neanderthal woman from Denisova Cave in the Altai Mountains. There, they reported a high coefficient of Neanderthal inbreeding, a finding that might tell us something about the transition to anatomical and cultural modernity affecting our own species. Prufer *et al.* showed that long runs of homozygosity in the Altai Neanderthal woman would be consistent with unions between double first cousins, an uncle and a niece, an aunt and a nephew, a grandfather and a granddaughter, or a grandmother and a grandson. Given typical hominin life spans in this time range, the final two (grandkin) scenarios are logical but unlikely possibilities. The other options have intriguing ramifications if we shift our perspective from one of biological mating to one involving kin concepts. The double first cousins scenario diagrammed by Prufer *et al.* (2014: their figure 3.3b) features two sisters who have married two brothers (i.e., same-sex siblings whose children would be *parallel cousins* should that logic have prevailed at the time). The same genetic complement for the Neanderthal female studied would be achieved, however, if the first cousins resulted from unions between a brother and a sister and another sister and a brother (i.e., opposite-sex siblings whose children were *cross-cousins*, should that logic have applied).

These findings have been dramatically reinforced by an instance of a Neanderthal family, quite literally, in Chagyrskaya Cave – a bison- and horse-hunting camp occupied 50,000 to 60,000 years ago in the western Altai of Siberia (Skov *et al.*, 2022). In sequencing eleven Neanderthal individuals from Chagyrskaya (as well as two from nearby Okladnikov Cave), Skov *et al.* (2022) concluded that, for Chagyrskaya, a father and daughter were represented. Two other adult males from the cave came from the same maternal lineage, while another male and female were second-degree relatives, likely cousins. All of the genomes of the Chagyrskaya Neanderthals had low diversity between maternal and paternal copies indicating that the interconnected population of breeding adults was low. The pattern of patrilocality existed here too, as males were very closely related, whereas females were not. The best-fitting scenarios suggested a community size of twenty individuals in which more than half of women were born elsewhere. The shared heteroplasmy between individuals Chagyrskaya C and D indicated that at least some females remained with the group they were born in; they would be closely related to other small group members. The presence of one baby tooth and two lightly worn permanent teeth belonging to the same adolescent male was intriguing, suggesting that these Neanderthals tended to either stay in place or return to the same place frequently.

Scenarios in which cousins could be involved are not innocent ones from the perspective of modern human kin semantics. Many kin systems make a distinction between cross- and parallel relatives. Societies preferring marriages involving the children of same-sex sibling sets are rare.[5] As we have seen already in Dravidian kin systems, the semantic mapping of cross–parallel distinctions precisely parallels discriminations between consanguines (blood relatives who may not be married) and affines (in-laws or spouses). When these kin distinctions hold, the children of same-sex sibling sets are terminological siblings themselves, invariably the subject of incest taboos. The children of opposite-sex sibling sets are distinguished as cross-cousins and are potential marriage partners.

Allen (1998, 2008) has argued that a simple tetradic form of cross–parallel distinctions, involving eight kin terms, constitutes an archetypal[6] kin system that anatomical moderns possessed upon leaving Africa. It could be that Neanderthal genetic findings in the 50,000–120,000-year time range reflect declining populations already facing significant exigencies, resulting in very close matings. On the other hand, we have consistently underestimated the cognitive capacities of Neanderthals for symbolic behavior including rock art, mortuary activities, and speech. Could it be that Neanderthals had the speech and cognitive capacities to manage simple kin systems like Allen's hypothetical tetradic configuration? If that were the case, then relatively close marriages of cross-cousins could reflect *purposeful* efforts to retain potential spouses within more highly endogamous populations (in low population density settings in which it might be uncertain where spouses could be found). Whatever the case may be (and continued aDNA recovery and analysis would expand upon currently tiny samples), at least some Neanderthal mating practices must have ranged even farther afield. In Denisova Cave, a hominin digit was discovered and the aDNA revealed a previously unknown population, called "Denisovan." Slon *et al.* (2018) discussed a Denisova Cave instance in which a first-generation female was the result of mating between a Denisovan father and a Neanderthal mother, indicating that occasional Neanderthal–Denisovan interbreeding did in fact occur.

Soffer (1994) cogently noted that Middle Paleolithic lifeways such as those favored by Neanderthals generally had proximity to rugged but altitudinally diverse habitats where varied resources were comparatively near each other. Neither Neanderthal nor Denisovan populations thrived in the cold and open environments

---

[5] Father's brother's daughter marriages in the Muslim world provide a notable exception.
[6] Archetypal in that logical semantic shifts in the eight tetradic terms can be used to generate virtually all semantic forms of kinship, whether classificatory or descriptive in nature.

of Siberia that would set the stage for Indigenous population movements into the Western hemisphere across the Bering Land Bridge. Such environments create scheduling and geographic conflicts that require considerable forward planning and fallback strategies. A variety of technological innovations, such as the capacity to produce finely tailored clothing and footwear, elaborate osseous (i.e., bone, antler, horn, and ivory) weaponry, and clever passive devices like snares were undoubtedly connected with the successful expansion of human populations into such regions, along with the cognitive capacities to manage logistical and scheduling issues in those open environments (e.g., Hoffecker and Hoffecker, 2017). It is nevertheless appropriate to ask if social factors might have had even more profound effects in contending with these late Pleistocene Siberian and Beringian environments.

There is a striking Upper Paleolithic contrast with our impression of seemingly patrilocal, endogamous Neanderthal settings. The remarkable Sunghir burials in Russia involved an adult male (Sunghir 1) and another two sub-adults (Sunghirs 2 and 3) interred head-to-head. These remains were covered in ochre and accompanied by rich grave goods including hundreds of ivory beads as well as spears, armbands, carvings, and Arctic fox canines. Associated, less complete human remains (Sunghirs 5 to 9) were radiocarbon dated to ages of ~34,000 years. Some of these individuals bear a close genetic relationship with individuals from the remarkable Kostenki populations inhabiting the Central Russian Plain mammoth bone structures ~24,000 years ago (e.g., Soffer, 1985, 1994; Pryor *et al.*, 2020). Intriguingly, Sikora *et al.* (2017) found that, despite their interment together, the Sunghir individuals were not closely related. The authors concluded that their findings were consistent with a social and population network of demes that preferentially mated within subgroups, *with exogamy and regular exchanges between demes* (Sikora *et al.*, 2017: 662, my emphasis).

Sikora *et al.* (2017: 663) went on to say that, were these findings to be representative of early Upper Paleolithic hunter-gatherers more generally, they would "reveal a social structure and cultural practices that *emphasized exogamy*" (my emphasis). Vanhaeren and d'Errico (2006) used patterning in forms of personal adornment across Aurignacian Europe (after 36,500 years ago) to suggest that an ethnolinguistic geography of Europe existed by then – a geography that would very likely be founded upon familiar hunter-gatherer marriage practices that would regulate kin-structured strategies surrounding relative degrees of endogamy, agamy, or exogamy.

Moving farther eastward in Siberia, the Yana RHS site has a record beginning 33,500–31,000 years ago (Pitulko *et al.*, 2013). It is situated above 71 degrees latitude, showing that modern human populations were capable of operating far to the north by then. The lithic industry had a relatively straightforward stone tool inventory involving flakes, but other aspects of the Yana record are remarkable. These include spectacular mammoth ivory and woolly rhinoceros horn weaponry, a variety of beads and other items of adornments, as well as needles and bird-bone needle cases – all indicative of sophisticated Arctic capacities. Two deciduous human teeth from Yana RHS belonged to a population that Sikora *et al.* (2019) termed Ancient North Siberians, with some East Asian admixture, but closer ties to West Eurasian populations at that time. The deciduous teeth were contemporaneous, but came from individuals who were not closely related. Once again, Sikora *et al.* (2019) found the two Yana individuals did not exhibit signatures of recent inbreeding and came from an effective-population size of up to 500 individuals. The authors concluded that this second finding "reinforced the view that wide-ranging mate exchange networks were present among Upper Paleolithic foragers across the pre-LGM landscape" (Sikora *et al.*, 2019: 184).

Here then is a case in which there probably *was* a salient social difference between archaic hominins (i.e., Neanderthals and Denisovans) and anatomical and cultural moderns (i.e., *Homo sapiens*), with Neanderthal populations either purposefully or through exigent circumstances retaining potential mates close at hand through endogamous practices. Those archaic hominin populations would be prone to both purely stochastic effects and any of the significant challenges that Pleistocene environments could feature. While there is important debate regarding the degree of Neanderthal carnivory, a high degree of reliance on fluctuating terrestrial game populations could be one such Pleistocene risk, as the Chagyskaya setting might suggest (cf. Jaouen *et al.*, 2022; Skov *et al.*, 2022). In contrast, sophisticated Upper Paleolithic societies across northern Eurasia apparently managed exogamous marriage practices. At this most elementary level, the "social technology" of exogamous marriage practices would have been critical to meeting the demographic challenges of high-latitude worlds. While there, in all likelihood, were strong incentives for modern human movement toward the Western hemisphere – namely diverse and abundant megafaunal game and other resources – ancestral Indigenous groups would have existed in extremely low-density populations, a topic to which we now turn.

3.7 Founding Indigenous Populations of the Americas

Any populations proceeding eastward from Siberia as the LGM approached were entering a vast cul-de-sac, defined by glacial geography, representing the edge of the known human world. There is every reason to keep these unusual conditions foremost in our minds: No other human populations lay ahead of those making this journey.

The onset of the LGM was severe enough to drive people southward in Siberia toward regions with less severe conditions: the Lake Baikal region, the Amur Basin, and the "Paleo-Sakhalin-Hokkaido-Kurile Peninsula" (then connected by lowered sea levels) (Kobe *et al.*, 2021; Buvit *et al.*, 2022). Descendant Ancient North Siberian populations included those such as the 24,000-year-old little boy interred at the Mal'ta site in the Lake Baikal area. In a site with slab-lined semi-subterranean dwellings, the ochre covered three-year-old was accompanied by sculpted ivory swan figures, beads, and a pendant. The Mal'ta child was the first instance of an ancient whole genome study with a bearing on ancient populations that have a distant but shared genetic relationship with Indigenous ancestors in the Americas (Raghavan *et al.*, 2014). The appearance of genetically distinct Indigenous ancestors came slightly later, when an Ancient North Siberian popula-tion received an influx of eastern Asian genetic heritage, perhaps reflected in the Diuktai archaeological culture, which became important in eastern Siberia some 20,000–22,000 years ago. It is in this time range that geneticists recognize the origins for founding populations of the Americas, namely as Indigenous ancestors became separated from other Eurasian populations.

Pathways into the Americas

Beginning with an Earth science perspective and routes leading to the Americas, the most recent reconstructions of pre-LGM sea levels and ice masses show that ~35,000 years ago, a narrower Bering Land Bridge linked Asia and the Americas, while massive Laurentide and Innuitian ice sheets occupied most of the Canadian Shield (Dalton *et al.*, 2022). It is unclear if passage along the Pacific Coast might have been possible in this time range (because of significant Cordilleran glaciation), but the remainder of western Canada could certainly have been entered from Beringia in this very early time range (a topic to which we will return). With the onset of the LGM, however, massive quantities of water were tied up in expanding ice sheets, creating a global drop in sea levels of about 120 meters. By roughly 23,000 years ago, almost all of Canada was covered in glacial ice that could become as thick as two to three kilometers; Cordilleran and Laurentide ice sheets expanded and coalesced along the eastern slopes of the Rockies. By the height of the LGM, the Bering Land Bridge became a vast, subcontinental region. It was part of greater Beringia, extending from the Verkhoyansk range in Siberia to the West and Alaska and northwestern Yukon in the East. Aridity in Beringia meant that these areas remained largely ice free.

This glacial geography controlled prospective late Pleistocene routes of entry to the Western hemisphere south of the ice masses, of which two would be available, with somewhat different timings. The fabled ice-free corridor (along the eastern slopes of the Rockies) linking eastern Beringia with the continental United States to the south – which was erroneously believed by many to have been the subject of decades of intensive research – was closed by coalesced ice throughout the height of the LGM until significant deglaciation began after ~16,000 years ago. Significant parts of the southern Cordilleran and Laurentide ice masses had detached from each other by 15,000 years, after which there was rapid retreat; by the onset of the Younger Dryas ~13,000 years ago, ice-free conditions prevailed as far east as the Cree Lake Moraine at the Alberta–Saskatchewan border (Norris *et al.*, 2021, 2022). A central constriction in the ice-free corridor, the Peace River Country of northwestern Alberta and northeastern British Columbia, deglaciated ~13,800 years ago (Clark *et al.*, 2022). The biotic habitability of that central constriction has been questioned, but there has for some time been clear evidence that bison and horse and even some treed vegetation was present there no later than 13,100 years ago, and quite possibly a few centuries earlier (as a consequence of the spread of fluted point technology, to which we will return) (cf. White and Mathewes, 1986; Weinstock *et al.*, 2005; Heintzman *et al.*, 2016; *contra* Pedersen *et al.*, 2016).

Glacial processes were critically important to a second pathway into the Americas, along the north Pacific Coast. It too was inaccessible at the height of the LGM, but massive Cordilleran ice created a "forebulge" effect that enhanced seaward glacial refugia such as Haida Gwaii and northern Vancouver Island. Current evidence indicates that, sometime after ~17,000 years ago, challenging though it might have been, Indigenous groups with some sea-faring capacity could have traversed the North American Pacific Rim until they were south of the ice masses and then entered unglaciated North America (Lesnek *et al.*, 2018, 2020; Hebda *et al.*, 2022). This pathway has been linked to a "kelp highway" hypothesis that stresses the rich near-shore resources of kelp forests along the North Pacific Rim (e.g., Erlandson *et al.*, 2007).

Genetic Evidence

Turning to the genetic evidence, we should first note how very few remains of the earliest humans have been recovered in Northeast Asia and the entire Western hemisphere. We must acknowledge the extent to which current understandings are based upon tiny sample sizes, projections of lineage differentiation from aDNA from more recent Indigenous burial populations, and modern population genomic studies. There are also genetic "ghosts" for which human remains with corroborating aDNA have not yet been recovered, but that are detectable in aDNA signals from other, known, samples. Significant surprises have already been noted and are sure to continue surfacing.

That said, genetic evidence suggests that, around 20,000 to 23,000 years ago, those Ancient North Siberian populations that experienced significant East Asian gene flow became two distinct, subsequent populations, one termed "Ancient Paleo-Siberians" and a second becoming the basal lineage whose descendants would cross Beringia and enter the Americas (see Willerslev and Meltzer, 2021, for a current synopsis). For the founding Indigenous population of the Americas, just where this isolation from Eurasian gene flow took place is not currently well understood. The "Beringian standstill" model has been influential over the last two decades – and it could be that the central Bering platform was the locus for that isolation for an interval of several thousand years (Tamm *et al.*, 2007). Popular though that idea has been, there is in fact uncertainty about where and when the basal American branch emerged. It may have been the consequence of a longer period of genetic isolation in eastern Beringia, but other studies suggest that the initial divergence may have taken place in Northeast Asia (such as the Amur Basin), that there may have been little or no Bering standstill (i.e., a much shorter interval of isolation on the Bering platform or, instead, a rapid crossing of Beringia), and that the initial divergence might have occurred after the onset of the LGM (e.g., Ning *et al.*, 2020; Mao *et al.*, 2021; Sun *et al.*, 2021; Willerslev and Meltzer, 2021; Buvit *et al.*, 2022). Y chromosome data in particular would suggest that there was only a brief Bering standstill, if one took place at all (cf. Wei *et al.*, 2018; Grugni *et al.*, 2019; Pinotti *et al.*, 2019).

One intriguing possibility would be that, while much of Northeast Asia was depopulated during the LGM, populations persisted in more mesic regions such as the Amur Basin and adjacent "Paleo-Sakhalin-Hokkaido-Kurile Peninsula" (then connected by lowered sea levels) (e.g., Buvit *et al.*, 2022). A demographic "push" back toward more challenging, higher latitude Northeast Asian and Beringian settings as the LGM ameliorated could conceivably have been the pulse that led to the divergence of basal populations for the Americas. While the earliest pottery in the entire world comes from farther south in China, very early instances of pottery begin to occur in the Amur Basin and Japanese archipelago after 16,000 years ago (Shoda *et al.*, 2020). Residue analysis of the stylistically distinct Osipovka pottery from the lower Amur, like contemporary Japanese pottery, had "aquatic" uses, particularly the processing of salmonids. In contrast, the Gromatukha pottery of the middle Amur was used in processing terrestrial rumin-ant fats. Shoda *et al.* (2020) wondered if Transbaikal and middle Amur populations were keen to extract maximum nutrition from carcasses by rendering bone grease during lean seasons, while lower Amur populations had focused on aquatic foods as an important alternative to depleted supplies of terrestrial resources during LGM conditions. One implication of the Amur situation could be that higher density, somewhat more sedentary, populations were already emerging. Growth and fissioning of these populations might have sent daughter groups toward Beringia, with attendant founder effects (cf. Mason, 2020; Tournebize *et al.*, 2022). Buvit *et al.* (2022) related a post-LGM human expansion back into Northeast Asia to the spread of microblade technology from these more southerly human refugia. That is expressed as the Diuktai technology, which made use of purposefully fragmented bifaces to prepare microblade cores and microblades – the Yubetsu technique, so named after its occurrence in the Paleo-Sakhalin-Hokkaido-Kurile Peninsula. The rapid expansion of Diuktai is coincident with Y chromosome data that would situate the earliest male Paleo-Indigenous ancestors in northern Asia at ca. 17,000 calibrated years before the present (cal BP) to 19,500 cal BP (Pinotti *et al.*, 2019).

It could be that we do know something about the *conditions* under which that isolation must have taken place. Hlusko *et al.* (2018) showed that variations in the fatty acid desaturase (FARS) gene cluster and the ectodysplasin A receptor (EDAR) gene variant 370A, which were widespread in Indigenous populations of the Americas, are intimately involved in the enhanced capacity to produce fatty acids (including vitamin D) in breast milk and in creating more complex ducting in the female breast. As we are all subtropical primates in our evolutionary history, vitamin D deficiency in high-latitude settings would be a truly fundamental challenge. Elevated FARS and EDAR variant frequencies arose as the height of the LGM passed ~19,000–20,000 years ago (Hlusko *et al.*, 2018; Mao *et al.*, 2021). The capacity to better produce and provide fatty acids in low ultraviolet (UV) environments would convey a significant adaptive advantage. There has been considerable emphasis in the recent literature on prospects for a coastal adaptation for the founding Indigenous population, anywhere from Hokkaido to the southern fringe of Beringia. For the microevolutionary trend, Hlusko *et al.* described, however, that it is very difficult to see how the requisite selective pressures could exist in coastal environments, with their marine mammal, various rock fish, Pacific cod, and shellfish resources providing one of the richest vitamin D settings imaginable. Instead, the EDAR and FARS findings would direct us to those same high-latitude settings that populations like those at Yana explored: today's northern Siberia and, perhaps even more likely, the vast Arctic plain exposed by lowered glacial period sea levels running north from Siberia and eastward across the Bering platform (Hoffecker *et al.*, 2022).

Genetic research also tells us that the founding population for Indigenous ancestors was small and resulted from population bottlenecking. Fagundes *et al.* (2018) modeled an effective founding population that may have been as small as 275–300 people, with the highest estimates ending at roughly 5,000 people (e.g., Kitchen *et al.*, 2008). From a social perspective, that must be another of the singular features of Indigenous arrival in the Americas – vast, uninhabited

Beringian and Western hemisphere landscapes and *extraordinarily* small human populations. No matter what form of kinship these populations had, many of its members would have been closely related, posing *from the outset* a paramount question: Where might one find a spouse?

However, further findings lead research: By ~20,000 to 21,000 years ago, the basal American branch had begun to diverge into separate lineages, none of them showing subsequent gene flow from Ancient Paleo-Siberian or other Northeast Asian populations. This initial divergence involved a poorly known "genetic ghost" (called unsampled population A [UPopA]), an Ancient Beringian population, and an ancestral Native American population, all of which crossed into North America. The deep divergence and limited gene flow between these three indicated to Willerslev and Meltzer (2021) that they probably did so in separate movements.

Diuktai technology, featuring Yubetsu-style microblade cores, makes its appearance in eastern Beringia in a small encampment at Swan Point, Alaska, just over 14,000 years ago, marking a sustained Indigenous presence in the Western hemisphere (Coutouly and Holmes, 2018; Surovell *et al.*, 2022). Authorship of the Swan Point record is currently not known, although a number of researchers suspect that the site's inhabitants came from the Ancient Beringian clade that had diverged at an early date from the founding Indigenous population of the Americas. The next genetic signpost we encounter, reflected in the 11,500-year-old Upward Sun River child burials of central Alaska, provides intriguing insights (Moreno-Mayar *et al.*, 2018a; Halffman *et al.*, 2020). Within a semisubterranean house, a ~three-year-old child was cremated in a pit. Contemporaneous with, but below that, hearth, two more infants were interred: One died shortly after birth, while a second was a late-term fetus. It was possible to recover aDNA from these two perinates, for which whole-genome sequencing revealed they were closely related, as female half siblings (with differing mtDNA C1b and B2 haplogroups) or first cousins. The Upward Sun River children (known as *Xach'itee'aanenh t'eede gaay* or sunrise child-girl [USR1] and *Yełkaanenh t'eede gaay* or dawn twilight child-girl [USR2]) provided the first indication that the founding population for the Americas split at an early date into an Ancient Beringian lineage (reflected in these Upward Sun River children) and a founding lineage for the remainder of the Americas. The fate of the Ancient Beringian lineage is not well understood but, about a millennium later, it disappeared entirely.

Although we are speaking of a single moment in time, two key inferences may be drawn from these Upward Sun River findings. Here we have a different scenario from Sunghir or Yana in which contemporaneous individuals at a site were not closely related, a situation indicative of exogamous marriage practices. The two Upward Sun River perinates had different mothers, but they were nevertheless closely related as first cousins or half siblings (Moreno-Mayar *et al.*, 2018a;

Halffman *et al.*, 2020). In at least this instance, it would appear that marriages were taking place close at hand, plausibly involving cross-cousin marriage or the levirate (in which a man is expected to marry his brother's widow). This would have been a sad time for the community involved, which would likely have been in the range of twenty-five to thirty-five persons, as was typical of Subarctic local groups. To lose three children so closely together would be a serious blow. With respect to this small, coresidential group, then, the boundary for acceptable marriages had been drawn inward.

After its separation from the Ancient Beringian lineage, further branching of the Ancient Native American lineage took place between 16,000 and 21,000 years ago when a "Big Bar" lineage[7] (known from a mid-Holocene burial in British Columbia) diverged (Willerslev and Meltzer, 2021). At about 15,700 years ago, a major split in the Ancient Native American lineage took place between northern Native American (NNA) and southern Native American (SNA) populations. The northern lineage was ancestral to Haida, Tsimshian, Salishan, Algonquian, Tlingit, and Dene (Athapaskan) peoples, while the southern lineage was ancestral to all other Indigenous populations of the southern United States and Central and South America. A number of geneticists believe the NNA–SNA differentiation took place south of the Canadian ice masses (e.g., Reich *et al.*, 2012; Rasmussen *et al.*, 2014; Raghavan *et al.*, 2015; Moreno-Mayar *et al.*, 2018b; Scheib *et al.*, 2018; Willerslev and Meltzer, 2021). This is by no means certain, however, and other genetic studies leave open the possibility that differentiation of the Ancient Native American population actually took place in Northeast Asia or perhaps in eastern Beringia (Achilli *et al.*, 2013; Sun *et al.*, 2021). Should that be the case, a very brief Bering standstill, if there was one at all, would be likely, so that the timing and potential use of available pathways into the Americas (a north Pacific coastal route versus the ice-free corridor) could be highly significant in the differentiation process for the basal Ancient Native American population.

In their synopsis, Willerslev and Meltzer (2021) noted that the relatively small size of the incoming population and the vast distances involved in initial movement throughout the Western hemisphere would have increased the chances of isolation and divergence, resulting in repeated splitting within the SNA lineage as it moved southward, leading to considerable ancestry variability in ancient South Americans. At a later point in time (as yet poorly defined), admixture of NNA and SNA lineages took place, with admixed populations also moving southward.

---

[7] The Big Bar lineage is one of the small sample "surprises" mentioned earlier – a branch previously unknown to geneticists, but revealed as more ancient genome sequencing took place.

Clovis and the Early Human History of the Americas

aDNA from a small number of early Indigenous individuals in North America (such as On-Your-Knees Cave, Spirit Cave, and Paisley Cave) and other key early populations in Central and South America (such as Hoyo Negro and Lagoa Santa) contributed to our understanding of the NNA–SNA lineage distinction. None was more influential than the roughly three-year-old Anzick boy from Montana, however. This child was buried with a rich assemblage of 125 Clovis tools, among them fluted projectile points, bifacial projectile point preforms, very large and thin bifaces, and bevelled elk antler rods (foreshaft implements intended to hold the fluted projectile points while inserted at the tip of a main spear or dart shaft). The artifacts were coated with ochre; the raw stone materials for the tools were aesthetically appealing, high-quality cherts, flaked by adept craftpersons. There was a clear conception of an afterlife, with the ochre, elk antler foreshafts, and other elements of the grave goods providing an echo of the Mal'ta child, to which the Anzick child was in fact distantly related. This Clovis child belonged to the SNA clade, relating it to all First Peoples from the southern United States through Central and South America. In a truly evocative way, the Anzick child belonged to a population whose descendants were most closely related to the Maya[8] who would of course found one the world's great civilizations.

The notion of "Clovis" had for many decades been pivotal in the archaeological understanding of the earliest time period in North America. There are two views of the age of Clovis assemblages: (1) a "short" chronology in which a carefully culled list of sites with the very best accelerator mass spectrometry (AMS) radiocarbon dates indicates a narrow time from ~13,100 to 12,800 cal BP (Waters and Stafford, 2007; Waters *et al.*, 2020) or (2) a "long" chronology, primarily based on the statistical argument that a larger list of high-quality dates will eventually expand the onset by two to three centuries (e.g., Prasciunas and Surovell, 2015). Even with a broader time range, the Clovis phenomenon was rather short lived.

The Clovis projectile point is the signature technology for this archaeological construct, although macroblades and bone, antler or ivory rods, weapon tips, and bevelled foreshafts also typically occur in Clovis assemblages. Clovis points have basal thinning (whereby one or more shorter flakes are driven from the point base) or fluting (whereby one or more lengthier flakes are driven toward the tip). Fluting running from the tip of a projectile point (leaving a deadly sharp edge at the tip) has been recognized in later Arctic Small Tool tradition assemblages in the High Arctic and in Neolithic Arabia. Nowhere else in the world (including Northeast Asia), however, was the base of the point thinned by channel flakes running toward the point tip as a means of facilitating hafting (i.e., binding within the foreshafts

---

[8] As well as to the South American Karatiana people.

mentioned previously) (Crassard *et al.*, 2020). Some research suggests that fluting allowed for predictable, "controlled" breakage at the point base during use as a weapon. This could have been important for highly mobile populations still exploring vast landscapes for required resources, including toolstone (e.g., Thomas *et al.*, 2017). Despite the foreshortened time frame for Clovis, this unusual idea of basal thinning or fluting eventually spanned the entire Western hemisphere, from the North Slope of Alaska to the Straits of Magellan, in just a few centuries. In South America, different stemmed or fishtailed points were a common early form but, in a number instances, these too have basal thinning or fluting. Basal thinning or outright fluting, then, is uniquely American, with a geographic scope akin to that of Venus figurines in Eurasia.

In the middle years of the twentieth century, there emerged the credible argument that the widespread appearance of Clovis was the first indication of the presence of Indigenous peoples. This idea held sway for a few decades, becoming inextricably linked with two other "fellow travellers." The age of Clovis suggested that it was the initial technology brought through the ice-free corridor route as deglaciation proceeded. Some researchers went farther, suggesting that this expanding migration wave into the Americas was the direct cause of many terminal Pleistocene megafaunal extinctions or extirpations. The term "Clovis First" has been applied to the notion that Clovis was the first Indigenous technology, that it arrived through the ice-free corridor, and that the demic expansion that followed would be implicated in terminal Pleistocene extinctions. None of these ideas was inextricably linked in logical terms, so that "Clovis First" was always an artificial construct.

The origins of Clovis and fluting technology have been highly sought after but remain uncertain even today. The tremendous density of fluted point occurrences, along with pre-Clovis assemblages at sites like Gault, Debra L. Friedkin, and Page-Ladson in the southern United States, is taken by many to suggest that the center of gravity for this development lay in those regions. From this point of origin, the idea of fluting would spread both north and south. This an appealing idea, but it is not the only perspective. It remains conceivable that the idea of fluting emerged in or near the corridor, expanding southward, as Buchanan and Hamilton (2009) argued on morphological grounds for a continental sample of fluted points. Wygal *et al.* (2022) pointed out that the earliest Western hemisphere instances of extensive ivory use for points, rods, and foreshafts occurred between 13,600 and 13,300 years ago at the Holzman site in interior Alaska, with significant ivory use at Swan Point and Broken Mammoth as well, extending to 14,100 years ago. Ivory, antler, and bone rods and foreshafts are another signature aspect of Clovis assemblages, so that it could be that early northern populations emerging from the corridor made the fluting innovation, which then spread rapidly. Schroedl (2021) used Clovis blade and biface cache discoveries to suggest that Clovis emanated from the Pacific

Northwest, south of the ice masses. Wherever the fluting idea actually emerged, it spread rapidly.

The early-period literature for the Americas is of course voluminous and need not detain us at length here (Waters, 2019, is a concise review). That said, by the last two decades of the twentieth century, serious faults appeared in the "Clovis First" scenario. There has been and will continue to be debate about some sites that would be regarded as pre-Clovis, that is, sites that would indicate Clovis could not have been the first expression of Indigenous presence. Without delving deeply into that literature, suffice it to say that sites such as Gault and Debra L. Friedkin in Texas, Page-Ladson in Florida, Paisley Cave in Oregon, and Monte Verde in Chile came to show that Indigenous ancestors were present by at least 14,000–14,500 years ago, predating Clovis by a significant margin. It also became apparent that some stemmed point assemblages, such as those of the Western Stemmed Point tradition in the Pacific Northwest, and elsewhere, were at least coeval with Clovis and, in some cases, earlier (e.g., Smith *et al.*, 2020).

In terms of the Clovis First conceptualization, it became clear that these pre-Clovis sites were occupied in a time frame when the ice-free corridor was not open early enough to explain an already established Indigenous presence. This led to a rapid intellectual shift toward the north Pacific Coast as the route that must have been followed in order to proceed south of the ice masses. For those who had favored a "blitzkrieg" notion of human involvement in Pleistocene extinctions, the deeper time frame began to suggest a longer period of human–megafauna inter-action, diminishing the prospects for human hunting being the single factor in the demise of species such as mammoths, mastodons, horses, camels, and ground sloths.

By the end of the fluted point time frame, there came to be a profound diversifi-cation of cultures and economies across the Western hemisphere as Indigenous peoples adapted to the Holocene environments that were being established. In North America, this would involve caribou-hunting populations in the Northeast near retreating ice masses as inferred for Bull Brook, Massachusetts; sophisticated hunting of bison chronospecies[9] on the Great Plains; an early focus on plant foods and fiber perishables in the Great Basin; maritime adaptations along the Pacific Coast; and even the later fluted point cemetery at the Sloan site of Arkansas, with its

---

[9] Bison survived the wave of late Pleistocene extinction but, in doing so, their ethology and morphology changed considerably from that of their steppe bison ancestors. Despite the enormous size of modern wood and plains bison, their Pleistocene ancestors were yet larger and armed with massive, deadly, flaring horns that suggest to paleontologists that at least the males were more aggressive and less gregarious animals than the modern forms. As Holocene changes progressed, bison diminished in size, evolving more tightly curving horns that would be less dangerous during the conflict that comes with rutting behavior, while becoming more gregarious. Quaternary paleontologists and archaeologists refer to these earlier forms as chronospecies such as *Bison antiquus* and *Bison occidentalis* – bison to be sure, but not yet looking or acting just as modern forms (*Bison bison*) would after some 6,000–7,000 years.

Dalton points and adzes for woodworking (Morse, 1997). Of course, in Central and South America, many more lifeways would emerge, eventually to include some of the world's great civilizations, the development of many domesticates of critical global food significance today (think of the maize, bean, and cucurbit triumvirate), and other expressions in an extraordinary array of human diversity.

Rapidly Shifting Perspectives

Until the fall of 2021, then, Earth science, genetic studies, and archaeological findings had converged on a relatively coherent (with several uncertainties as noted previously), if broad, picture of the initial peopling of the Americas (e.g., Waters, 2019; Willerslev and Meltzer, 2021). That is, the basal population for the Americas took shape some 20,000–23,000 years ago, with rapid separation of an Ancient Beringian branch, then Big Bar, and, finally, an Ancient Native American lineage. The latter split into NNA and SNA lineages sometime between 15,000 and 17,000 years ago. That separation may have taken place in Beringia or Northeast Asia, although a number of authors felt it took place south of the North American ice masses. The "molecular clock" dating, coupled with established archaeological records south of the ice masses in the 14,000–15,000 age range, indicated that a north Pacific coastal route must have been followed by these initial Indigenous groups. Soon thereafter, at least the SNA lineage members would be responsible for not just Clovis, but the Western Stemmed Point tradition, and some preceding technologies emerging 13,000–14,000 years ago. It was at this time, however, that dramatic findings from White Sands National Park, New Mexico, were announced. A series of human trackways, stratigraphically constrained, bracketed by *Ruppia* seed layers, and interspersed with trackways of Pleistocene ichnotaxa (such as mammoths, dire wolves, and ground sloths) were dated (Bennett *et al.*, 2021). The seeds impressed in human footprints produced calibrated radiocarbon ages between ~21,000 and 23,000 years ago, results supported by U-series dating.

Of all the sites in the Western hemisphere that might document a pre-LGM human presence, White Sands is currently the most compelling. The dramatic nature of these results, in particular the age of the *Ruppia* seeds, attracted close scrutiny (e.g., Haynes, 2022; Oviatt *et al.*, 2022; Pigati *et al.*, 2022). More recently, however, those previous ages have been affirmed by radiocarbon dating of terrestrial pine pollen from the relevant strata as well as optically stimulated luminescence ages (Pigati *et al.*, 2023). The four independent means of dating the White Sands footprints remain in close agreement, providing conclusive evidence of human presence. Should continued analyses and other promising sites lead to widespread acceptance of much earlier ages, then the relatively tidy picture noted previously would, to a degree, unravel. One key implication of the White Sands findings would be that human populations were *already* present in North America during the LGM, well in advance of the 15,000–17,000-year-old "molecular clock" and other archaeological site dates that have emerged, raising many questions. As noted at the outset of this section, a Bering Land Bridge existed prior to the LGM and there would be *no* geographic impediment whatsoever to movement through an ice-free interior of western North America, changing our perception of prospective routes. Another series of questions would involve intervening time periods, between roughly 21,000–23,000 years ago and 15,000 years ago. Given the scale of development activity in North America and numerous archaeological investigations, could there truly be a 5,000–8,000-year gap in which founding populations had succeeded, but left no trace? Are the genetic findings and their inferred dates in serious error or is there another possibility, seldom alluded to, but occasionally suggested? Not every settlement effort will succeed, and it is possible that a failed effort of decades or even a few centuries could leave a detectable archaeological record, but have no surviving descendants (Meltzer, 1989, 2002, 2003, 2021; Ives, 2015; Potter *et al.*, 2018; Fiedel, 2022). In that case, the White Sands results would hold, but the remainder of that "coherent" picture would too, with the ultimate founding populations entering the Western hemisphere in that post-LGM time frame.

This is an opportune time to see if scientific and Indigenous knowledge could flow together to illuminate our understanding of how Indigenous peoples entered the Americas, in one of the most extraordinary chapters of human history. Socially informed "thought models" can be useful in inflecting our understanding of what, for the time being, is an unsettled picture. Before attempting that task, however, some probing of assumptions would clearly be in order.

3.8 Warranting Some Key Modeling Assumptions

Trautmann (2001: 283) wrote that the deep history that lies between "the end of the last ice age and the beginning of the Victorian era, is not thickly populated by anthropologists." One reason for this would be that it is no simple matter to support realistic assumptions about kinship in this deep time frame. It is certainly legitimate to ask how it could be that kin systems, understood from so much more recent intervals, could possibly be relevant to the late Pleistocene worlds explored and settled by Indigenous ancestors. Yet we have seen that kin-structured principles of group formation were essential in creating social frameworks that allowed the successful reproduction of Dene societies throughout several millennia of Subarctic history. To provide a foundation for evaluating the thought modeling that follows, it is useful to consider ways in which some key variables in Subarctic life might resemble some aspects of late Pleistocene life. From there, we can move to historical and logical reasons why some kin models are more likely than others, even for the distant, Pleistocene, past.

As Kelly and Todd (1988) pointed out, the situation for founding Indigenous populations was novel by any recent human standard. At some point, the Beringian "cul-de-sac" mentioned earlier (along with the entire Western hemisphere) had no human populations, so there must have been extended circumstances in which human population densities were vanishingly low. It strains credulity to think that founding Indigenous populations could have been anything other than acutely aware of that fact and its consequences for societal reproduction. Although one must be cautious of the colonial impacts created by epidemic disease, a great deal of fur trade and ethnographic literature would also suggest that Indigenous population densities in the boreal forest have typically been some of the lowest in the world, even in comparison with the severely arid environments of central Australia and southern Africa. Subarctic figures of one person for every 200–400 km^2 are common; therefore, in pragmatic terms, were we to look for a comparator, the Subarctic would be one logical cultural area to consult.

Given the enormous distances involved in traversing Beringia and then entering an unoccupied continent, we can also be assured that the earliest Indigenous populations were highly mobile. Although there are complexities to interpretation (particularly with respect to the impact of early exchange networks), the far-flung instances of "exotic" toolstones such as obsidian, Knife River Flint, and a variety of high-quality cherts have long suggested to early-period specialists that the first Indigenous populations of the Americas regularly travelled widely. In his careful treatment of this subject, Ellis (2011) assessed distances from toolstone sources for the fluted point populations in northeastern North America in sorting out a high degree of seasonal mobility from likely instances of trade or resupply through special task groups. His work was informed by Subarctic ethnographic examples, most notably with respect to Dene-speaking Ethen-eldèli (caribou eaters) who followed or intercepted barren-ground caribou, reflecting the extreme end of the known seasonal range for pedestrian human travel.

I am by no means suggesting that the Holocene boreal forest ecosystem would be directly comparable to the predominantly open and dry Arctic steppe-tundra biomes through which the earliest Indigenous ancestors moved. While they may not have been uniformly distributed, the megafaunal resources of the Arctic steppe tundra, or "mammoth steppe" as it has been termed, were at times both diverse and abundant relative to the more limited faunal resources of closed boreal forest settings. There is in fact a sound case to be made that the advent of the closed boreal forest ended faunal interchange between eastern Beringia and lower latitude North America, as the expanding ice-free corridor first created a steppe-tundra "tail" extending across Canada and the northern tier of states that then vanished (e.g., Guthrie, 1990; MacDonald and McLeod, 1996; Pedersen *et al.*, 2016). There nevertheless remains one parallel between the two worlds. The prevalence of mutations in the EDAR and FADS gene complexes in ancestral Indigenous populations (noted previously) indicates that the founding Indigenous populations for the Americas may at some point have been heavily reliant on high-latitude terrestrial mammal resources as their principal food source, as were later boreal forest peoples (Hlusko *et al.*, 2018).

Without mistaking Holocene Subarctic circumstances for those of the terminal Pleistocene, then, it is the case that ancestral Indigenous populations likely did resemble Subarctic societies in these respects: They featured very low population densities and high degrees of mobility, and relied upon terrestrial mammal resources.

This brings us back to the subject of kinship and Dr. Tallbear's remark with which we began this chapter. A narrow focus on subsistence and economy cannot account for that other dimension of human life: the paramount need to reproduce a human society through subsequent generations. Although there have been suggestions that founding Indigenous populations would have been unaware of these larger circumstances, and that early episodes of settlement would have proceeded through simple demic expansion, it is truly difficult to believe that the first Indigenous ancestors were unaware that the Western hemisphere world before them was *entirely* empty – by simple logic, a world of phenomenally low human population densities. To accomplish what these ancestors so clearly did, one can only imagine that these were vibrant societies, keenly aware of the challenges that would accompany the issue of where children could find marriage partners. As we have seen already, it is well within the capacities of various group-forming principles for societies to address precisely where the socio-geographic boundary for incest would lie, thereby influencing the scope of where legitimate marriage partners could be found.

If we are to be more specific about the semantic content of late Paleolithic kin systems, there is an additional matter of logic. Variability in kin systems is not limitless, and human minds have fixed upon a smaller number of alternatives. Of these alternatives, kin systems that make cross–parallel distinctions are in fact common throughout the world, and it would not be surprising on those grounds alone to find that this kind of semantic framework was a feature of ancestral Indigenous kin systems at the very outset of the human history of the Americas. There would also be the perspective on the tetradic origins of all human kinship, as mentioned earlier. Allen (1998, 2008: 112) suggested that tetradic kin systems or their derivatives may have dispersed from sub-Saharan Africa with culturally and anatomically modern human populations. That is conceivable, and kin systems making cross–parallel distinctions might have remained widely current in the late

Paleolithic Siberian world, as ancestral Paleo-Indigenous populations were organized in Northeast Asia or Beringia.

Moving forward in time, the rigor of the comparative method in historical linguistics does allow for the reconstruction of prototypical kin systems and their semantic form. It is true that such methods do not work reliably beyond a mid-Holocene time frame, perhaps 4,000–5,000 years ago. Yet, just a bit of probing on this front reveals that Proto-Dene-Eyak, Proto-Algonquian, Proto-Numic, and Proto-Mayan all featured cross–parallel distinctions mapping cross-relatives as affines in Dravidian or Kariera form (Eggan, 1955a, b; Hockett, 1964; Krauss, 1977; Wheeler, 1982; Ives, 1990, 1998; Hage, 2003; Hage *et al.*, 2004). Thus, one could argue that, in roughly mid-Holocene time, kin systems of just the sort we have been considering were quite widespread in the Americas. Their more ancient antecedents could reasonably be expected to have had related structural properties.

While to this point I have focused upon cross–parallel distinctions, as these are related to Dravidian crossness, a number of scholars have suggested that there are historical relationships among Dravidian, Iroquoian, and Crow-Omaha systems, with progressive transformations among these different kin systems (e.g., Ives, 1998; Trautmann and Barnes, 1998; Ives *et al.*, 2010; Whiteley and McConvell, 2021). The precise mapping of the Dravidian cross–parallel categories is the more straightforward one because of its logical connection to bilateral cross-cousin marriage. Even so, leading scholars in kin studies have viewed the several modalities in the dimension of crossness (Australian, Dravidian, Iroquoian, Yafar, Kuman, and Crow-Omaha) not as completely different, but rather as "dialectical" variation in a "language" of crossness, which in every case has an affinal implication (see various papers in Godelier *et al.*, 1998; Whiteley and McConvell, 2021). These "species" of crossness have a vast spread in both North and South America that may reflect the transformation of Dravidian-like systems toward the more complex alliance regimes we see with Iroquoian and Crow-Omaha systems. Some form of cross–parallel reckoning may lie at the base of historical developments in *very* many language families in the Americas. Again, with little effort at compiling numbers, we find that cross–parallel distinctions are known for Inuit, Tsimshian, Haida, Tlingit, Eyak, Athapaskan, Algonquian, Siouan, Iroquoian, Uto-Aztecan, Mayan, Chibchan, Carib, Panoan, Yanomamö, Tukunoan, and Jivaroan (Spier, 1925; Godelier *et al.*, 1998). Purely in logical and distributional terms, then, there is a strong *prima facie* case that the first kin systems to enter the Western hemisphere were classificatory in Morgan's sense, and that they made cross–parallel distinctions (cf. Ives, 1998, 2015).

Finally, there are traces of these forms of social organization in the early archaeological records of both Eurasia and North America. We have been speaking of elementary kin structures with essentially binary marital exchanges and, as

Lévi-Strauss (1963) noted, such societies have a tendency to map their social structure onto settlements (see also Means, 2007). This is especially prevalent in societies with dual organization, where the two halves of a society are reflected in linear or circular arrangements of settlements with bilateral or concentric symmetry. Bilateral or symmetrical cross-cousin marriage commonly accompanies such spatial organization of settlements; kin terminologies with cross–parallel distinctions are typical in these settings.

Late Paleolithic sites in Ukraine and on the Central Russian Plain have just such settlement features, with both arrangements of circular structures and "long house-like" dwellings with rows of central hearths (see the plans in Klein, 1973 and Soffer, 1985). In some instances, these sites have produced split image art,[10] which frequently accompanies dual organization in concentric or other forms (Lévi-Strauss, 1963). Later in time, the 13,000-year-old Ushki site of the Kamchatka Peninsula featured small, semisubterranean dwellings in its component 7 layer. These tend to occur in pairs facing each other, as we might expect in simple expressions of dual organization (Coutouly and Ponkratova, 2016; see the plans in Dikov and Titov, 1984: 77). In North America, there is one spectacular instance at the ~12,500-year-old Bull Brook site in Massachusetts. As Robinson *et al.* (2009) have documented, the site appears to be a large fall or winter aggregation locale situated at a strategic location with respect to communal caribou hunting. It consisted of thirty-two artifact-rich "hot spots" arrayed in a ring structure, for which refitting studies revealed contemporaneous occupation of the various loci. This ring also has concentric differentiation of functional activities. Among the well-chosen ethnographic analogies that Robinson *et al.* (2009) cited are the large group camps that Slobodin (1962) documented for the Dene-speaking Teetl'it Gwich'in, where exogamous sibs were situated on opposite sides of a circle, with more aged dependents arranged behind each of the family encampments within the circle. Noble's Pond and a few other Paleo-Indigenous sites have suggestive traces of such patterning, but the precise circumstances in which we can trace this are rare (Seeman, 1994). Were one to diagram a site in which there were profound conceptions of duality, linked to cross–parallel distinctions in kin terminologies and broader social entities (such as clans or moieties), the result would be very much like Bull Brook.

While we cannot say with certainty what form kin relations would take in specific instances in these distant time ranges, all of these warranting factors make clear that it would, at a minimum, be essential to model variables influenced by the semantic patterning and related group-forming principles that we have been

---

[10] The painted mammoth skull from Mezerich illustrated by Soffer (1985: 78, figure 2.73), for example, has a design that curves somewhat, but otherwise has nearly perfect symmetry: Almost every design element to the left of the midline is mirrored to the right.

discussing. In reality, there is a strong case to be made that models manipulating these specific semantic forms ought to be the *preferred* starting point for any such exercise.

3.9 Constructing a Useful Thought Model for Earliest Paleo-Indigenous Worlds

One should first conceptualize the social model and then evaluate its quantitative implications.

(Brown et al.*, 2007)*

My purpose here is to meet the initial objective that Brown and others have specified. That requires parsing the key modeling variables that ought to be incorporated in more sophisticated quantitative modeling – variables that we must expect the first Indigenous ancestors to have been keenly aware given the late Pleistocene circumstances in Beringia and the entire Western hemisphere. My purpose is not to complete a sophisticated mathematical modeling process, an ultimately desirable task that I would leave to others with the necessary skills. Beyond this, I will conclude by making some preliminary assessments of whether or not ideas generated by such "thought modeling" have the capacity to generate useful, testable hypotheses for ongoing research, and whether modeling predictions bear informative resemblances to our current knowledge of the earliest human record in the Americas.

Earlier modeling efforts, such as those of Moore and Moseley (Moore, 2001; Moore and Moseley, 2001) and Anderson and Gillam (2000, 2001) took as their starting point a variety of small initial group sizes. This was appropriate from a comparative ethnographic perspective. Two decades later, such a starting point has been affirmed by the genetic studies noted earlier. The effective size of the founding population for the Americas may have been as small as ~300 persons. The founding population might therefore have consisted of several hundred individuals amounting to the equivalent of one or a few regional groups or macrobands. It is difficult to conceive of any form of human kinship in which populations of this size would not have relatively high quotients of relatedness in the first place. Constraints over marriage choice would therefore have existed from the inception of the founding population. To inject a quantum of traditional knowledge into our considerations, we could also say that relatedness would involve sibling sets. We have no indication in later northeastern Siberian archaeological records (Diuktai in particular) that coresident groups numbering in the hundreds were spending sustained periods of time together. We would also infer that dispersed phases of the subsistence settlement system with smaller coresident groups must have occurred, in which sibling sets would have figured prominently.

The pivotal model variable would be to specify alternative states with respect to internalized or externalized marriages, where different approaches would situate coresident groups on the spectrum of endogamy versus exogamy within that initial population. Just how endogamous or exogamous a founding population might be would be determined by conscious human decisions that were structured by kin semantics. It is essential that this be regarded as a variable with alternative degrees of expression. Some treatments have made this key variable a constant; for instance, in his modeling, Moore (Moore, 2001; Moore and Moseley, 2001) stipulated that all members of ego's generation would be terminological siblings, necessitating coresident group exogamy. While it is true that this situation is common, we have seen that it is by no means necessary. When cross-cousin marriage is allowed or favored, there will in fact be potential affines in ego's generation. Beyond this, there are a variety of customs often associated with cross–parallel distinctions that have significant capacity to mediate stochastic demographic effects. These include polygamy, whereby a spouse can have more than one partner; the levirate, whereby a woman is expected to marry the brother of her deceased husband; and the sororate, whereby a man may be married to a woman and her sisters. Each of these customs, distinct from Western conceptions of monogamous conjugal pairs, can, to a degree, rectify sex ratio imbalances caused by stochastic birth and death events.

In related work, I have suggested that the Beaver and Slavey Dene group-forming principles can be abstracted into more generalized *local group growth* and *local group alliance* modes of socioeconomic reproduction (Ives, 1990, 1998, 2015). Societies applying these kin principles make deliberate choices about *where* the sociogeographic locus of the incest taboo will be situated. Local group growth systems feature unlike-sex sibling cores and make extensive cross–parallel kin distinctions. Their marriage practices can range from agamy (i.e., they are relatively neutral regarding internal or external marriages) to actively promoting endogamy. In the latter case, the consequence will be inwardly focused alliance patterns. Coresident local groups or microbands will occur in a wider range of sizes, with differing economic capacities that would range from broad spectrum foraging to more complex ventures like communal hunting. Should they experience economic and demographic success, they can have significant intergenerational duration, growing from modest sizes into large bilateral kindreds. Alternatively, however, this strategy can lead to coresident groups that could be seriously affected by stochastic and other negative factors with the potential to result in their demise. In so many words, the local group growth pathway leads to coresident groups that are "implosive" and characteristically restrict the external circulation of individuals. These patterns do have detectable material and genetic correlates for which archaeological records can be investigated in more incisive ways (Table 3.2).

Table 3.2 *Characteristics and implications of a local group growth framework*

Local group growth	Genetic and archaeological and correlates
Agamous to endogamous marriage practices	Contemporaneous individuals that aDNA shows are closely related
Settlement mode: prone to growth and fission with limited external ties	"Leapfrog" or "outpost" patterns potentially lacking landscape connectivity
Implosive: restricted circulation of individuals	Enhanced potential for cultural drift and style "enclaves"
Unsuccessful local groups: prone to stochastic and environmental risks that could lead to dissolution of the coresident groups	Discontinuous but detectable archaeological records
Successful local groups: sporadic, but of potentially longer intergenerational duration	Longer term range occupancy with focal landscape signatures
Economic accommodation: through population growth	Foraging coexists with sporadic communal game hunting

Local group alliance systems, on the other hand, have like-sex sibling cores or mixed-sex sibling cores that apply kin terminologies in which zero generation crossness is extinguished, either of which will result in a profound emphasis upon exogamy (Ives, 1998). The upshot of this is a situation in which local groups "pump out" personnel within and beyond regional marriage isolates. Local group alliance systems yield microbands with narrower size ranges and short intergenerational duration. They seek political and economic accommodations to external environments by linking microbands together through their extensive alliance networks. While small coresident groups may be fluid and cycle into and out of existence, larger regional populations are able to persist more readily in the face of stochastic or other environmental factors. The genetic and material correlates of local group alliance systems are given in Table 3.3.

These are two simplified modeling poles, for which there is predictable logical variability. Just within the Subarctic Dene world, there are a number of variant factors that influence the precise form that developmental processes and patterns of group fissioning may take. For exogamous local group alliance systems, these would include the relative concentration (e.g., a group of brothers marry a group of sisters) or dispersal of affinal ties (e.g., brothers and sisters marry spouses from a variety of other groups), customs in the care of the elderly (i.e., in the founding generation), the degree to which affinal or sibling working relationships are privileged, affective sentiments toward egalitarian or more structured social environments, and so forth (Ives, 1990, 1998; Stevenson, 1997).

Table 3.3 *Characteristics and implications of a local group alliance framework*

Local group alliance (exogamous)	Genetic and archaeological and correlates
Exogamous marriage practices	Contemporaneous individuals that aDNA shows are less closely related
Settlement mode: repositioning of "webs" of individuals or local groups across range	"Matrix" or "string of pearls" patterns, with stronger landscape connectivity
Pumping out: wide circulation of individuals	Diminished potential for cultural drift with stylistic variability diffused through and beyond immediate region
Successful local groups: tend to dissolve and reassemble through elapsing generations	Long-term range occupancy shifts across a larger landscape
Sustained contact within and between regional marriage isolates: local groups cycle but regional groups are less prone to stochastic and environmental risks	Continuous and geographically contiguous archaeological records for regional marriage isolates
Economic accommodation: through allying groups	Foraging with consistent seasonal capacity for communal game hunting

The impetus to shift this locus outside the local group is strong because it fosters the creation of a *web* of kin ties providing local groups with important options when the exigencies of Subarctic or Great Basin life inflicted hardship. Recall the situation that Dene or Innu groups using treeline intercept strategies would find themselves in at times when caribou failed to materialize along a regular migration route. The capacity to call upon relatives elsewhere would provide the kind of "safety net" that Whallon (2006) has noted in his work with late Upper Paleolithic and Mesolithic societies in Germany. Of course, in a late Pleistocene world with phenomenally low population densities, regular contact with and the maintenance of alliance ties among a series of coresident groups would be an effective hedge against the stochastic challenges that very small groups would otherwise face.

A strategy focusing upon endogamy is not without its attractive qualities, however. When local groups experience success within the course of a generation, their membership grows through both endogamy and recruitment. The idealized, tiny unlike-sex group with which we began could not undertake communal game hunting. A group that had grown through both internal and external recruitment could explore that option independently. At the phenomenally low human population densities that accompanied early Paleo-Indigenous colonization, such advantages would scarcely have gone unnoticed in settings in which finding a marriage partner could have posed extraordinary challenges. Deliberately positioning the socio-geographic locus at which marriages could occur *inside* the local group would have obvious advantages. The disadvantage is that groups following these developmental processes more consistently forego opportunities to forge external alliances and are comparatively isolated from surrounding populations whose larger numbers would mediate stochastic and other potential limiting factors.

A More Overtly Social Approach to the Peopling of the Western Hemisphere

The "social contouring" of the evidence we have reviewed can be summarized as follows. Between 40,000 and 50,000 years ago, anatomically and culturally modern people began using high-latitude environments that had largely eluded archaic hominins like Neanderthals and Denisovans. By ~33,000 years ago, we see at Sunghir and Yana exogamous Ancient North Eurasian populations with sophisticated northern adaptations. In addition, at sites such as Dolní Věstonice, Mezerich, and Kostenki, extending from Moravia to the Ukraine and Central Russian Plain, we see spatial structuring of sites including elaborate mammoth bone constructions that would be, as Soffer (1985) inferred, the very likely consequence of corporate kin groups that undoubtedly had complex social arrangements (perhaps more akin to those of North America's Northwest Coast than those of many other foraging or hunter-gatherer groups). The social sphere of influence for these societies apparently extended as far east as the Baikal region, where at sites like Mal'ta, we see slab-lined semisubterranean dwellings, the elaborate Mal'ta child burial, and even figurines reminiscent of Venus figurines farther to the west.

Ancient North Eurasians largely abandoned northern Siberia as the severe effects of the LGM set in. East Asian populations did persist through the LGM in less severe settings such as the Amur Basin and adjacent Sakhalin and Hokkaido, where early pottery traditions appeared by 16,000 years ago. This general region appears to be the "hearth" from which a sophisticated microblade technology emanated, expanding into Northeast Asia as the Diuktai tradition about 20,000 years ago. It is possible that somewhat better settled, higher density populations in this region provided a renewed demographic push into higher latitude Northeast Asian and Beringian regions.

The founding Indigenous population for the Americas took shape in this larger context. The genetic evidence indicates that there was a small founding group in which – no matter in what form kin relations might have been cast – people must already have been relatively closely related to each other. Diuktai technology makes a definitive appearance at Swan Point, Alaska, just over 14,000 years ago. Generally speaking, post-LGM archaeological sites in the Americas share facets of a Northeast Asian cultural substrate: high precision stone tool making; ivory, antler, and bone rods and foreshafts; a symbolic use of red ochre; and mixed hunting and fishing economies with a capacity to hunt Pleistocene megafauna. Yet, this early time range in eastern Beringia and early sites in the Western hemisphere have a much fainter archaeological signature. Gone are large settlements with elaborate structures, sophisticated and at times copious items of adornment, and so forth. The next genetic locus for social information then comes from the Upward Sun River children, who were later members of that Ancient Beringian lineage. Unlike, the Upper Paleolithic Eurasian indications of exogamous social networks, the Upward Sun River children were as closely related to each other as half siblings or first cousins, suggesting that this would have been an endogamous coresident group. The ancient Beringian lineage would vanish roughly a millennium later, with eastern Beringia and, indeed, nearly all of Canada coming to be inhabited instead by the northern lineage branch of the founding Indigenous population.

While Swan Point provides for a remarkably early record in eastern Beringia, evidence has steadily accumulated for a prior Indigenous presence south of the ice masses. As of yet, we have no genetic indicators of familial relatedness from aDNA studies. We do know that there were "settled-in" appearing, pre-Clovis populations, greater than 14,000 years of age, at places such as the Gault-Friedkin and Idaho localities in Texas, the Aucilla River in Florida, interior Washington (Paisley Cave and Cooper's Ferry), and South America (e.g., Monte Verde). Penecontemporaneous to contemporaneous archaeological expressions (notably Clovis and the Western Stemmed Point tradition) were used by southern clade Indigenous ancestors who were nevertheless members of the same, rather than different, biological populations.

3.10 Research Agendas that Could Flow from the Thought Model

An immediate test of any model would be an assessment of whether or not its predictions correspond in important ways with known circumstances. To conclude, I will suggest items for a research agenda that could arise from the thought model while also making an assessment of model predictions against currently known genetic and archaeological records. The purpose of laying out such an agenda is *not* to suggest that specific kin systems from the past can be detected, but rather to *inflect* our understanding of early archaeological records by purposefully evaluating the societal context in which these archaeological records were likely to have formed.

Discontinuities in Archaeological Records

The foremost objective would be to explore genetic data for indications of the degree of relatedness between contemporaneous individuals, as we have seen for Altai Cave, Sunghir, Yana, and Upward Sun River. For any discoveries in which early-period, contemporaneous individuals come to light, it will be informative to determine whether endogamy, agamy, or exogamy prevailed. It is perhaps conceivable that further development of bioinformation statistics may also allow conclusions in this vein based on rare, individual Paleo-Indigenous human remains. As we have seen, internalized versus externalized marriage practices should not be misconstrued as kinship itself. On the other hand, that distinction is pivotal with regard to elementary decisions affecting the principles of group formation that can be developed from a given semantic framework.

Genetic relatedness will of course remain important in determining broader forms of human relationship. As already noted in connection with the extraordinary White Sands[11] discoveries, we do need to bear in mind that not every effort at settlement will succeed. A small founding population, systematically retaining spouses in a local group growth strategy, could establish a population nucleus that had initial success, but for the variety of reasons noted in the thought model, would eventually fail. The geoarchaeological odds of discovering such a circumstance may not be large, but this is both a conceivable and a potentially detectable outcome.

While one should not overread a single record, the Upward Sun River children reveal an interesting pattern. In a seasonal fishing and hunting encampment, three perinates died. Two of those children had different mothers but were nevertheless very closely related – precisely the pattern one would predict for a local group growth scenario in which unfortunate events could unfold. The Ancient Beringian lineage to which the children belonged persisted for about a millennium more as aDNA from Alaska's Trail Creek individual shows. On the other hand, the entire Ancient Beringian lineage *did* eventually come to an end without further issue, suggesting that it must ultimately have been affected by events like these. It is by no means speculative, therefore, to think that detectable archaeological records of several human generations could be created by early populations that were unable to persist in the longer term.

This certainly provides food for thought with respect to the remarkable White Sands (New Mexico) footprints, the enigmatic Bluefish Caves (Yukon) record, sites like the recently reported Hartley Mammoth locality in New Mexico, and other earlier Western hemisphere sites. There are some convincing cut marks on Bluefish Cave bones that begin 23,000 years ago and extend throughout the LGM (suggestive of human butchery), but they provide a rather faint signature (Bourgeon *et al.*,

---

[11] Footprints are among the rarest but most evocative finds that archaeological records can provide. Many of the fooprints reported for White Sands were of children and adolescents. Bennett *et al.* (2021) provide a number of reasons why these age groups might be overrepresented, but, on the face of it, the White Sands population at this time was not struggling. On the other hand, one female adult made a trip in one direction, carrying a child that was put down briefly from time to time, after which the journey resumed. The same person returned without the child; perhaps some form of duress, such as an illness, might have caused this trip (National Park Service, 2021).

2017; Bourgeon and Burke, 2021). Burin and microblade technology is also present, but the extent to which it is associated with LGM-aged bone is not clearly established. The Bluefish Caves would be situated at the far northeastern extremity of that Beringian cul-de-sac and would have been occupied at a time when many populations had retreated southward from northern Siberia. Under those circumstances, some tentative and impermanent probing by populations at the periphery of the entire human range might be all that could be reasonably expected.[12]

Should human aDNA evidence emerge that links these 20,000–40,000-year-old sites to the founding Indigenous population, a *very* substantial reworking of genetic and archaeological data (the "tidy picture" mentioned earlier) would be required. It could be, however, that the White Sands and other early populations would *not* be genetically related to either the NNA or SNA lineages that did establish a sustained and continuous archaeological record throughout North America from ~16,000 years onward (cf. Ives, 2015; Potter *et al.*, 2018; Fiedel, 2022). In this case, we would be speaking of archaeologically detectable colonizing efforts that were to a greater or lesser degree unsuccessful.

In August 2022, findings from the 37,000-year-old Hartley Mammoth locality in New Mexico were published (Rowe *et al.*, 2022). The authors present thoroughly developed evidence of human butchery of mammoths that was accomplished with fractured bone tools and a simple stone tool inventory. This locality would join other early sites that appear to reflect mammoth butchery (see Holen and Holen, 2013, with references to sites like Lovewell and La Sena). It is possible that there will be new chapters to be written in this very early time period, which may have involved technologies more heavily reliant on osseous raw materials and simple lithic inventories (contrasting with well-developed late or Upper Paleolithic lithic assemblages characteristic of the post-16,000 year interval). Rowe *et al.* (2022) accept that founding northern and southern lineages came to dominate in North America after 16,000 years ago, but raise the possibility that archaeological findings from this very early time range could reflect the presence of ghost population Y that would vanish in North America, but persisted long enough to leave a genetic signature in southern lineage South American populations.

The most likely pathway into the Americas during the pre-LGM time frame would then be the vast area between the Rockies and Laurentide ice far to the east.[13] However, given that entire, detectable lineages like the Ancient Beringian one (and perhaps North American instances of ghost populations like Y) did disappear,

---

[12] Intriguingly, Morlan (2003) noted that between 40,000 and 50,000 years ago, new taphonomic processes began to affect eastern Beringian bone assemblages, with green bone breakage and other modifications that *could* be attributed to human intervention (see also Bourgeon (2021) regarding Bluefish Cave mammoth bone fractures that could be of human origin).

[13] The massive impacts of subsequent late Wisconsinan glaciation across Canada would of course not be helpful for search strategies concerning that possibility.

the possibilities of genuine discontinuities in early archaeological records must not be discounted. This in part raises critical geoarchaeological and "search image" challenges; relevant sites may be deeply buried and have less readily recognizable archaeological signatures. We would nevertheless be remiss not to take note that vanishingly low population densities, stochastic birth and death events, and highly endogamous marriage practices could be genuine contributors to the scattered and apparently discontinuous record one would predict from a social perspective. There could be a detectable human presence preceding the current understanding of the founding American lineage by as much as 20,000 years. Whether or not population Y existed long enough for its genetic detection in South American southern lineage representatives, we can certainly conceive of earlier colonization events (with as yet unknown genetic signatures) that did not lead to a lasting human presence.

Patterns of Settlement

Records that yield such ancient genomic data will always be rare, so it is important to consider how modeled variables would be expected to influence other parts of the archaeological record. However findings relevant to the pre-LGM and LGM intervals proceed, when we turn to the time frame in which the northern and southern lineage founders became dominant in the Americas (after 16,000 years ago), one would still expect detectable consequences for purposeful human choices in group-forming principles. Fluted point distributions in the Western hemisphere are interesting to consider in this regard. Figure 3.5 illustrates fluted point densities for combined data from The Paleoindian Database of the Americas and the Western Canadian Fluted Points Database.[14] Anderson and Gillam (2000, 2001) and Moore and Moseley (2001) exchanged perspectives with respect to these distributions. Anderson and Gillam proposed that, in terms of microband–macroband dynamics, two alternative settlement patterns could be predicted from their modeling efforts. In the "string of pearls" alternative, in which ranges for microbands circulating within a macroband had a circular diameter of 400 km, they expected a contiguous series of ranges extending down either lineal north Pacific coastal or ice-free corridor routes (hence the "string of pearls") and then expanding across North America. The alternative "leap frog" pattern involved noncontiguous, spatially discrete episodes of expansion. Anderson and Gillam (2000) argued that the pattern

[14] There are two schools of thought regarding this information, which does need to be interpreted with caution. Both US and Canadian data sources involved avocational collections, as well as local, regional, and national museum collections and locational data. Some authors have suggested that too many biasing factors influence these data: These would include the prevalence of agriculture, collecting intensity, proximity to modern population centers, and environmental productivity (e.g., Prasciunas, 2011). Buchanan (2003) considered similar biasing factors but reached the conclusion that the patchiness in these density distributions had genuine, underlying ancient causes. Here, I adopt the latter perspective, in significant measure because the observed patterning derives from thousands of fluted point instances that I feel reflects a sort of "brute force" sampling.

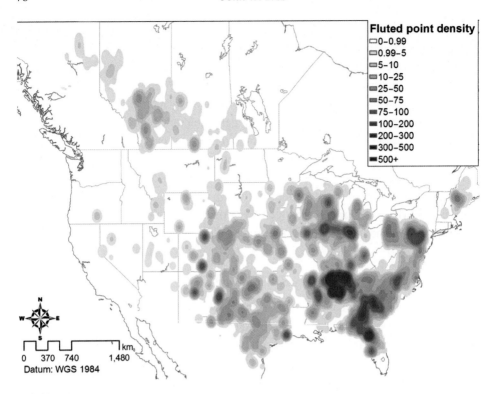

Figure 3.5 A heat map created from density isopleths for fluted point occurrences in Canada and the United States. (Heat map prepared from the Western Canadian Fluted Points Database and The Paleoindian Database of the Americas; data by Kisha Supernant; see also Ives *et al.*, 2013, 2019.)

evident in fluted point densities is consistent with a leap frog style of colonization and not the string of pearls alternative.

Moore and Moseley (2001) responded that only the string of pearls alternative could have been the case, but there were two primary difficulties with their scenario. It began with the assumption that the small population aggregates with which they began modeling had members who were not initially related to each other. As we have seen, this is not realistic. There would have been high degrees of relationship in the small founding population scenarios that geneticists project, so the difficulties of finding a marriageable partner already existed when the founding Indigenous population became isolated from other Northeast Asian populations. Perhaps by virtue of Moore's (2001) extensive experience with Cheyenne kinship principles, Moore and Moseley's (2001) modeling further stipulated that zero-generation crossness would be extinguished in favor of recognizing all as siblings, meaning that, within a generation or two, all members of succeeding generations would be too closely related to each other for marriages to take place in any case.

The ethnographic and thought model provisions reviewed here show that that too is not realistic. Dravidian kin semantics are capable of generating group-forming principles in which highly endogamous marriage practices are the desired outcome. Those same kin semantics are often associated with customs such as polygamy, the levirate, and the sororate that can mediate stochastic imbalances in sex ratios. Human agency is very much at issue here and, as we have seen, there would be many incentives for early Indigenous populations to make conscious choices about the social alternatives available for dealing the late Pleistocene Beringian and Western hemisphere circumstances. Contingent upon the empirical validity of this density distribution, it is quite possible that the underlying patchiness resulted from the frequent choice of the local group growth pathway and its tendency to facilitate population growth and subsequent fissioning, whereby daughter groups may actually be propelled away from the parent population. That pathway would seem distinctly possible for immediate pre-Clovis and early Clovis populations.

Steele (2009) reviewed literature that incorporated more realistic factors like anisotropic environmental gradients and the Allee effect. The Allee effect posits that low-density populations face genuine constraints over successful reproduction; a colonization event may fail if it does not satisfactorily transcend certain threshold values and spatial configurations (Lewis and Kareiva, 1993). Steele found that stochastic factors are likely to be influential in the evolution of a front driven by dispersal behavior of these more complex kinds, concluding that the consequence "may be *a chaotic-seeming series of outbreaks of secondary dispersal foci ahead of the main front*" (Steele, 2009: 131; my emphasis). Steele (2009: 129) and Hazelwood and Steele (2003) also observed that the archaeological response variable for colonization events was *not* first arrival time but rather the cumulative occupancy signature when the archaeological signal is the time-averaged density of discarded artifacts. These quite different modeling efforts also point toward the local group growth alternative as an underlying social cause for such patterning. Incorporating thought model variables in more sophisticated quantitative models holds significant potential for key insights.

Demic Expansion versus Diffusion

One of the significant issues that will continue to be important in understanding how Western hemisphere human history unfolded is our understanding of the fluted point phenomenon. As with Steele's conclusions outlined previously, the predominant view of the Clovis world would be one in which demic expansion into uninhabited environments took place. In this view, the widespread appearance of fluting technology would be the consequence of rapid population growth and fissioning as populations established new regions of land use.[15] It is very likely that high rates of mobility and demic expansion were indeed factors. Yet, it is equally likely that this cannot be the whole story. A more socially oriented interpretation would take into account the prospect for the communication of ideas across preexisting populations.

The alternative perspective would be that new social settings developed once long-term populations existed in the Americas. With sustained presence in eastern Beringia from just prior to 14,000 years ago, plus a number of convincing sites south of the ice masses in the 13,500- to 15,000-year-old time frame, there was a millennium or more in which early northern and southern lineage populations could grow and expand prior to Clovis along a local group growth pathway. Clovis and the broader fluted point phenomenon might then be construed as a demographic and cultural point of "take off," in which relatively isolated, preexisting populations that had experienced success came into more regular contact with each other (e.g., Gruhn and Bryan, 2011; Ives, 2015; Meltzer, 2021; Surovell *et al.*, 2022). From this point of view, fluted point distributions are informative about not just the Clovis phenomenon, but the demographic conditions immediately *before* the advent and spread of fluting technology.

Whether we conceive of leap frog or string of pearls patterns of settlement, this would mean that ideas (technological or otherwise) could diffuse rapidly among already established populations. The speed with which the unusual idea of basal thinning or fluting reached both the North Slope of Alaska and Tierra del Fuego would strongly suggest that more than one factor was at work – ideas could be transmitted by cultural diffusion much more rapidly than through demic expansion. It is speculative to be certain, but Clovis technology and the fluting idea may well have been accompanied by a number of other ideas. Perhaps Clovis reflected the "opening out" of early populations that had, of necessity, previously focused on a more endogamous, local group growth lifeway, but began shifting into a more outward-looking, exogamous local group alliance framework in which both people and ideas began to circulate more widely.

One avenue to explore lies in instances in which a material culture "tell" may be informative of interactions. The provenance of the spectacular Fenn cache artifacts, believed to come from the northeastern Utah-Wyoming border area, has been the subject of some discussion (Frison and Bradley, 1999). The dozens of artifacts include large, carefully thinned bifaces, Clovis points, and bifacial preforms for

---

[15] Here I reiterate that Cannon and Meltzer (2022) recently applied optimal foraging modeling leading them to the conclusion that high rates of mobility connected with the pursuit of previously unhunted large game animals in the Clovis time frame were insufficient to explain the rapid expansion of fluted point technology and that other demographic and social factors must have been involved.

those points, all made from aesthetically pleasing, high-quality toolstones. If the assemblage truly reflects a Paleo-Indigenous cache, two aspects of the assemblage can be singled out. One of the Fenn cache Clovis points is virtually identical to the "lobed" bases of fluted points from the Colby mammoth kill in Wyoming, sharing a basal form not replicated elsewhere in the Clovis world. There is also a carefully made "crescent," an artifact form that is widespread in California and the Great Basin and is considered by some as possibly used in taking waterfowl (Moss and Erlandson, 2013). The obsidian artifacts in the Fenn cache came from the Malad source in southern Idaho (Hughes, 1990). It could be that the Fenn cache reflected contributions from groups with different technological traditions, with the crescent being typical of the Western Stemmed Point tradition, intermingled with classic Clovis forms. While the Northwest is the heartland of the Western Stemmed Point tradition, there certainly is extensive geographic overlap of fluted points as well as another spectacular Clovis find, the Richey–Roberts cache from the Wenatchee area of Washington (Gramly, 1993). The interdigitation of fluted point finds and the longstanding stemmed point tradition (which for earlier sites predate and then are contemporaneous with Clovis) would suggest that there were opportunities for interactions between two different cultural traditions of shared genetic origin (see also Beck and Jones, 2010; Rhode *et al.*, 2022).

Looking northward, there has been an unfortunate tendency among many archaeologists, and certainly media outlets, to play a "zero sum" game with respect to pathways into the Americas. If the ice-free corridor was not available in an early enough time frame to explain sites south of the Late Wisconsinan ice masses, and a north Pacific coastal route was available, then the significance of the ice-free corridor recedes into irrelevance in this way of thinking. Yet, when we know there are distinct northern and southern founding lineages that became genetically isolated from each other, the presence of two pathways into the Americas (with slightly different time frames for availability) may be highly relevant to the processes through which the Americas became fully inhabited (Achilli *et al.*, 2013; Potter *et al.*, 2018; Waters, 2019; Willerslev and Meltzer, 2021).

The massive coalescence of Laurentide and Cordilleran ice along Alberta's eastern slopes would have made that region uninhabitable during the LGM, until such time as those ice masses parted just after 15,000 years ago (Margold *et al.*, 2019). This is one region of the Americas where we can be certain that no one was present until physical deglaciation took place and there was baseline biotic habitability. By 14,000 years ago, however, there were established Indigenous populations in eastern Beringia as well as south of the ice masses. The southern funnel of the ice-free corridor became inhabited 13,450 years ago, when horse and camel hunters took down their prey at the Wally's Beach site in the St. Mary Reservoir of southern Alberta (Waters *et al.*, 2015; Devièse *et al.*, 2018). Roughly 500 years later, the Anzick child, with its classic Clovis assemblage, was buried not far to the south, in Montana. In recent years, it has become apparent that the Alaskan instances of fluting or basal thinning are noticeably younger than the principal Clovis time range, and are typically ~12,500 years of age (Smith and Goebel, 2018; Buvit *et al.*, 2019). This has led a number of investigators to believe that the idea of fluting spread northward through the deglaciating corridor region (see also Ives *et al.*, 2013, 2019). The predominant reasoning would be that this too involved demic expansion, but there are grounds for thinking that there were Indigenous ancestors in the corridor before the onset of Clovis (notably, the complete predominance of local or regional raw materials for corridor fluted points plus the age of the Wally's Beach finds; see Ives, 2006; Ives *et al.*, 2013, 2019, regarding fluted points and raw materials in western Canada).

There are some detectable shifts in the nature of basal thinning or fluting strategies in early projectile points from the ice-free corridor and eastern Beringia. While fluting strategies south of the ice masses tend to involve one or two channel flakes to achieve their objective, in a number of corridor, Alaskan, and Yukon points, we see instead multiple, parallel to subparallel basal thinning flakes that are reminiscent of the microblade and burin technologies common in eastern Beringia, when fluted and basally thinned points appear there (roughly two millennia after Swan Point). These differences are illustrated in Figure 3.6. It is possible that we are seeing highly skilled stone tool artisans from northern settings basally thinning points in a way that is familiar to them, achieving that objective in a manner different from skilled artisans to the south. This is a hypothetical idea in the fluted point time range that merits further testing (see also Smith, 2019). This idea ceases to be hypothetical about 12,000 years ago when actual microblade technology does show up at Charlie Lake Cave in northeastern British Columbia and Vermilion Lakes in southwestern Alberta (see Figure 3.6) (Fedje *et al.*, 1995; Driver *et al.*, 1996; see Heintzman *et al.*, 2016, for the most recent Charlie Lake Cave dates). Whether that presence involved the movement of people or ideas, these are northern influences appearing far to the south. This heightens my suspicion that the deglaciating corridor may well have been a massive confluence zone in which people and ideas circulated (Ives *et al.*, 2019). High degrees of exogamy with intermarriage among different groups would certainly foster such circumstances, allowing the transmission of an idea like fluting.

"Exotic" Toolstones: High Seasonal Mobility or Early Exchange Systems?

A recurrent theme in Paleo-Indigenous studies has involved raw stone materials or toolstones that were used to make lithic assemblages, particularly projectile points. Clovis caches and fluted points more generally are notable for the high-quality,

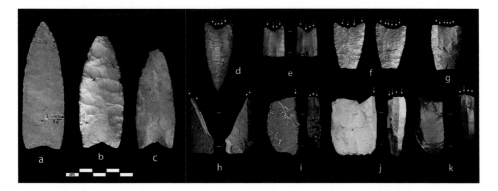

Figure 3.6 "Classic" fluted points (a–c) from the ice-free corridor where one or two broad channel flakes were removed to thin the base; Alberta and Yukon examples of points (d–g) where basal thinning has been accomplished with a series of finer parallel to subparallel flakes; a dihedral burin and three southern Alberta microblade cores (h–k) showing the similarity of burin spalls and microblade core fluted faces to the basal thinning technique seen in (d–g). Photo credits: (a) Anzick, Montana fluted point (University of Alberta archival photo); (b) Dyck collection fluted point found in a construction trench in Lethbridge, Alberta; (c) Clearwater Pass Clovis point, Banff National Park (photographs [b] and [c] courtesy of the Royal Alberta Museum); (d) Donaldson collection fluted point, Eaglesham area, northwestern Alberta; (e) fluted point from the Dog Creek site (NcVi-3), Yukon, closely resembling a number of Alaska specimens (Esdale et al., 2001; photograph courtesy of the Canadian Museum of History); (f) basally thinned point from the Grande cache area (photographs courtesy of Matt Rawluk, Elizabeth Robertson, and Kyle Belanger), northwestern Alberta; (g) basally thinned point from the Oyen area, southeastern Alberta; (h) dihedral burin from the Dog Creek site (NcVi-3), Yukon (photograph courtesy of the Canadian Museum of History); (i) biface fragment used as a microblade core, Vermilion lakes (photograph courtesy of Gwyn Langemann, Parks Canada); (j, k) High River (Cody Complex) microblade cores (photographs courtesy of Martin Magne).

aesthetically appealing toolstones that were frequently employed. There were no doubt several reasons for this. Fluted point makers were the first widespread inhabitants of the Americas, and they had initial access to many of the prime, never-before-used raw stone sources. In some cases, this would include recently deglaciated areas that may have had greater degrees of surficial exposure as revegetation began. Whatever the case, it would seem that Indigenous ancestors in this time frame were adept prospectors and that, beyond this, there were highly skilled artisans who often took an aesthetic interest in the attractive raw materials with which they worked, well beyond the functional requirements of the artifact.

Bearing in mind that this was a pedestrian era, these raw materials were often transported considerable distances (200–600 km) where they then appear in bison or mammoth kills, caches, and campsites. This has generally been attributed to high rates of mobility, whether seasonal, residential, or connected with a task group (with the objective of the activity being to secure the toolstone). These strategies appear to have been subject to change through time. Ellis (2011), for instance, was able to show that in the Great Lakes and the Northeast region more generally, toolstones for earlier Clovis assemblages were acquired from significantly greater distances than that of post-Clovis foragers. Those post-Clovis foragers were beginning to undergo the economic and cultural regionalization upon which the diversity of Western hemisphere societies would be founded. Boulanger *et al.* (2015) went on to use geochemical evidence acquired from neutron activation analysis of stone flaking debris from the Paleo Crossing site, a 12,900-year-old Clovis camp in northeastern Ohio: The majority of stone raw material at the Paleo Crossing originated from the Wyandotte chert source area in Harrison County, Indiana, a straight-line distance of 450–510 km. These authors attributed these patterns to geographically widespread hunter-gatherer social networks linked to a colonization phase of fluted point expansion. Loebel (2009) provided a specific example from the Withington site (Clovis/Gainey) in Wisconsin with an endscraper made from Cobden chert. The chert came from a source 585 km to the south in Illinois. Loebel (2009: 243) suggested it might signal the exogamous movement of a female band member from a neighboring group that more routinely ranged to the south.

While there has been a common expectation in fluted point studies that movement of exotic toolstones over large distances arose from high degrees of mobility, there are important questions to be asked in this connection, like authors such as Speth *et al.* (2013) and Amick (2017) have done. While a settling-in process would require exploration to find suitable raw materials, once they were know, distant toolstone sources could be sought out for ceremonial purposes, rites of passage, and epic voyaging situations. Patterned use of stone tool sources requires specific examination in each case. This research was funded by the National Partk Service under CESU Master Cooperative Agreement P16AC00003. The Great Lakes and the Northeast region pattern reported by Ellis (2011) unfolded as ice retreated around the Champlain Sea. The ice-free corridor constitutes another major region that deglaciated and then saw human expansion, in a slightly earlier time frame. Here, however, the situation was decidedly different. The Western Canadian Fluted Points Database points are made almost exclusively of local and regional raw materials (97 percent of them are quartzites, siltstones, mudstones, pyrometamorphics, miscellaneous cherts, and others; see Ives, 2006, 2015, as well as Ives *et al.*, 2013, 2019, for the discussion that follows). There is a single instance of a Knife River Flint fluted point from within the ice-free corridor, and a probable obsidian example sourced to Timber Butte, Idaho. Logically, one would expect that, if the appearance of fluting technology followed from a straightforward demic expansion to the Northwest (or structured seasonal use of the deglaciating corridor region from points to the south), there would be a significant lithic "founder effect" as stone tools prepared from more abundant, high-quality raw materials in the adjacent northern tier of states entered tool kits and were then discarded at locations in present-day Alberta. This is clearly not the case. In fact, many of these fluted points are at the end of their use lives, having been broken, re-sharpened, and discarded at locations suggestive of dispersed phases of a subsistence settlement system.

Perhaps rare "exotic" discards from an initial settlement pattern were simply swamped once fluted point populations were well established inside the expanding corridor. This seems unlikely, especially when there are other viable explanations. It is important to remember that the butchered camel and horses in the St. Mary Reservoir have been dated to 13,450 years of age by the best available methods. Fluted points are nearby, but not directly associated with those kills. The weaponry used in the kills is not known but, with such secure dating, we must conclude that these were pre-Clovis hunters or, at the very least, the earliest Clovis representatives. We must also recall that eastern Beringia had become continuously occupied by just over 14,000 years ago. Those populations could certainly be considered to be preadapted to the deglaciating "tail" of the mammoth steppe or Arctic steppe tundra extending from Alaska and the Yukon toward the Great Lakes. These were also lower latitude settings with rich, newly exposed landscapes potentially creating high-quality forage for game animals that may well have attracted human exploration from the north as much as from the south (e.g., Turner *et al.*, 1999; Ives, 2006). There is, therefore, a significant prospect that Indigenous populations were already present in parts of the corridor when the notion of fluting arose. Northern populations would have traversed hundreds of kilometers of terrain in which only locally or regionally available toolstones could be accessed.

The data indicate that this pattern holds for the remainder of the fluted point interval and that too is intriguing. As noted earlier and in many other publications, exotic toolstones were quite often transported over great distances during the fluted point time frame in the co-terminus United States. This is linked by many researchers to geographically widespread hunter-gatherer social networks. One implication of this would be that, unlike many other Clovis situations, the corridor fluted point makers were relatively isolated from other fluted point-using groups below the continental ice masses. This is of significant interest from the perspective of genetic findings, in which there is a clear separation of northern and southern lineages of the founding Indigenous population. If, as noted earlier, that separation took place at an early date south of the North American ice masses, it might mean that the northern lineage entered the rapidly deglaciating southern funnel of the ice-free corridor after 15,000 years ago and became relatively isolated there. If that separation had taken place in Northeast Asia or Beringia, it would suggest that a southward movement through the deglaciating corridor could have been a contributing factor to the isolation process. The divergence of northern and southern lineages is truly one of the singular features of the Western hemisphere record, for which a deeper understanding would be most welcome.

Perhaps the unusual (microblade core-like) basal thinning described earlier reflects that isolation breaking down through both demic expansion and diffusion. Although it requires more exploration, it should also be noted that various stemmed points present both in the St. Mary Reservoir and in Alberta more generally are difficult to distinguish from Western Stemmed Point tradition examples (i.e., Haskett, Parman, Cougar Mountain, Lind Coulee, Windust, and Lake Mojave varieties in the Pacific Northwest) or, in other cases, Alaskan stemmed points (Sluiceway and other oblanceolates) (Ives, 2021a, b). Both the Western Stemmed Point and Alaskan examples come from traditions that span roughly the 13,000- to 8,000-year interval, although it is likely that many of the Alberta instances pertain to the post-fluted point era. Their presence nevertheless suggests other vectors of interaction in a subsequent "cultural confluence" setting.

It is at this point that another phenomenon decidedly out of phase with the remainder of the North American record takes place, in which economies, cultural spheres, and toolstone use were generally becoming more regionalized. The succeeding Cody complex, well known from the Plains and adjacent regions, takes shape initially with a stemmed point style known as "Alberta," but is quickly accompanied by elegant Scottsbluff and Eden dart or spear tips that span ~11,500 to 9,000 years of age. Cody complex sites involve both substantial kills of evolving bison chronospecies and generalized hunting in foothills and montane settings. Cody complex points and an unusual, asymmetric form of knife (the Cody knife) are common in Alberta throughout what was once the corridor region, although Laurentide ice was by then far to the east, beyond Glacial Lake Agassiz (Figure 3.7 shows examples).

Of the more than 1,000 documented specimens, the pattern of raw material use can be determined for more than 400 specimens (Dawe, 2013). It is the obverse of the corridor fluted point interval: More than 40 percent of Alberta points and Cody knives are made of Knife River Flint, an "exotic" toolstone (a beautiful honey- to chocolate-colored silicified lignite, much preferred by both modern and ancient stone tool knappers) with sources in Dunn and Mercer counties along the Missouri in North Dakota (Kristensen *et al.*, 2018). Ahler (1986: 105) estimated that Knife River Flint was used for the production of roughly 640 million tools or cores, examples of which have been found over an area 3.7 million square kilometers in interior North America. In the corridor region, Knife River Flint artifacts are found in all time periods, roughly as commonly as obsidian from Idaho and Oregon sources. These rarer instances are characteristic of "down-the-line" exchange

Figure 3.7 A cache or tool kit of heavily patinated Knife River Flint artifacts from the Wally's Beach site, St. Mary Reservoir, southern Alberta, including two Alberta points (a, b), seven end scrapers (c, middle row), and six fragments or flakes (d, lower row). To the right are weakly or unpatinated Knife River Flint Alberta points from the Edmonton area (e, f), an Alberta point from the Peace Country of northwestern Alberta (g) (Ogrodnick collection), and a Cody knife from the Brazeau Reservoir, west central Alberta (h). (Photograph (g) courtesy of Todd Kristensen and (h) courtesy of Courtney Lakevold; all of the other photographed artifacts are courtesy of the Royal Alberta Museum.)

practices, whereby the frequency of valued items drops sharply in a "distance-decay" pattern as one moves away from a source area.

The Cody complex pattern in western Canada is so striking because it is quite literally the case that in looking at collections from Alberta that, for every four Cody complex diagnostics one sees, one or two of those items will be made of Knife River Flint,[16] a material prized throughout the millennia, but never to the extent we see with the Cody complex. These figures pertain specifically to the Cody complex diagnostics, but there are cache and other instances in which associated artifacts are also of Knife River Flint (see Figure 3.7 for an example from the Wally's Beach site, St. Mary Reservoir, in southern Alberta and other Knife River Flint Cody complex artifacts from Alberta). In extreme southeastern Alberta or southern Saskatchewan, these frequencies might be explained by high degrees of early-period mobility and

[16] Dawe's (2013) Cody complex figures for a sample of 475 artifacts for which the toolstone could be accurately determined were as follows: 43.8 percent Knife River Flint for Cody knives, 42.7 percent Knife River Flint for Alberta points, 25.6 percent Knife River Flint for Scottsbluff points, and 21.9 percent Knife River Flint for Eden points.

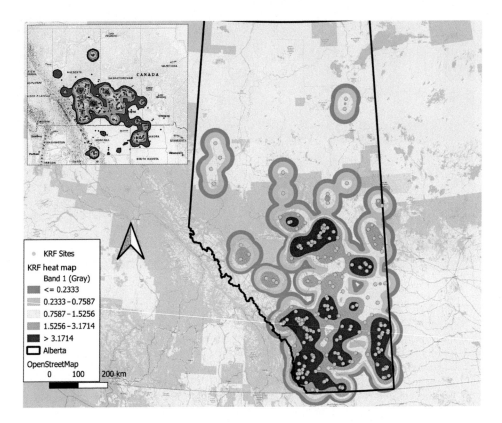

Figure 3.8 A heat map of Knife River Flint artifact densities for sites in Alberta. The inset in the upper left shows the unusual concentration of Alberta, Scottsbluff, and Eden points as well as Cody knives in western Canada and the northern tier of states, including the Knife River Flint source area in North Dakota. (Arc GIS maps prepared by Cody Sharphead with data from the Archaeological Survey, Historical Resources Division, Alberta Culture, the Saskatchewan Heritage Conservation Branch, the Historic Resources Branch, Manitoba Sport Culture and Heritage, Montana State Historic Preservation Office, and North Dakota State Historic Preservation Office.)

proximity to the North Dakota source area. However, proceeding as far away as the St. Mary Reservoir, straight-line distances to the source area stretch to 800 kilometers (as shown in the heat maps in Figure 3.8). This pattern holds into north central Alberta, where we are now 1,000–1,200 km in straight-line distance. Cody complex Knife River Flint artifacts are even present in northwestern Alberta and northeastern British Columbia, at distances of 1,500–1,600 km from the North Dakota sources.

The straight-line distances are hardly likely to have been followed and, therefore, the extraordinary travel requirements simply cannot be explained by high degrees of seasonal mobility. Other social factors must have been involved. Even for just Cody complex points and knives, episodic forms of transport, such as young men undertaking voyages to the Knife River Flint source area as a rite of passage, seem inadequate to explain the phenomenon. Root (1997) and Root *et al.* (2013) described specific instances of systematic overproduction of Knife River Flint bifacial preforms within the source area. Skilled Cody complex artisans made dozens of preforms that would outstrip individual needs over the course of a year, strongly suggesting to him that there was a degree of role specialization taking place, with surpluses being produced for exchange purposes. It is difficult to conceive of such an extensive interaction sphere in which high levels of exchange in material goods were taking place, without comparable levels of social inter-action. Wide circulation of spouses for outward-looking, exogamous societies would certainly foster the conditions required for this remarkable, late Paleo-Indigenous phenomenon, just prior to the onset of the regionalized lithic assem-blages typical of the Holocene Archaic.

Taking a societal perspective also causes us to look at other aspects of the early archaeological record in different ways. Environmental change in the Cody time frame would certainly be the order of the day, as major North American biomes took shape in the early Holocene. For a time, however, the former corridor region may have offered a unique set of opportunities, with a post-glacial landscape that remained well watered despite severe warming trends, as well as a mixture of grasslands and gallery forests that did not yet have their full complement of tree and other species. This is well illustrated by the Fletcher site in southern Alberta, where what is today dry prairie once featured a large, freshwater lake, beside which Cody complex hunters ambushed several large bison (see Vickers and Beaudoin, 1989). Many of today's large alkali lakes in southern and central Alberta likely were freshwater then; to the north, the closed boreal forest was only beginning to take shape. In these terms, one group of late Paleo-Indigenous societies may have been responding to unique lifeway opportunities that would be a final echo of the possibilities that had existed in the preceding two millennia (Ives *et al.*, 2014).

Genetic Processes and Events and the Archaeological Record

A societal perspective is also helpful for reconnoitering when the archaeological record might indicate that specific genetic processes and events were likely to take place. Moreno-Mayar *et al.* (2018b) and Scheib *et al.* (2018) both concluded that there was admixture between some members of the founding northern and southern lineage populations, after which there was southward movement of those admixed populations in North and Central America. They situate that admixture in either the Pacific Northwest or the Great Lakes regions. While that is possible, the corridor region of which we have been speaking must surely be a candidate region to consider. The time frame in which that admixture took place is currently uncertain, but it is possible to suggest temporal thresholds when such an admixture would be likely. Perhaps that is the message that the Wally's Beach site and the Anzick child convey: that northern lineage peoples were already in the corridor, in proximity to the Anzick child population, which was of southern lineage heritage (as were Western Stemmed Point tradition members). If for any reason genetic evidence comes to show that that northern and southern clade admixture took place some-what later in time, the Cody complex circumstances would definitely have featured conditions that would have promoted admixture.

Reinterpreting Other Forms of Information

Should others find them of value, the local group growth and local group alliance characteristics in Tables 3.2 and 3.3 could be used to generate ideas and perhaps testable hypotheses with archaeological and genetic data. Here I will suggest two alternatives that could be more fully explored.

For the first of these, there has been debate about whether early Paleo-Indigenous societies were specialists in hunting terminal Pleistocene megafauna or were more generalized foragers. On the one hand, there does seems to have been a Western hemisphere predilection for hunting proboscideans including mammoth, mastodons, and gomphotheres, a pattern sometimes thought to be archetypal for Clovis (e.g., see Surovell and Waguespack, 2008, for one discussion). On the other hand, that pattern is occasionally contrasted with findings such as those from the Pennsylvania Clovis site Shawnee-Minisink, with its evidence of fish and plant remains (including hawthorn seeds and hickory nutshell fragments), suggestive of more generalized foraging (Gingerich, 2011). The thought model would encourage us to think along longitudinal lines in these respects. Societies with microbands following local group growth pathways would be constrained to broad-spectrum foraging practices in the early stages of their developmental processes. Communal large-game hunting, whether of proboscideans, bison, horses, or caribou, is invariably situated in a social context. It requires a larger number of people to execute some type of drive, intercept, or surround strategy and, similarly, a larger number of people to process the meat, hides, and other products of a successful hunt. When such small coresident groups experienced demographic success, they would literally grow into a size range in which the labor needs of communal hunting could be met directly by the local group. We thus predict episodic, irregular approaches to communal game hunting in this alternative, with the options conditioned by where a coresident group would be in its developmental process.

Local group alliance systems are equally capable of communal hunting, but they achieve this in a different way (see Ives, 1990, for a fuller treatment). The social context for communal game hunting resides in the web of exogamous ties shared among local groups that would facilitate sustained and regular interaction. By linking themselves together in large seasonal gatherings, predominantly exogamous groups could engage in systematic, relatively regular communal hunting. Storck (2006) suggested that small sets of activity loci, such as those seen at encampments like the Thedford II site in southwestern Ontario, were created by small coresidential groups. The larger array of activity loci at Bull Brook was to him the outcome of several coresident groups coalescing for communal caribou hunting.

Pioneering populations need not be characterized as megafaunal specialists at the expense of broader spectrum foraging or vice versa. Instead, an intergenerational perspective on group-forming principles lets us grasp when and how these different tactics would be applied during phases in the life histories of small hunter-gatherer populations. What appear to be episodes of intensive mammoth hunting in the San Pedro valley of Arizona at sites such as Murray Springs, Naco, and Lehner Ranch or the Colby locality in Wyoming can thus be seen in a different light, as can the probable caribou intercept strategy revealed at Bull Brook.

A second productive area of analysis concerns the milieu in which learning and transmission of ideas would take place. In terms of material culture, face-to-face interaction would be the primary means of learning. Gradients in the rate of endogamy through exogamy will condition the degree of circulation of personnel in a society, and this in turn could be expected to have impacts on the transmission of knowledge about material culture. In the local group growth alternative, there is more limited circulation of people in the wider regional society precisely because of endogamous retention of personnel. When there are high rates of endogamy, we might logically expect the formation of style enclaves as material culture drift or innovation takes place. Because exogamy systematically promotes the circulation of people more widely beyond their natal groups, one would expect a greater degree of homogeneity to arise for larger spheres of material culture.

The thought model would suggest ideas for exploring cultural transmission in variable but intimate circumstances. Savishinsky (1974) noted that Hare (Sahtu Dene or North Slavey people) felt that sled dogs should be hitched either as groups of brothers or as groups of sisters because they had established working relationships from their birth and early family life, allowing for greater harmony. While certainly a reflection of detailed knowledge of sled dogs, this was also perhaps a projection of human values, arising from the widespread Mackenzie Basin idea of uniting same-sex sibling sets. In those human instances, same-sex sibling working relationships are privileged, and it would be groups of brothers and groups of sisters (not simply one's parents) that provided role models and instruction for young men and women. In other Dene cultural settings, it is actually the same-sex affinal working relationship that is the more privileged one. McClellan (1975) described for the southern Yukon Tutchone how a man's brother-in-law would be a critical work partner in life and, similarly for a woman, her sister-in-law.

In the last two decades, sophisticated geometric morphometric studies of fluted and stemmed points from large areas of North America have become common. These studies allow the analysis of patterns in shape and other attributes. In most cases, the approach has been linked to models of cultural transmission guided by neo-evolutionary principles, using fluted point attributes as cultural proxies for processes analogous to the biological effects of drift and selective pressures. O'Brien (2019) noted that attributes such as basal configuration or length–width ratios for both regional and continental samples of fluted points produced what appear to be conflicting results. Sholts *et al.* (2012, 2017) and Gingerich *et al.* (2014) applied an array of analyses to attributes including traditional metrics and ratios, flake scar angles, and 3D model-based flaking pattern contouring to samples of fluted points from the Plains and southwestern regions as well as the northeastern USS. Building on this work, O'Brien (2019) concluded that fluted point shapes were prone to processes more like drift, with greater interregional time transgressive variance. Fluted point flaking patterns, of greater structural significance to the artifact, were more broadly shared across North America as the consequence of conforming to ancestral tool-making processes. O'Brien agreed with Sholts *et al.* (2012, 2017) that earlier Clovis populations maintained fairly close social ties, perhaps through regularly congregating at important toolstone sources, thereby limiting interregional variance in flaking patterns on Clovis points. As fluted point populations "settled in" and began to have less contact with each other, shape began to vary considerably. These are topics that could be explored in greater detail at major toolstone sources and at geographic scales that would mirror microband or regional marriage isolate ranges (rather than at more continental scales).

Sholts *et al.* (2012) also proposed that flake scar contours had considerable power for differentiating the techniques of individual knappers (reflecting aspects like handedness and other individual idiosyncrasies) when applied to a mixture of modern replicas made by experts and ancient fluted points. While a degree of inference is necessarily involved, topics at the more intimate scale mentioned previously may also be approachable in the archaeological record. Amick (2004) described the McNine assemblage, a cache of Western Stemmed Point tradition obsidian points (typed as Parman) and preforms from a Nevada cave. He voiced the intriguing idea that a number of flaking platforms on points and preforms "were isolated and prepared but not removed," and that they "may have been left intentionally to serve as static representations of the process of Parman point manufacture"

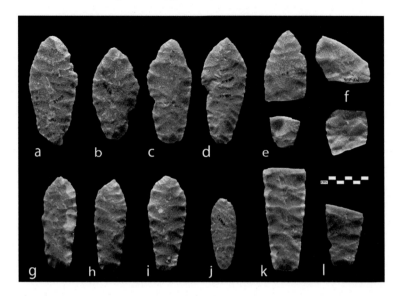

Figure 3.9 Late-stage stemmed point preforms and points of northern quartzite from the Poohkay collection of northwestern Alberta's Peace River country. Note the unfinished flake facets at the tips of (a), (h), and (i), in contrast with the more finished tips in (d) and (e). The bases of (a), (c), and (e) are irregular compared with other examples, such as (k). There are well-defined flaking platforms along the edges of the preforms that could be used for further, medially directed flaking on many of the specimens. While the Poohkay specimens remain large, they were being expertly thinned. Specimen (j) is virtually finished, however, and is much thicker than the other preforms (that would become yet thinner as they were finished). Specimen (j) is unlike the other preforms, but is indistinguishable from Sluiceway points typical of the north slope of Alaska. (Photographs courtesy of the Royal Alberta Museum.)

(Amick, 2004: 139). This may have resulted from a situation involving ritual (the artifacts appeared to be coated with red ochre), but one had didactic intent, for instructional purposes. Figure 3.9 illustrates a cache of that stemmed point and stemmed point preforms from the Peace Country of northwestern Alberta. These beautiful vitreous quartzite artifacts were recovered from a discrete landform in a spatially restricted area.[17] They may reflect a tool kit lost while being transported, but might alternatively have been a votive offering connected with ritual or ceremonial

---

[17] The recovery circumstances for these artifacts, the Poohkay cache, did not allow for dating. On morphological grounds, one would expect them to fall in the 11,000- to 12,000-year time range, although they could conceivably be either earlier or later than this estimate. The more finished artifact of Figure 3.9(j) is not the end goal of the preforms, which had already been thinned to a greater degree than this example. In fact, this much thicker but more finished artifact is indistinguishable from Sluiceway points known from the North Slope of Alaska. This may be another instance of an artifact with more distant connections being incorporated in a cache as noted in an earlier section.

activity. Like the McNine artifacts, note the mosaic of unfinished attributes present in this assemblage. Some of the preforms have well-finished tips, but others retain unaltered facets that would need to be sharpened for there to be a useful tip. Similarly, some bases are squared off as would be intended for the finished point, whereas others are unfinished. The edges of the preforms also have a number of isolated, prepared platforms that would be needed for the final flaking that would take the preform to the finished projectile point form. Perhaps these were simply a mosaic of tasks that had yet to be completed when the artifacts were deposited, but it is also possible that these *were* static representations of remaining steps in manufacture, as Amick wondered, of pedagogical value for master artisans instructing novices.

In a related vein, Lohse (2010: 170–1) assessed how levels of skill might be reflected in Clovis macroblade manufacture. For macroblade cores, he discriminated between the work of master craftspersons, the work of those of moderate skill levels, and the work of those with low levels of skill. In the final category were some instances of egregious knapping errors, along with other instances in which appropriate blade manufacture strategies were being employed, but the individual had not mastered the execution of those strategies. One can readily imagine circumstances in which novices attempted to emulate more skilled individuals or were being guided by those more skilled individuals, with variable results. These are all domains in which the thought modeling process can provide meaningful, underlying structure from which hypotheses could be framed for testing with increasingly powerful analytical methods.

Conclusions

To break down the silos between disciplines and subdisciplines, Menéndez *et al.* (2022) advocated the use of interdisciplinary strategies in the study of human expansion into the Americas. First among their objectives was the need to design research in ways that would incorporate Indigenous perspectives. My hope in providing this chapter is not to insist that the thought model developed here must be correct in all its details or that it would be the only viable approach to modeling exercises. I would certainly encourage other, socially informed modeling approaches that could be developed from different conceptual bases and would definitely advocate that the key variables discussed here be explored in other efforts. That is because it is so difficult to conceive of scenarios in which Indigenous ancestors would be unaware of the social context for their lives in a late Pleistocene world or that these ancestors lacked agency in creating the means to enter the Americas. Until the dramatic Polynesian expansion across the South Pacific, the manner in which Indigenous ancestors arrived in the Western hemisphere and then proceeded to thrive for the next several hundred generations, truly a time immemorial, would stand out as perhaps the most astonishing of all modern human journeys. Applying a social lens to this deeper time frame reinvests Indigenous creativity in our research, as Tallbear, Pitblado, and Menendez *et al.* seek.

It is true that the approach I have taken could be called a "hybrid" of ideas from traditional life and Western (anthropological and other) scholarship. I would stress, however, that the wellspring for this hybrid approach lay in accepting the wisdom of traditional Dene thought about kinship and ways to form and sustain local and regional groups. In this connection, it is helpful to distinguish between two modes of thinking. One might be termed "dimensional," letting us harken back to the sibling cores with which we began. If we take a strictly empirical approach to dimensions of variability, then those idealized like- and unlike-sex sibling cores reveal virtually nothing of consequential difference. To move beyond ambiguous assertions about the importance of ancient social life, our modeling must be infused with specific cultural content. A "generative" approach can draw us closer to Indigenous perspectives because it is concerned with how and why the sibling cores come into existence. There is a strong case that any conceptual or quantitative modeling of Paleo-Indigenous demography and kin systems can be enlivened by including alternatives in which cross–parallel distinctions were an important feature of social life. The process of animating our research in this way can be an integral factor in generating new and productive lines of thought, with testable hypotheses, for archaeology and anthropology, genetics, earth scientists leading focused geoarchaeological search strategies, and even oceanographers considering a North Pacific pathway into the Americas (e.g., Royer and Finney, 2020).

In teaching these concepts, I find it helpful to use an example involving bannock, which became a traditionally important food in the Canadian North and elsewhere. I ask students to describe in the most concise way possible an item of baking like a cookie or a bannock. Students will invariably – from a Western perspective – spend a great deal of time listing the tangible attributes of a bannock. What is its shape or texture? Does it have berries in it? How does it taste? After time and not a little prompting, the odd voice will stumble upon the idea of ingredients – could one indicate what ingredients were used? After more prompting about whether knowing the ingredients would be sufficient to make bannock, a student will eventually hit upon the idea of a *recipe*. If we wanted to know more about making good bannock, we would not engage chemists or physicists. To take a generative approach, we would learn from a good cook or chef *how* to make bannock, that is, what are the ingredients and how are they combined? In so many ways, if we pay attention to Indigenous perspectives on what coresident groups were, we can become party to "recipes" that provided long-lasting, holistic solutions to a wide array of social, political, and economic challenges – up to and including the ancient settlement of the entire Western hemisphere.

References

Achilli, A., Perego, U. A., Lancioni, H., *et al.* (2013). Reconciling migration models to the Americas with the variation of North American native mitogenomes. *Proceedings of the National Academy of Sciences*, **110**(35), 14308–13.

Ahler, S. A. (1986). *Knife River Flint Quarries: Excavations at Site 32DU508*. Bismarck, ND: State Historical Society of North Dakota.

Allen, N. J. (1998). The prehistory of Dravidian-type kin terminologies. In M. Godelier, T. R. Trautmann, and F. Fat, eds., *Transformations of Kinship*. Washington, DC, and London: Smithsonian Institution Press, pp. 314–41.

Allen, N. J. (2008). Tetradic theory and the origin of human kinship systems. In N. J. Allen, H. Calan, R. Dunbar, and W. James, eds., *Early Human Kinship: From Sex to Social Reproduction*. Oxford, UK: Royal Anthropological Institute/Blackwell Publishing, pp. 96–112.

Amick, D. S. (2004). A possible ritual cache of Great Basin Stemmed Bifaces from the Terminal Pleistocene-Early Holocene occupation of NW Nevada, USA. *Lithic Technology*, **29**(4), 119–45.

Amick, D. S. (2017). Evolving views on the Pleistocene colonization of North America. *Quaternary International*, **431**, 125–51.

Ammerman, A. J. (1975). Late Pleistocene population dynamics: an alternative view. *Human Ecology*, **3**(4), 219–33.

Anderson, D. G. and Gillam, J. C. (2000). Paleoindian colonization of the Americas: implications from an examination of physiography, demography, and artifact distribution. *American Antiquity*, **65**(1), 43–66.

Anderson, D. G. and Gillam, J. C. (2001). Paleoindian interaction and mating networks: reply to Moore and Moseley. *American Antiquity*, **66**(3), 526–9.

Asch, M. I. (1980). Steps toward the analysis of Athapaskan social organization. *Arctic Anthropology*, **17**(2), 46–51.

Asch, M. I. (1988). *Kinship and the Drum Dance in a Northern Dene Community*. The Circumpolar Research Series. Edmonton, AB: The Boreal Institute for Northern Studies and Academic Printing and Publishing.

Asch, M. I. (1998). Kinship and Dravidianate logic: some implications for understanding power, politics and social life in a northern Dene community. In M. Godelier, T. R. Trautmann, and F. T. S. Fat, eds., *Transformations of Kinship, the Round Table: Dravidian, Iroquois and Crow-Omaha Kinship Systems*. Washington, DC, and London: Smithsonian Institution Press, pp. 140–9.

Beck, C. and Jones, G. T. (2010). Clovis and Western Stemmed: population migration and the meeting of two technologies in the Intermountain West. *American Antiquity*, **75**(1), 81–116.

Bennett, M. R., Bustos, D., Pigati, J. S., *et al.* (2021). Evidence of humans in North America during the Last Glacial Maximum. *Science*, **373**(6562), 1528–31.

Binford, L. R. (1962). Archaeology as anthropology. *American Antiquity*, **28**, 217–25.

Binford, L. R. (2001). *Constructing Frames of Reference: An Analytical Method for Archaeological Theory Building Using Hunter-gatherer and Environmental Data Sets*. Berkeley, CA: University of California Press.

Boulanger, M. T., Buchanan, B., O'Brien, M. J., *et al.* (2015). Neutron activation analysis of 12,900-year-old stone artifacts confirms 450–510+ km Clovis tool-stone acquisition at Paleo Crossing (33ME274), northeast Ohio, U.S.A. *Journal of Archaeological Science*, **53**, 550–8.

Bourgeon, L. (2021). Revisiting the mammoth bone modifications from Bluefish Caves (YT, Canada). *Journal of Archaeological Science: Reports*, **37**, 102969.

Bourgeon, L. and Burke, A. (2021). Horse exploitation by Beringian hunters during the Last Glacial Maximum. *Quaternary Science Reviews*, **269**, 107140.

Bourgeon, L., Burke, A., Higham, T., and Hart, J. P. (2017). Earliest human presence in North America dated to the Last Glacial Maximum: new radiocarbon dates from Bluefish Caves, Canada. *PLOS One*, **12**(1), e0169486.

Bradley, B. A. and Collins, M. B. (2013). Imagining Clovis as a revitalization movement. In K. E. Graf, C. V. Ketron, and M. R. Waters, eds., *Paleoamerican Odyssey*. College Station, TX: Peopling of the Americas Publications, Center for the Study of the First Americans, Texas A & M University, pp. 247–55.

Brown, C. T., Liebovitch, L. S., and Glendon, R. (2007). Lévy flights in Dobe Ju/'hoansi foraging patterns. *Human Ecology*, **35**(1), 129–38.

Buchanan, B. (2003). The effects of sample bias on Paleoindian Fluted Point recovery in the United States. *North American Archaeologist*, **24**(4), 311–38.

Buchanan, B. and Hamilton M. J. (2009). A formal test of the origin of variation in North American early Paleoindian points. *American Antiquity*, **74**(2), 279–98.

Buvit, I., Rasic, J. T., and Izuho, M. (2022). Archaeological evidence shows widespread human depopulation of Last Glacial Maximum Northeast Asia. *Archaeological and Anthropological Sciences*, **14**(7), 1–11.

Buvit I., Rasic J. T., Kuehn S. R., and Hedman, W. H. (2019). Fluted projectile points in a stratified context at the Raven Bluff site document a late arrival of Paleoindian technology in northwest Alaska. *Geoarchaeology*, **34**(1), 3–14.

Cannon, M. D. and Meltzer, D. J. (2022). Forager mobility, landscape learning and the peopling of Late Pleistocene North America. *Journal of Anthropological Archaeology*, **65**, 101398.

Clark, J., Carlson, A. E., Reyes, A. V., *et al.* (2022). The age of the opening of the Ice-Free Corridor and implications for the peopling of the Americas. *Proceedings of the National Academy of Sciences*, **119**(14), e2118558119.

Coutouly, Y. A. G. and Holmes, C. E. (2018). The microblade industry from Swan Point Cultural Zone 4b: technological and cultural implications from the earliest human occupation in Alaska. *American Antiquity*, **83**(4), 735–52.

Coutouly, Y. A. G. and Ponkratova, I. Y. (2016). The Late Pleistocene microblade component of Ushki Lake (Kamchatka, Russian Far East). *PaleoAmerica*, **2**(4), 303–31.

Crassard, R., Charpentier, V., McCorriston, J., *et al.* (2020). Fluted point technology in Neolithic Arabia: an independent invention far from the Americas. *PLOS One*, **15**(8), e0236314.

Dalton, A. S., Stokes, C. R., and Batchelor, C. L. (2022). Evolution of the Laurentide and Innuitian ice sheets prior to the Last Glacial Maximum (115 ka to 25 ka). *Earth-Science Reviews*, **224**, 103875.

Damas, D. (1969). *Contributions to Anthropology: Band Societies*. Bulletin 228, Anthropological Series 84. Ottawa, ON: National Museum of Canada.

Dawe, R. J. (2013). A review of the Cody Complex in Alberta. In E. J. Knell and M. P. Muñiz, eds., *Paleoindian Lifeways of the Cody Complex*. Salt Lake City, UT: University of Utah Press, pp. 144–87.

Devièse, T., Stafford Jr, T. W., Waters, M. R., *et al.* (2018). Increasing accuracy for the radiocarbon dating of sites occupied by the first Americans. *Quaternary Science Reviews*, **198**, 171–80.

Dikov, N. N. and Titov, E. E. (1984). Problems of the stratification and periodization of the Ushki sites. *Arctic Anthropology*, **21**(2), 69–80.

Driver, J. C., Handly, M., Fladmark, K. R., *et al.* (1996). Stratigraphy, radiocarbon dating and culture history of Charlie Lake Cave, British Columbia. *Arctic*, **49**(3), 265–77.

Dumont, L. (1953). The Dravidian kinship terminology as an expression of marriage. *Man* (New Series), **54**, 34–9.

Eggan, F. (1955a). The Cheyenne and Arapaho kinship system. In F. Eggan, ed., *Social Anthropology of North American Tribes*. Chicago, IL: University of Chicago Press, pp. 35–95.

Eggan, F. (1955b). Social anthropology: methods and results. In F. Eggan, ed., *Social Anthropology of North American Tribes*. Chicago, IL: University of Chicago Press, pp. 485–551.

Eggan, F. (1980). Shoshone kinship structures and their significance for anthropological theory. *Journal of the Steward Anthropological Society*, **11**(2), 165–93.

Ellis, C. (2011). Measuring Paleoindian range mobility and land-use in the Great Lakes/ Northeast. *Journal of Anthropological Archaeology*, **30**(3), 385–401.

Erlandson, J. M., Graham, M. H., Bourque, B. J., et al. (2007). The kelp highway hypothesis: marine ecology, the coastal migration theory, and the peopling of the Americas. *Journal of Island Coastal Archaeology*, **2**(2), 161–74.

Esdale, J. A., Le Blanc, R. J. and Cinq-Mars, J. (2001). Periglacial geoarchaeology at the Dog Creek Site, northern Yukon. *Geoarchaeology: An International Journal*, **16**(2), 151–76.

Fagundes, N. J., Tagliani-Ribeiro, A., Rubicz, R., et al. (2018). How strong was the bottleneck associated to the peopling of the Americas? New insights from multilocus sequence data. *Genetics and Molecular Biology*, **41**(1, suppl), 206–14.

Fedje, D. W., White, J. M., Wilson, M. C., et al. (1995). Vermilion Lakes site: adaptations and environments in the Canadian Rockies during the latest Pleistocene and early Holocene. *American Antiquity*, **60** (1), 81–108.

Fiedel, S. J. (2022). Initial human colonization of the Americas, redux. *Radiocarbon*, **64**(4), 845–97.

Frison, G. C. and Bradley, B. A. (1999). *The Fenn Cache: Clovis Weapons & Tools*. Santa Fe, NM: One Horse Land & Cattle Company.

Gannon, M. I. (2019). The knotty question of when humans made the Americas home. *Sapiens Anthropology Magazine*. www.sapiens.org/archaeology/native-american-migration/.

Gingerich, J. A. M. (2011). Down to seeds and stones: a new look at the subsistence remains from Shawnee-Minisink. *American Antiquity*, **76**(1), 127–44.

Gingerich, J. A. M., Sholts, S. B., Wärmländer, S. K. T. S. and Stanford, D. (2014). Fluted point manufacture in eastern North America: an assessment of form and technology using traditional metrics and 3D digital morphometrics. *World Archaeology*, **46**(1), 101–22.

Goddard, P. E. (1916). The Beaver Indians. *Anthropological Papers of the American Museum of Natural History*, **10**, 202–93.

Godelier, M., Trautmann, T. R., and Fat, F. E. T. S. (1998). *Transformations of Kinship*. Washington, DC, and London: Smithsonian Institution Press.

Gramly, R. M. (1993). *The Richey Clovis Cache*. Santa Clara, CA: Persimmon Press.

Grugni, V., Raveane, A., Ongaro, L., et al. (2019). Analysis of the human Y-chromosome haplogroup Q characterizes ancient population movements in Eurasia and the Americas. *BMC Biology*, **17**(1), 1–14.

Gruhn, R. and Bryan, A. (2011). A current view of the initial peopling of the Americas. In D. Vialou, ed., *Peuplements et Préhistoire en Amériques*. Paris: Éditions du Comité des Travaux Historiques et Scientific, pp. 17–30.

Guthrie, R. D. (1990). *Frozen Fauna of the Mammoth Steppe: The Story of Blue Babe*. Chicago, IL: University of Chicago Press.

Hage, P. (2003). The ancient Maya kinship system. *Journal of Anthropological Research*, **59**(1), 5–21.

Hage, P., Milicic, B., Mixco, M., and Nichols, M. J. P. (2004). The proto-Numic kinship system. *Journal of Anthropological Research*, **60**(3), 359–77.

Halffman, C. M., Potter, B. A., McKinney, H. J., *et al.* (2020). Ancient Beringian paleodiets revealed through multiproxy stable isotope analyses. *Science Advances*, **6**(36), eabc1968.

Hamilton, M. J., Milne, B. T., Walker, R. S., Burger, O., and Brown, J. H. (2007). The complex structure of hunter-gatherer social networks. *Proceedings of the Royal Society B*, **274**, 2195–202.

Haynes, C. V. (2022). Evidence of humans at White Sands National Park during the Last Glacial Maximum could actually be for Clovis people. *PaleoAmerica*, **8**(2), 95–8.

Hazelwood, L. and Steele, J. (2003). Colonizing new landscapes: archaeological detectability of the first phase. In M. Rockman and J. Steele, eds., *Colonization of Unfamiliar Landscapes: The Archaeology of Adaptation*. New York: Routledge, pp. 203–21.

Hebda, C. F.G., McLaren, D., Mackie, Q., *et al.* (2022). Late Pleistocene palaeoenviron-ments and a possible glacial refugium on northern Vancouver Island, Canada: evidence for the viability of early human settlement on the northwest coast of North America. *Quaternary Science Reviews*, **279**, 107388.

Heintzman, P. D., Froese, D., Ives, J. W., *et al.* (2016). Bison phylogeography constrains dispersal and viability of the Ice Free Corridor in western Canada. *Proceedings of the National Academy of Sciences*, **113**(29), 8057–63.

Helm, J. (1961). *The Lynx Point People: The Dynamics of a Northern Athapaskan Band*. Bulletin 176, Anthropological Series 53. Ottawa, ON: National Museum of Canada.

Helm, J. (1965). Bilaterality in the socio-territorial organization of the Arctic Drainage Dene. *Ethnology*, **4**(4), 361–85.

Highway, T. (2022). *Permanent Astonishment: Growing Up Cree in the Land of Snow and Sky*. Toronto, ON: Anchor Canada.

Hill, K. R., Walker, R. S., Božičević, M., *et al.* (2011). Co-residence patterns in hunter-gatherer societies show unique human social structure. *Science*, **331**(6022), 1286–9.

Hlusko, L. J., Carlson, J. P., Chaplin, G., *et al.* (2018). Environmental selection during the last ice age on the mother-to-infant transmission of vitamin D and fatty acids through breast milk. *Proceedings of the National Academy of Sciences*, **115**(19), E4426–32.

Hockett, C. F. (1964). The Proto Central Algonkian kinship system. In W. H. Goodenough, eds., *Explorations in Cultural Anthropology*. Toronto, ON: McGraw-Hill Book Company, pp. 239–57.

Hoffecker, J. F. and Hoffecker, I. T. (2017). Technological complexity and the global dispersal of modern humans. *Evolutionary Anthropology: Issues, News, and Reviews*, **26**(6), 285–99.

Hoffecker, J., Pitulko, V., and Pavlova, E. (2022). Beringia and the settlement of the Western Hemisphere. *Vestnik of Saint Petersburg University. History*, **67**(3): 882–909. https://doi.org/10.21638/spbu02.2022.313.

Holen, S. R. and Holen, K. (2013). The mammoth steppe hypothesis: the Middle Wisconsin (Oxygen Isotope Stage 3) peopling of North America. In K. E. Graf, C. V. Ketron, and M. R. Waters, eds., *Paleoamerican Odyssey*. College Station, TX: Peopling of the Americas Publications, Center for the Study of the First Americans, Texas A & M University, pp. 429–44.

Hughes, R. E. (1990). X-ray fluorescence analysis of obsidian from the Fenn Cache, Wyoming. Unpublished manuscript in possession of the author.

Ives, J. W. (1990). *A Theory of Northern Athapaskan Prehistory*. Boulder, CO, and Calgary, AB: Westview Press/University of Calgary Press.

Ives, J. W. (1998). Developmental processes in the pre-contact history of Athapaskan, Algonquian and Numic kin systems. In M. Godelier, T. R. Trautmann, and F. Fat, eds., *Transformations of Kinship*. Washington, DC, and London: Smithsonian Institution Press, pp. 94–139.

Ives, J. W. (2006). 13,001 years ago – human beginnings in Alberta. In M. Payne, D. Wetherell, and C. Cavanaugh, eds., *Alberta Formed – Alberta Transformed*, Vol. 1. Calgary/Edmonton, AB: University of Calgary/University of Alberta Presses, pp. 1–34.

Ives, J. W. (2015). Kinship, demography and paleoindian modes of colonization: some western Canadian perspectives. In M. D. Frachetti and R. N., Spengler III, eds., *Mobility and Ancient Society in Asia and the Americas*. Cham, Switzerland: Springer, pp. 127–56.

Ives, J. W. (2021a). A Canadian perspective on later Paleoindian technocomplexes and emerging genetic data. Paper presented to the 86th Annual Meeting, Society of American Archaeology, April 15–17, 2021.

Ives, J. W. (2021b). Stemmed points and the ice-free corridor. Paper presented to the Canadian Archaeological Association Virtual Meetings, May 5–7, 2021.

Ives, J. W. (2022). Seeking congruency – search images, archaeological records, and Apachean origins. In J. W. Ives and J. C. Janetski, eds., *Holes in Our Moccasins, Holes in Our Stories. New Insights into Apachean Origins from the Promontory, Franktown, and Dismal River Archaeological Records*. Salt Lake City, UT: University of Utah Press, pp. 27–42.

Ives, J. W. (submitted). Intersecting worlds in Denendeh: reflections on nascent anthropology, the fur trade and Dene kinship. Paper submitted for University of Alberta Press volume celebrating the work of Michael I. Asch.

Ives, J. W., Froese, D., Collins, M., and Brock, F. (2014). Radiocarbon and protein analyses indicate an Early Holocene age for the bone rod from Grenfell, Saskatchewan, Canada. *American Antiquity* **79**(4), 782–93.

Ives, J. W., Froese, D., Supernant, K., and Yanicki, G. (2013). Vectors, vestiges and Valhallas – rethinking the corridor. In K. E. Graf, C. V. Ketron, and M. R. Waters, eds., *Paleoamerican Odyssey*. College Station, TX: Peopling of the Americas Publications, Center for the Study of the First Americans, Texas A & M University, pp. 149–69.

Ives, J. W., Rice, S., and Vajda, E. J. (2010). Dene-Yeniseian and processes of deep change in kin terminologies. *Anthropological Papers of the University of Alaska*, **5**, 223–56.

Ives, J. W., Yanicki, G., Lakevold, C., and Supernant, K. (2019). Confluences: fluted points in the Ice-Free Corridor. *PaleoAmerica*, **5**(2), 143–56.

Jaouen, K., Villalba-Moucoc, V., Smith, G. M., *et al.* (2022). A Neandertal dietary conundrum: Insights provided by tooth enamel Zn isotopes from Gabasa, Spain. *Proceedings of the National Academy of Sciences*, **119**(43), e2109315119.

Johnston, G. (1982). Organizational structure and scalar stress. In C. Renfrew, M. Rowlands, and B. Segrave, eds., *Theory and Explanation in Archaeology*. New York, NY: Academic Press, pp. 389–421.

Kelly, R. L. and Todd, L. C. (1988). Coming into the country: early Paleoindian hunting and mobility. *American Antiquity*, **53**(2), 231–44.

Kitchen, A., Miyamoto, M. M. and Mulligan, C. J. (2008). A three-stage colonization model for the peopling of the Americas. *PLOS One*, **3**(2), e1596.

Klein, R. G. (1973). *Ice-Age Hunters of the Ukraine*. Chicago, IL: University of Chicago Press.

Kobe, F., Leipe, C., Shchetnikov, A. A., *et al.* (2021). Not herbs and forbs alone: pollen-based evidence for the presence of boreal trees and shrubs in Cis-Baikal (Eastern Siberia) derived from the Last Glacial Maximum sediment of Lake Ochaul. *Journal of Quaternary Science*, **37**(5), 868–83.

Krauss, M. E. (1977). The Proto-Athabaskan and Eyak kinship term system. Unpublished manuscript, Alaska Native Language Center Archive. www.uaf.edu/anla/record.php? identifier=CA961K1977b.

Kristensen, T., Moffat, E., Duke, J. M., *et al.* (2018). Identifying Knife River flint in Alberta: a silicified lignite toolstone from North Dakota. *Archaeological Survey of Alberta Occasional Paper*, **38**, 1–24.

Lalueza-Fox, C., Rosas, A., Estalrrich, A., *et al.* (2011). Genetic evidence for patrilocal mating behavior among Neandertal groups. *Proceedings of the National Academy of Sciences*, **108**(1), 250–3.

Lee, R. B. and DeVore, I. (1966). *Man the Hunter*. Chicago, IL: Aldine Publishing Company.

Lesnek, A. J., Briner, J. P., Baichtal, J. F., and Lyles, A. S. (2020). New constraints on the last deglaciation of the Cordilleran Ice Sheet in coastal Southeast Alaska. *Quaternary Research*, **96**, 140–60.

Lesnek, A. J., Briner, J. P., Lindqvist, C., Baichtal, J. F., and Heaton, T. H. (2018). Deglaciation of the Pacific coastal corridor directly preceded the human colonization of the Americas. *Science Advances*, **4**(5), eaar5040.

Levi-Strauss, C. (1963). *Structural Anthropology*. Boston, MA: Beacon Press.

Levi-Strauss, C. (1969 [orig. 1949]). *The Elementary Structures of Kinship*. New York, Basic Books.

Lewis, M. A. and Kareiva, P. (1993). Allee dynamics and the spread of invading organisms. *Theoretical Population Biology*, **43**(2), 141–58.

Loebel, T, J. (2009). Withington (47Gt158): A Clovis/Gainey campsite in Grant County, Wisconsin. *Midcontinental Journal of Archaeology*, **34**(2), 223–48.

Lohse, J. C. (2010). Evidence for learning and skill transmission in Clovis blade production and core maintenance. In B. A. Bradley, M. B. Collins, A. Hemmings, M. Shoberg, and J. C. Lohse, eds. *Clovis Technology*, Archaeological Series 17. Ann Arbor, MI: International Monographs in Prehistory, pp. 157–76.

MacDonald, G. M. and McLeod, T. K. (1996). The Holocene closing of the "Ice-Free" Corridor: a biogeographical perspective. *Quaternary International*, **32**, 87–95.

MacNeish, J. H. (1960). Kin terms of the Arctic drainage Dene: Hare, Slavey, Chipewyan. *American Anthropologist*, **62**(2), 279–95.

Mao, X., Zhang, H., Qiao, S., *et al.* (2021). The deep population history of northern East Asia from the late Pleistocene to the Holocene. *Cell*, **184**(12), 3256–66.

Margold, M., Gosse, J. C., Hidy, A. J., *et al.* (2019). Beryllium-10 dating of the Foothills Erratics Train in Alberta, Canada, indicates detachment of the Laurentide Ice Sheet from the Rocky Mountains at ∼15 ka. *Quaternary Research*, **92**(2), 1–14.

Mason, O. K. (2020). The Thule migrations as an analog for the early peopling of the Americas: evaluating scenarios of overkill, trade, climate forcing, and scalar stress. *PaleoAmerica*, **6**(4), 308–56.

McClellan, C. (1975). *My Old People Say: An Ethnographic Survey of Southern Yukon Territory*, Publications in Ethnology Number 6. Ottawa, ON: National Museum of Man.

Means, B. K. (2007). *Circular Villages of the Monongahela Tradition*. Tuscaloosa, AL: University of Alabama Press.

Meltzer, D. J. (1989). Why don't we know when the first people came to North America? *American Antiquity*, **54**, 471–90.

Meltzer, D. J. (2002). What do you do when no one's been there before? Thoughts on the exploration and colonization of new lands. In N. Jablonski, eds., *The First Americans: The Pleistocene Colonization of the New World*. San Francisco, CA: California Academy of Sciences. University of California Press, pp. 27–58.

Meltzer, D. J. (2003). Lessons in landscape learning. In M. Rockman and J. Steele, eds., *The Colonization of Unfamiliar Landscapes: The Archaeology of Adaptation*. London: Routledge, pp. 222–41.

Meltzer, D. J. (2021). *First Peoples in a New World: Populating Ice Age America*, 2nd ed. Cambridge, UK: University of Cambridge Press.

Menéndez, L. P., Paul, K. S., de la Fuente, C., *et al.* (2022). Towards an interdisciplinary perspective for the study of human expansions and biocultural diversity in the Americas. *Evolutionary Anthropology*, **31**(2), 62–8.

Moore, J. H. (2001). Evaluating five models of human colonization. *American Anthropologist*, **103**(2), 395–408.

Moore, J. H. and Moseley, M. E. (2001). How many frogs does it take to leap around the Americas? Comments on Anderson and Gillam. *American Antiquity*, **66**(3), 526–9.

Moreno-Mayar, J. V., Potter, B. A., Vinner, L., *et al.* (2018a). Terminal Pleistocene Alaskan genome reveals first founding population of Native Americans. *Nature*, **553**(7687), 203–7.

Moreno-Mayar, J. V., Vinner, L., de Barros Damgaard, P., *et al.* (2018b). Early human dispersals within the Americas. *Science*, **362**(6419), eaav2621.

Morgan, L. H. (1966). *Systems of Consanguinity and Affinity of the Human Family*. Smithsonian Contributions to Knowledge 218 (reprint of the 1871 edition). Oosterhaut, Netherlands: Anthropological Publications.

Morlan, R. E. (2003). Current perspectives on the Pleistocene archaeology of eastern Beringia. *Quaternary Research*, **60**(1), 123–32.

Morse, D. F. (1997). *Sloan: A Paleoindian Cemetery in Arkansas*. Washington, DC: Smithsonian Institution Press.

Moss, M. and Erlandson, J. (2013). Waterfowl and lunate crescents in western North America: the archaeology of the Pacific Flyway. *Journal of World Prehistory*, **26**, 173–211.

National Park Service. (2021). White Sands National Park footprints offer a glimpse into the perilous life of a prehistoric mom and child. US National Parks Service. www.travel-experience-live.com/white-sands-national-park-footprints-fossilized-human-tracks/ (last accessed August 9, 2022).

Ning, C., Fernandes, D., Changmai, P., *et al.* (2020). The genomic formation of First American ancestors in East and Northeast Asia. *BioRxiv*, https://doi.org/10.1101/2020.10.12.336628.

Norris, S. L., Garcia-Castellanos, D., Jansen, J. D., *et al.* (2021). Catastrophic drainage from the northwestern outlet of glacial Lake Agassiz during the Younger Dryas. *Geophysical Research Letters*, **48**(15), e2021GL093919.

Norris, S. L., Tarasov, L., Monteath, A. J., *et al.* (2022). Rapid retreat of the southwestern Laurentide Ice Sheet during the Bølling-Allerød interval. *Geology*, **50**(4), 417–21.

O'Brien, M. J. (2019). More on Clovis learning: individual-level processes aggregate to form population-level patterns, *PaleoAmerica*, **5**(2), 157–68.

Oviatt, C. G., Madsen D. B., Rhode, D., and Davis, L. G. (2022). A critical assessment of claims that human footprints in the Lake Otero basin, New Mexico date to the Last Glacial Maximum. *Quaternary Research*, **111**, 138–47.

Pedersen, M. W., Ruter, A., Schweger, C., *et al.* (2016). Postglacial viability and colonization in North America's ice-free corridor. *Nature*, **537**(7618), 45–9.

Pigati, J. S., Springer, K. B., Holliday, V. T., *et al.* (2022). Reply to "Evidence for Humans at White Sands National Park during the Last Glacial Maximum could actually be or Clovis people ~13,000 Years Ago" by C. Vance Haynes, Jr. *PaleoAmerica*, **8**(2), 99–101.

Pigati, J. S., Springer, K. B., Honke, J. S., *et al.* (2023). Independent age estimates resolve the controversy of ancient human footprints at White Sands. *Science*, **382**(6666), 1–3.

Pinotti, T., Bergström, A., Geppert, M., *et al.* (2019). Y chromosome sequences reveal a short Beringian standstill, rapid expansion, and early population structure of Native American founders. *Current Biology*, **29**(1), 149–57.

Pitblado, B. L. (2021). On rehumanizing Pleistocene people of the Western Hemisphere. *American Antiquity*, **87**(2), 217–35.

Pitulko, V., Nikolskiy, P. A., Basilyan, A., and Pavlova, E. Y. (2013). Human habitation in Arctic western Beringia prior to the LGM. In K. E. Graf, C. V. Ketron, and M. R. Waters, eds., *Paleoamerican Odyssey*. College Station, TX: Texas A&M University Press, pp. 13–44.

Pitulko, V. V., Tikhonov, A. N., Pavlova, E. Y., *et al.* (2016). Early human presence in the Arctic: evidence from 45,000-year-old mammoth remains. *Science* **351**(6270), 260–3.

Potter, B. A., Baichtal, J. F., Beaudoin, A. B., *et al.* (2018). Current evidence allows multiple models for the peopling of the Americas. *Science Advances*, **4**(8), eaat5473.

Prasciunas, M. M. (2011). Mapping Clovis: projectile points, behavior, and bias. *American Antiquity*, **76**(1), 107–26.

Prasciunas, M. and Surovell, T. A. (2015). Reevaluating the duration of Clovis: the problem of non-representative radiocarbon dates age estimates for the duration of the Clovis complex. In A. M. Smallwood and T. A. Jennings, eds., *Clovis: On the Edge of a New Understanding*. College Station, TX: Texas A&M University Press, pp. 21–35.

Prüfer, K., Racimo, F., Patterson, N., *et al.* (2014). The complete genome sequence of a Neanderthal from the Altai Mountains. *Nature*, **505**(7481), 43–9.

Pryor, A. J. E, Beresford-Jones, D. G., Dudin, A. E., *et al.* (2020). The chronology and function of a new circular mammoth-bone structure at Kostenki 11. *Antiquity*, **94** (374), 323–41.

Raghavan, M., Skoglund, P., Graf, K. E., *et al.* (2014). Upper Palaeolithic Siberian genome reveals dual ancestry of Native Americans. *Nature*, **505**(7481), 87–91.

Raghavan, M., Steinrücken, M., Harris, K., *et al.* (2015). Genomic evidence for the Pleistocene and recent population history of Native Americans. *Science* **349**(6250), aab3884-1–10.

Rasmussen, M., Anzick, S. L., Waters, M. R., *et al.* (2014). The genome of a late Pleistocene human from a Clovis burial site in western Montana. *Nature* **506**(7487), 225–9.

Reich, D., Patterson, N., Campbell, D., *et al.* (2012). Reconstructing Native American population history. *Nature* **488**(7411), 370–4.

Rhode, D., Smith, G. M., Dillingham, E., Kingrey, H. U., and George, N. D. (2022). The Nye Canyon Paleo site: an upper montane mixed fluted point, Clovis blade, and Western Stemmed Tradition assemblage in western Nevada. *PaleoAmerica*, **8**(2), 115–29.

Ridington, R. (1968a). The environmental context of Beaver Indian behavior. Unpublished PhD dissertation, Department of Anthropology, Harvard University.

Ridington, R. (1968b). The medicine fight: an instrument of political process among the Beaver Indians. *American Anthropologist*, **70**(6), 1152–60.

Ridington, R. (1969). Kin categories versus kin groups: a two section system without sections. *Ethnology*, **8**(4), 460–7.

Robinson, B. S., Ort, J. C., Eldridge, W. A., Burke, A. L., and Pelletier, B. G. (2009). Paleoindian aggregation and social context at Bull Brook. *American Antiquity*, **74**(3), 423–47.

Root, M. J. (1997). Production for exchange at the Knife River flint quarries, North Dakota. *Lithic Technology*, **22**(1), 33–50.

Root, M. J., Knell, E. J., and Taylor, J. (2013). Cody Complex land use in western North Dakota and southern Saskatchewan. In E. J. Knell and M. P. Muñiz, eds., *Paleoindian Lifeways of the Cody Complex*. Salt Lake City, UT: University of Utah Press, pp. 121–43.

Rowe, T. B., Stafford Jr, T. W., Fisher, D. C., *et al.* (2022). Human occupation of the North American Colorado Plateau 37,000 years ago. *Frontiers in Ecology and Evolution*, **10**, 534.

Royer, T. C. and Finney, B. (2020). An oceanographic perspective on early human migrations to the Americas. *Oceanography*, **33**(1), 32–41.

Savishinsky, J. S. (1974). *The Trail of the Hare: Life and Stress in an Arctic Community*. New York, NY: Gordon and Breach.

Scheib, C. L., Li, H., Desai, T., *et al.* (2018). Ancient human parallel lineages within North America contributed to a coastal expansion. *Science*, **360**(6392), 1024–7.

Schroedl, A. R. (2021). The geographic origin of Clovis technology: insights from Clovis biface caches. *Plains Anthropologist*, **66**(258), 120–48.

Seeman, M. F. (1994). Intercluster lithic patterning at Nobles Pond: a case for "disembedded" procurement among Early Paleoindian societies. *American Antiquity*, **59**(2), 273–88.

Shoda, S., Lucquin, A., Yanshina, O., *et al.* (2020). Late Glacial hunter-gatherer pottery in the Russian Far East: indications of diversity in origins and use. *Quaternary Science Reviews*, **229**, 106124.

Sholts, S. B., Gingerich, J. A. M., Schlager, S., Stanford, D. J., and Wärmländer, S. K. T. S. (2017). Tracing social interactions in Pleistocene North America via 3D model analysis of stone tool asymmetry. *PLOS One*, **12**(7), e0179933.

Sholts, S. B., Stanford, D. J., Flores, L. M., and Wärmländer, S. K. T. S. (2012). Flake scar patterns of Clovis points analyzed with a new digital morphometrics approach: evidence for direct transmission of technological knowledge across early North America. *Journal of Archaeological Science*, **39**(9), 3018–26.

Sikora, M., Pitulko, V. V., Sousa, V. C., *et al.* (2019). The population history of northeastern Siberia since the Pleistocene. *Nature*, **570**(7760), 182–8.

Sikora, M., Seguin-Orlando, A., Sousa, V. C., *et al.* (2017). Ancient genomes show social and reproductive behavior of early Upper Paleolithic foragers. *Science*, **358**(6363), 659–62.

Skov, L., Peyrégne, S., Popli, D., *et al.* (2022). Genetic insights into the social organization of Neanderthals. *Nature*, **610**(7932), 519–25.

Slimak, L., Zanolli, C., Higham, T., *et al.* (2022). Modern human incursion into Neanderthal territories 54,000 years ago at Mandrin, France. *Science Advances*, **8**(6), eabj9496.

Slobodin, R. (1962). *Band organization of the Peel River Kutchin*. Bulletin No. 179, Anthropological Series No. 55. Ottawa, ON: National Museum of Canada.

Slon, V., Mafessoni, F., Vernot, B., *et al.* (2018). The genome of the offspring of a Neanderthal mother and a Denisovan father. *Nature*, **561**(7721), 113–16.

Smith, G. M., Duke, D., Jenkins, D. L., *et al.* (2020). The Western stemmed tradition: problems and prospects in Paleoindian archaeology in the Intermountain West. *PaleoAmerica*, **6**(1), 23–42.

Smith, H. L. (2019). The manufacture of Northern Fluted Points: a production sequence hypothesis. *PaleoAmerica* **5**(2), 169–80.

Smith H. L. and Goebel, T. (2018). Origins and spread of fluted-point technology in the Canadian ice-free corridor and eastern Beringia. *Proceedings of the National Academy of Sciences*, **115**(16), 4116–21.

Soffer, O. (1985). *The Upper Paleolithic of the Central Russian Plain*. Orlando, FL: Academic Press.

Soffer, O. (1994). Ancestral lifeways in Eurasia – the Middle and Upper Paleolithic records. In M. H. Nitecki and D. V. Nitecki, eds., *Origins of Anatomically Modern Humans*. New York, NY: Plenum Press, pp. 101–19.

Speth, J. D. (2010). *The Paleoanthropology and Archaeology of Big-Game Hunting: Protein, Fat or Politics?* New York, NY: Springer.

Speth, J., Newlander, K., White, A., Lemke, A., and Anderson, L. (2013). Early Paleoindian big-game hunting in North America: provisioning or politics? *Quaternary International*, **285**, 111–39.

Speth, J. D. and Spielmann, K. A. (1983). Energy source, protein metabolism, and hunter-gatherer subsistence strategies. *Journal of Anthropological Archaeology*, **2** (1), 1–31.

Spier, L. (1925). The distribution of kinship systems in North America. *University of Washington Publications in Anthropology*, 1(2), 69–88.

Steele, J. (2009). Human dispersals: mathematical models and the archaeological record. *Human Biology*, **81**(2–3), 121–40.

Stevenson, M. (1997). *Inuit, Whalers, and Cultural Persistence: Structure in Cumberland Sound and Central Inuit Social Organization*. Oxford, UK: Oxford University Press.

Storck, P. L. (2006). *Journey to the Ice Age: Discovering an Ancient World*. Vancouver, BC: UBC Press.

Sun, J., Ma, P. C., Cheng, H. Z., *et al.* (2021). Post-last glacial maximum expansion of Y-chromosome haplogroup C2a-L1373 in northern Asia and its implications for the origin of Native Americans. *American Journal of Physical Anthropology*, **174**(2), 363–74.

Surovell, T. A., Allaun, S. A., Crass, B. A., *et al.* (2022). Late date of human arrival to North America: continental scale differences in stratigraphic integrity of pre-13,000 BP archaeological sites. *PLOS One*, **17**(4), e0264092.

Surovell, T. A. and Waguespack, N. M. (2008). How many elephant kills are 14? Clovis mammoth and mastodon kills in context. *Quaternary International*, **191**(1), 82–97.

Tamm, E., Kivisild, T., Reidla, M., *et al.* (2007). Beringian standstill and spread of Native American founders. *PLOS One*, **2**(9), e829.

Thomas, K. A., Story, B. A., Eren, M. I., *et al.* (2017). Explaining the origin of fluting in North American Pleistocene weaponry. *Journal of Archaeological Science*, **81**, 23–30.

Tjon Sie Fat, F. E. (1998). On the formal analysis of "Dravidian," "Iroquois," and "Generational" varieties as nearly associative combinations. In M. Godelier, T. R. Trautmann, and F. E. T. S. Fat, eds., *Transformations of Kinship*. Washington, DC, and London: Smithsonian Institution Press, pp. 59–93.

Tournebize, R., Chu, G., and Moorjani, P. (2022). Reconstructing the history of founder events using genome-wide patterns of allele sharing across individuals. *PLOS Genetics*, **18**(6), e1010243.

Trautmann, T. R. (1981). *Dravidian Kinship*. New York, NY: Cambridge University Press.

Trautmann, T. R. (2001). The whole history of kinship terminology in three chapters. *Anthropological Theory*, **1**(2), 268–87.

Trautmann, T. R. and Barnes, R. H. (1998). "Dravidian," "Iroquois," and "Crow-Omaha" in North American perspective. In M. Godelier, T. R. Trautmann, and F. E. T. S. Fat, eds., *Transformations of Kinship*. Washington, DC: Smithsonian Institution Press, pp. 27–58.

Turner, M. D., Zeller, E. J., Dreschoff, G. A., and Turner, J. C. (1999). Impact of ice-related plant nutrients on glacial margin environments. In R. Bonnichsen and K. L. Turnmire, eds., *Ice Age Peoples of North America: Environments, Origins, and Adaptations*. Corvallis, OR: Oregon State University Press for the Center for the Study of the First Americans.

Vanhaeren, M. and d'Errico, F. (2006). Aurignacian ethno-linguistic geography of Europe revealed by personal ornaments. *Journal of Archaeological Science*, **33**(8), 1105–28.

Vickers, J. R. and Beaudoin, A. B, (1989). A limiting AMS date for the Cody Complex occupation at the Fletcher Site, Alberta, Canada. *Plains Anthropologist*, **34**(125), 261–4.

Waters, M. R. (2019). Late Pleistocene exploration and settlement of the Americas by modern humans. *Science* **365**(6449), eaat5447.

Waters, M. R. and Stafford Jr, T. W. (2007). Redefining the age of Clovis: implications for the peopling of the Americas. *Science*, **315**(5815), 1122–6.

Waters, M. R., Stafford Jr, T. W., and Carlson, D. L. (2020). The age of Clovis – 13,050 to 12,750 cal yr B.P. *Science Advances*, **6**(43), eaaz0455.

Waters, M. R., Stafford Jr, T. W., Kooyman, B., and Hills, L. V. (2015). Late Pleistocene horse and camel hunting at the southern margin of the Ice-Free Corridor: reassessing the age of Wally's Beach, Canada. *Proceedings of the National Academy of Sciences*, **112**(14), 4263–7.

Wei, L-H. , Wang, L-X. , Wen, S-Q. , *et al.* (2018). Paternal origin of Paleo-Indians in Siberia: insights from Y-chromosome sequences. *European Journal of Human Genetics*, **26**(11), 1687–96.

Weinstock, J., Willerslev, E., Sher, A., *et al.* (2005). Evolution, systematics, and phylogeography of Pleistocene horses in the New World: a molecular perspective. *PLOS Biology*, **3**(8), e241.

Weiss, K. M. (1973). *Demographic Models for Anthropology*, Memoir 27. Washington, DC: Society for American Archaeology.

Whallon, R. (1989). Elements of cultural change in the later Paleolithic. In P. Mellars and C. Stringer, eds., *The Human Revolution: Behavioural and Biological Perspectives on the Origins of Modern Humans*. Edinburgh, UK: Edinburgh University Press, pp. 433–54.

Whallon, R. (2006). Social networks and information: non-"utilitarian" mobility among hunter-gatherers. *Journal of Anthropological Archaeology*, **25**(2), 259–70.

Wheeler, C. J. (1982). An inquiry into the Proto-Algonquian system of social classification and marriage: a possible system of symmetric prescriptive alliance in a Lake Forest Archaic Culture during the third millennium BC. *Journal of the Anthropological Society of Oxford*, **13**(2), 165–74.

White, J. M. and Mathewes, R. W. (1986). Postglacial vegetation and climatic change in the upper Peace River district, Alberta. *Canadian Journal of Botany*, **64**(10), 2305–18.

Whiteley, P. and McConvell, P. (2021). How Crow-Omaha skewing spreads. *Journal of Anthropological Research*, **77**(4), 483–519.

Willerslev, E. and Meltzer, D. J. (2021). Peopling of the Americas as inferred from ancient genomics. *Nature*, **594**(7863), 356–64.

Wobst, H. M. (1974). Boundary conditions of Paleolithic social systems: a simulation approach. *American Antiquity*, **39**(2), 147–78.

Wygal, B. T., Krasinski, K. E., Holmes, C. E., Crass, B. A. and Smith, K. M. (2022). Mammoth ivory rods in eastern Beringia: earliest in North America. *American Antiquity*, **87**(1), 9–79.

4

Native American Science in a Living Universe: A Paiute Perspective

RICHARD W. STOFFLE, RICHARD ARNOLD, AND KATHLEEN VAN VLACK

4.1 Introduction

Native Americans study their environment for many reasons but importantly because it is alive and speaking with them. They have always lived in a living university, so there has never been a time when they have not been seeking information from the environment, testing their interpretations of what they observe, and formulating conservation behaviors that positively reflect what they and their environment want from each other. Whether this process is termed Native science or culturally sensitive natural interactions, it is a pattern that is typical of the Native American people, who have participated in more than 136 ethnographic studies conducted by the authors. Paiute science is illustrated using ethnographic findings from forty-four field studies and the lived experiences of one of the authors.

This essay is about how Native Americans in North America have learned about the universe in which they have lived since the Creation, that is, what lawyers call time immemorial or what today's Western scientists consider as at least 23,000 years ago. This chapter argues that, because Native Americans have lived in and interacted with nature for great periods, they have learned about natural elements and their interrelationships with each other and humans. The resulting traditional ecological knowledge (or TEK) has become the foundation for Native culture, religion, and society itself, which includes a relationship with all natural elements. In a recent ethnographic study at Canyonlands National Park, a representative from the Navajo Nation described himself as a member of the five-finger people, which distinguishes but does not separate humans from all of their relatives around them, including plants, animals, water, minerals, mountains, and the wind (Stoffle *et al.*, 2017). Most Native people also consider the sun, moon, and stars as living and sentient relatives.

This chapter is also about the interface between Western science and Native science (Lewis, 1989). Especially important is the interpretation of natural resources and the processes that influence them. An epistemological divide occurs when there is a belief-based debate about the existence of either the natural resources themselves or the processes that influence them, and how those differences figure into both education and the management of natural resources.

The analysis in this chapter is split into four sections. First, there is the case of Vine Deloria, Jr., and his arguments for the existence and accuracy of Native science. The second section presents a research-based diachronic model of human environmental learning. The third section contains examples of Paiute science, ranging from examples that are easy to understand to others that are more complex. Finally, the fourth section sets out a discussion of why Paiutes, and by implication other Native American people, are scientists who have a body of actionable knowledge.

It is important to clarify that this analysis neither attempts to test Native TEK nor argues that it should replace alternative knowledge derived from Western science. The analysis is focused on what Native people know, especially those people living in the United States Southwest and the Great Basin who are today called Paiutes and by their traditional name *Nungwu*. The analysis is based on fifty years of field research and, in the case of coauthor Richard Arnold, traditional training in Southern Paiute history and culture. Findings from other research-based studies are used where appropriate.

4.2 Vine Deloria, Jr., and Native Science

Vine Deloria, Jr., was, for most of his career, recognized worldwide as a reliable spokesperson regarding Native American culture and history. When he spoke, as he did throughout North America, or when he wrote (having published more than twenty books and hundreds of published essays), others took him seriously because he was known to convey the truth. Vine Deloria, Jr., was a member of the Yankton and Standing Rock Sioux Tribes and received degrees from Kent State University (Ohio), Iowa State University, and the Lutheran School of Theology in Chicago, and received his Juris Doctor degree from the University of Colorado Law School, Boulder. He studied geology for two years at the Colorado School of Mines before entering the US Marine Corps. He was an episcopal archdeacon who intended to be a priest like his father and grandfather, but he gravitated toward law and positions in universities, where he taught and researched Native American cultural and policy issues. He was elected Executive Director of the National Congress of American Indians and was a Board Member of the National Museum of the American Indian. His sister was also a famous anthropologist.

During the last decade of his life, from about 1995 to 2005, he received funding from the US National Institutes for Humanities and from private donors to conduct

Native American science seminars. The seminars were sponsored locally by the American Indian Science and Engineering Society. These seminars were organized around topics that are commonly understood to be old knowledge by many Native groups, but importantly addressed topics of special interest to Deloria. The last of these, held in a round room in the Indian Pueblo Cultural Center in Albuquerque, New Mexico, in 2005, was focused on volcanos and their role in Native life. Two of the authors of this essay were among the twenty-five or so multigenerational participants in this seminar, including elder representatives from Hawai'i, Alaska, the Pacific Northwest, the Mount Shasta region in California, the Hopi Tribe, and the Paiutes of Arizona and Owens Valley, California. Richard Arnold was invited to explain Paiute cultural attachments with volcanos. The seminars were conducted in a traditional fashion with noted experts speaking about a common topic and others listening. No notes or tape recorders were permitted – just speech, songs, and memory.

Clearly, Vine Deloria, Jr., was a credentialed and honored scholar when he was confronted by other scholars for writing a book on Native knowledge of nature. Some scholars were surprised when he wrote *Red Earth, White Lies: Native Americans and the Myth of Scientific Fact* (Deloria, Jr., 1995), a book with the thesis that everything that recognized traditional Native experts say about the natural world is true and derived from science. Deloria maintains in *Red Earth* (Deloria, Jr., 1995: 60) that "corrective measures must be taken to eliminate scientific misconceptions about Indians, their culture, and their past" and that "there needs to be a way that Indian traditions can contribute to the understanding of scientific beliefs at enough specific points so that the Indian traditions will be taken seriously as valid bodies of knowledge."

Red Earth recounts Native truths that largely are unfathomable in Western non-Native epistemology, for example descriptions of people in the distant past who were much bigger or much smaller than people today and accounts of a volcanic mountain in Oregon that exploded, leaving three distinct peaks known as the Three Sisters who commemorate the event. He argues against two common science theories that directly conflict with Native TEK. These are the Bering Strait migration theory and the Pleistocene overkill theory. The contemporary political implications here are obvious. The Bering Strait migration theory suggests that Indians did not originate in the Americas, which can be read as the latest in a long history of colonial attempts to invalidate Indian land claims, while the Pleistocene overkill theory blames Paleo-Indians for mass extinctions of ancient flora and fauna, which can be interpreted as an effort to discredit Indian environmental practice. Both theories remain controversial now, twenty-five years later; however, the Pleistocene overkill theory is largely discredited. The time and path of arrival of Native people in North America is documented by scientists as probably occurring along the now-underwater Pacific continental shelf and more than 23,000 years ago. This certainly is a sufficient period for a people to assert that they were created in these lands; however, Indian people continue to stipulate they were created at specific places in holy lands in North America and did not come from Asia.

Because of his stature as an expert, as soon as *Red Earth* was published, scientists and scholars from many different disciplines set about debunking his book. Scholars of Native culture were critical (Barker, 1997), neutral (Mohawk, 1996), and applauded its thesis and importance (Treat, 1997). All recognized that the book's purpose was to increase the consideration of Native science as capable of producing confident findings, even if this was by creating a polemic. The motivations of science critics certainly varied, but many simply believed that there is no *Native science*, so the essays in the book were not possible. Some wanted to diminish a prominent scholar with a reputation of usually being correct and a trusted spokesperson about Native knowledge. Attacks on *Red Earth* were believed by Deloria himself to be an extension of an old process of intellectually distancing Native Americans, or what he called attacks on intellectual self-determination and sovereignty (Deloria, 1998; Hernandez-Avila and Varese, 1999).

Later, when Deloria and Stoffle were working together on a monograph entitled *Native Sacred Sites and the Department of Defense* (Deloria and Stoffle, 1998), Deloria explained that *Red Earth* was designed as a polemic, although it was not just rhetorical because he and the cultural groups discussed in the book all believe in the truth of what their elders say about nature, especially natural events from the distant past. Much of the case material discussed in the book would become the foundation for his Native American science seminars. It was as though he was further ground-truthing his *Red Earth* with new elders from many societies.

4.3 Learning from the Land: A Diachronic Model

One way to understand human adaptation is by using a diachronic model of human adaptation that begins with the hypothetical arrival of a group of people to a new place and observes their initial adaptations to this ecosystem. The following text derives from a chapter (Stoffle *et al.*, 2003: 97–114) in a book entitled *Nature Across Cultures: Views of Nature and the Environment in Non-Western Cultures*, edited by Helaine Selin and Anne Kalland. All of the environmental coadaptations in each of the environmental learning phases have been studied by anthropologists, including the authors of this chapter (consult the published essay for specific case references). The model (Figure 4.1) has continued to have heuristic value for demonstrating environmental learning continuity, namely from early arrivals until they are a people who have lived in a specific area for thousands of years. The model discusses how environmental knowledge increases in complexity with more understanding of trophic levels and species interactions. It also describes why tested environmental knowledge is increasingly viewed as sacred.

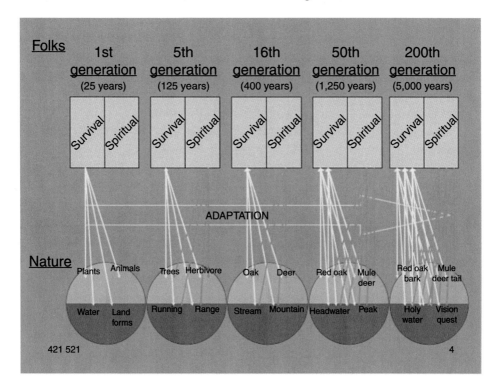

Figure 4.1 Diachronic learning model.

Eventually, the hypothetical group of newly arriving people begins to make certain modifications in their environment. The model then moves forward in time over generations to observe how these people become fully adapted and how the place becomes the center of their lives both physically and spiritually. The model is informed by the direct analysis of actual human adaptation and resulting environmental changes. A few cautionary points need to be made before proceeding. In this discussion, we are scaling up, that is, we are increasing the variables of space, time, and complexity of typical human ecological analysis. We are moving beyond what is confidently known in order to build a diachronic model with heuristic value for situating and perhaps guiding how we think about these issues. Most human–nature studies have a narrow time frame and focus on only a few species interactions. "Very little ecology deals with any processes that last more than a few years, involve more than a handful of species, and cover an area of more than a few hectares" (Pimm, 1991: xi). Nonetheless, Pimm builds food-web and temporal variability ecological models from studies with narrow time frames and with a strict species or ecological focus.

Social and natural scientists have had few opportunities for long-term joint research projects; thus, we tend not to understand each other's theories, methods, variables, and analytical techniques. As a consequence, many studies of how humans adapt to and change their environments are conducted by social scientists who know the human side of the equation. Carl Sauer (1925: 46; 1956) defined "cultural landscapes" as natural areas modified by a cultural group, because culture is the agent, nature is the medium, and a landscape is the result. Sauer was instrumental in organizing the first large-scale evaluation of what *has* happened and what *is happening* to the Earth under man's impress. The resulting book, entitled *Man's Role in Changing the Face of the Earth* (Thomas, 1956), contains chapters from all fields of knowledge regarding man's capacity to transform his physical-biological environment and upon his cumulative and irreversible alterations of the Earth. Most authors view humans as having had an impact on the Earth but not understanding what they were doing and thus being unable to control or mitigate their impacts. This science view persisted for decades.

The science of ecology continued to be dominated by the findings of natural scientists, while social science findings tended to be recognized only if they concurred and were supported by natural science studies. Studies by natural scientists tended to look at the nature impacts of humans as being caused by people who were simple consumers of the environment and lacked an understanding of what they did to it. In this theory, humans are perceived as without actionable environmental knowledge and thus lacking the data to modify their behavior and conserve nature; something like the basic predator–prey models in natural science. This premise was very much present in *Man's Role* and continued in anthropology until the debate caused by Rappaport (1968) in *Pigs for The Ancestors: Ritual Ecology of a New Guinea People*. Rappaport's findings – that the people understood the ecological implications of their warfare and related practices – were challenged by Andrew Vayda (1976), a prominent cultural ecologist, who said that people don't understand what they do, especially the people of New Guinea. The resulting debate made *Pigs* a classic and the most cited book in cultural ecology. In the second edition, Rappaport (1984) directly dealt with the debate (Dwyer, 1985); however, the debate was continued in modified form by Vayda and others (Vayda *et al.*, 2004).

After *Man's Role* was published, social scientists expanded their research on the interface of humans and nature, but often did so while lacking a full understanding of the natural systems. Similarly, biologists, marine ecologists, and nature ecologists focused on the topics discussed in *Man's Role,* but they often lacked a background in social and cultural systems. More recently, however, a more balanced analysis of the human–nature divide interface has been provided by ecologists like Turner *et al.* (2000), Davidson-Hunt (2006), Miller and Davidson-Hunt (2010), Berkes (2018), and Nabhan (2018), to name just a few.

The current analysis describes and explains the human dimensions of ecology based on a foundation of Native science. The model presented here is intended to sequence types of human–nature coadaptations so that we can begin to imagine the cultural implications of learning about nature and the resulting adaptive behavior. For example, when the Southern Paiute Indians say the Creator made them in Las Vegas Wash below their origin mountain, it can be viewed as the end result of thousands of years of living in this area rather than some inherent aspect of their culture or a collective social construction. There are many peoples around the world who have embedded portions of cultural landscapes into their lives and livelihoods and their definition of self. Our model suggests that it takes great periods of time to produce embeddedness in cultural landscapes.

Environmental Learning: Newcomers

How does environmental learning occur? The first generation of people living in a new environment actually learns much from observing native plants and animals. This is a period from arrival until 35 years later. Families observe what is around them and what happens when they use the environment. How can animals and plants teach? Plants tell the observant viewer about rainfall and the subsurface distribution of water. Longer term weather and climatic stories are told based on whether plants do or do not grow near or in the intermittent streams, which are subject to both unpredictable and cyclical El Niño-type catastrophic flood events. Animals move between econiches according to weather shifts and availability of food resources.

Why can a culture learn more from mistakes than from successes? It may be that, when they fail, they have reached a critical limit. On the other hand, they may succeed and never know why because they are working within the limits of the resource being used. People learn by hurting nature and seeing the undesired consequences. When they kill too many large mammals, they learn not to repeat those behaviors because they rely on that resource.

The first generation makes mistakes, learns how not to cause unwanted damage, and then tells family, friends, and perhaps the community. Some first-generation lessons remain at the family level and thus do not become cultural.

Second Generation: "Do What Parents Tell You to Do"

The second generation can build upon the lessons learned by the first generation. The study of biotic resources continues from about 35 years in one place until 125 years later. The first generation has scrambled to survive; the second generation begins to develop theories of how to learn about the environment. Lessons come from events: Perhaps a person follows a bee to water, person builds their home near a dry wash only to be flooded, or observes an animal healing itself with special white mud. The second generation will develop theories of how data move between people and the world and becomes useful for short-term survival and longer term adaptation.

The second generation receives these lessons from living people who can be questioned about the event. Knowledge comes from deeper understandings of nonintuitive aspects of the world. Few of these are expected in the first two generations, but they lay the foundation for later environmental learning.

We see at this stage what is called "adaptive behavior." Bennett (1969: 11) defines adaptive behavior as coping mechanisms or ways of dealing with people and resources in order to attain goals and solve problems. Our emphasis here is on patterns of behavior: problem-solving, decision-making, consuming or not consuming, inventing, innovating, migrating, and staying. Bennett documented how different people at different times constructed unique cultures in a single ecosystem in central Canada.

Fifth Generation: "Do What Great Grandparents Tell You to Do"

The fifth generation (occurs by approximately 125 years) is special. They have a firm information base about the environment and know how to use the land without hurting it. Families have experienced birth, life, and death in the new homeland, and these events emotionally connect the people with specific places. They have pushed a number of environmental boundaries, made mistakes, and changed their behavior. By the fifth generation, various aspects of successful and sustainable behavior have been put into place.

Environmental learning seems to be directly related to the amount and kinds of natural resource scarcity, especially when this occurs in annual and decennial cycles. People mostly learn to hunt and gather when basic resources are regularly available. It is not until the limits of the resource are reached that they have an opportunity to learn what parameters drive the system. It is at exactly these moments that people decide to modify their environmental resource use patterns in order not to drive other resources into extinction. By this time, they have reduced the possibility of Hardin's (1968) tragedy of the commons by developing community-based cooperative resource rules (McCay and Acheson, 1987). If they fail to adopt resource-conservation procedures and to build these into the rules by which they govern themselves, then the carrying capacity of the environment will be reduced. Often, the people will be forced to move to another ecological zone.

Learning from plants and animals continues during this generation, but the lessons are less intuitive and probably about natural processes that are less accessible to human view. Plants often serve as calendars, like the rabbitbrush (*Chrysothamnus nauseosus*) that blooms when it is time to go into the mountains to harvest pine nuts (*Pinus monophylla*). Smaller animals and insects provide information about pollination, food webs, and ecosystems that make up landscapes (Nabhan, 1997). Beaver dams keep the ecosystem from eroding and retain water during arid seasons, so the people minimize their consumptive use of beaver. People learn to stay away from water sources because it disturbs the animals. Also proactive is the planting of certain species along or diagonally across small streams to assist the beaver's dam efforts, which reduce erosion while at the same time retaining the water and moving it to places along the streams where it is used by other species. They often set aside areas or preserves for animal reproduction. Proactive adaptive behavior will be normal by this time. This might include selectively increasing the yields of certain species by burning or pruning. By creating a specialized patch ecology that does not occur naturally, the carrying capacity of the land will increase (Lewis, 1982; Lewis and Ferguson, 1999; Stewart *et al.*, 2009). Connell (1978) found that intermediate natural disturbances in ecosystems can have positive impacts on biodiversity and biocomplexity. These are intermediate in terms of scale and frequency of occurrence. Do traditional people use their knowledge of ecosystems and consciously make intermediate human changes that have positive benefits?

The people begin to have more complex and successful adaptive strategies. Bennett (1969: 14) defines these as the patterns formed by the many separate adjustments that people devise in order to obtain and use resources and to solve the immediate problems confronting them. The rules of adaptation become culturally embedded. Bennett (1969: 16) notes that, as time passes, the many separate adjustments that have become patterned as strategies can also "enter into culture." As repetitive patterns of actions, the people come view them as traditions – behavior defined as "right" or "good." These embedded traditions form part of a group's cultural style. The extent to which adaptive behavior becomes culturally sanctioned varies according to the demands placed on the society by various external factors. If the major changes in the natural system are recognized and occur in predictable cycles, there will be a tendency for a given strategic regime to become "sacred."

The population grows as the fit between technology and resources increases the carrying capacity of the land. Basic issues of village life, including the need for seasonal movements of the community or community members, exist.

Sixteenth Generation: "We Just Do It Like This"

There is a point in the development of a society when the origins of ancestral lessons are vaguely remembered but firmly established as the correct way to treat the environment. This is a time (perhaps by the sixteenth generation or approximately 400 years) when lessons are taught and maintained as general principles.

The name of the person who originally learned the lesson probably is not attached to it. People teach their children to behave in certain ways because "it is how we [the members of this community] do it." If you follow "our ways," you will always have food, shelter, and health, and the environment will be in balance.

By now, the people have amassed sufficient data from the natural environment to begin to develop knowledge about what is happening around them. They recognize the pollination of plants by certain species. They have learned many lessons about medicine. They have seen sufficient climate variability that they know with some confidence what the cycles provide in terms of opportunities and threats. There are still other forces that are unclear to them but appear to be variables within the system. The rules for engaging the environment are now well tested and those rules, which have repeatedly proven useful for maintaining and improving the productivity of the environment, are increasingly defined as sacred and not subject to debate.

The population grows, so there are more people to organize, to teach about "how we live here," and to make regulations for. The natural resources of the environment are further stressed. New boundaries are reached and perhaps broken. Certain long-term weather cycles and climatic events have occurred a few times, and there is the recognition that one must prepare for events that may not come within the lifetime of a generation.

Fiftieth Generation: Environmental Ethics Defined and Sanctioned by Supernatural Forces

The model now skips to the fiftieth generation, which begins about 1,250 years after the people's arrival. People have lived so long in this ecosystem that they have clear adaptive strategies that have survived a wide variety of temporal and biotic shifts in the ecosystem. Over the period of 1,250 years, many natural resource changes will also occur. New plants will come into the ecosystem and old ones will become extinct. Streams will flow more or less depending on weather cycles, climate change, and patterns of use by beavers and people.

By this time, they have embedded these adaptive strategies into their culture. They have moved key values into the realm of the supernatural, with both the lessons and the sanctions being supernaturally defined and thus protected against being interrogated by new generations. There is a confluence between science and religion, as the scientific findings of past generations have been recognized as essential to life and moved to the realm of the sacred and thus beyond human control. Despite their useful adaptive strategies, people still have things to learn as scientists. They are still watching and responding to changes in nature. Parents still observe, realize that something is happening, adjust their behavior, and tell their children; this is the continuity of first-generation learning. And the children follow the new behaviors to avoid the mistakes made before.

200th Generation: "We Were Created Here"

After 200 generations (approximately 5,000 years) in the same ecosystem, many kinds of complex human adaptations can be expected. Only a few biotic features will remain unchanged. Climate changes reduce or even eliminate most fauna and flora and certainly radically alter the food webs (Grayson, 1993). Even the topography changes. Sea levels will rise and fall. Volcanoes will build up the land, and erosion will tear down and transform it. Long wet periods will alternate with long dry periods. What does it mean to have adaptive strategies, which define, organize, and maintain a human group's adaptations to an environment, if this environment itself changes?

Those cultural groups who persist in structure and function through a 5,000-year period will not only remain but become increasingly central in the culture. For example, mountain peaks and ranges should remain the same and draw rain from the sky. This is what is called the "sky-island function" (Crowley and Link, 1989). Mountain ranges are central in the lives of people dependent on related resources. For example, Baboquivari Peak and Mountains feature centrally in the lives of the Tohono O'odham people (Toupal, 2001) and are primary rain makers in southern Arizona. Biotic and biological evolutionary forces cause some changes in the environment, while growth in human population size and technological innovations create others. The people know how to listen to the environment and change their behavior.

Discussion

If we believe that humans will learn about and adapt to the environment in which they live, then what do we need to know about this process? Key here is the time that a people have continuously lived in one place (Nabhan, 1997: 2). In general, it is assumed that people begin learning as soon as they arrive in a place (Figure 4.2). Such knowledge is often termed "local knowledge," and it may be useful in terms of developing proper environmental behavior within a generation. This kind of knowledge, however, references a kind of knowledge, rather than a place. Place-based local knowledge can be of any complexity and the term "local" is often used in another way by the United Nations Educational, Scientific and Cultural Organization (UNESCO), namely as part of the term "local and indigenous knowledge systems" (LINKS) in reference to where kinds of advanced knowledge are grounded (https://en.unesco.org/links).

To move from simple observations to deeper ecological understandings of food webs and trophic levels will take many generations. The hypothetical model used in this analysis (Stoffle *et al.*, 2003) is diachronic in order to

Access to nature				
		High	Medium	Low
Potential for knowledge	High	x		
	Medium		x	
	Low			x

Figure 4.2 Access to nature and the potential for knowledge.

present a coadaptation view of learning through time and it is illustrated by cases from around the world. It argues that, within five generations, people who continuously live in the same ecological system will acquire deep ecology understandings. This is the first point in time when the term "traditional knowledge" (applied to either culture or people) should be used. The arguments for this are that, after this period, (1) the people know something significant about the ecosystem functions, (2) they have developed various use strategies to both gain from and protect the ecosystem, and (3) they have experienced more than 100 years of environmental perturbations through which they can understand the resiliency of their adaptive strategies.

Indigenous knowledge is a third and deeper level of knowledge. Berkes (2018) uses the term "sacred knowledge" for this deeper knowledge because so much of it and the ethics it guides are defined in religious terms. Such knowledge can be illustrated by the people of Bali, who accurately understand underground hydrological systems and see these as being controlled by a water goddess (Lansing, 1991). By definition, the paths to these forms of Indigenous knowledge are perceived as sacred and thus resist simple explanation. Such knowledge is often believed to have been taught by the supernatural.

Environmental learning is generally understood to occur more often and to become more detailed when the natural elements being observed are most accessible to more people who would learn about them (Berkes and Turner, 2006; Turner and Berkes, 2006). The foundation of this premise is that people can see what they do to their environment and adjust their behaviors to protect their resources. This premise, however, seems problematic when Native knowledge is documented and analyzed. Plants and insects are generally well understood after a few generations, but so is the ecology of ancient Bristlecone Pines, which are 4,500-year-old trees located near timber line and Snow Fleas or springtails – *Hypogastrura harveyi* and *Hypogastrura nivicol* (Lin *et al.*, 2007) – found on the extreme tops of Nevada mountains (Stoffle *et al.*, 2004). Red ants are observable, but knowing how to consume them alive by the cup full to stimulate visions as they bite the vision seeker's stomach is a distant form of knowledge (Groark, 1996). Minerals like salt and red paint are understood, but so are the effects of radioactive minerals (Stoffle and Arnold, 2003). Hydrological systems and their sustainable conservation relationships with beavers are understood, but so are the patterns of underground rivers that are observed by dropping baskets through sinkholes and watching for them to emerge in another sinkhole tens of miles away.

Oceans are a barrier to the observer, but Native seaweed gatherers learned to conserve and stimulate underwater seabeds of seaweeds along the Pacific Coast (Turner, 2016). Native people learned to harvest seaweed from rocky intertidal zones and to accomplish this task in a sustainable way, so it stimulated the seaweed beds and made them healthier (Turner, 2003; Turner *et al.*, 2013). Elsewhere along the coast, rocks were rolled to the shore to make walled structures called clam gardens (Deur *et al.*, 2015). These stones formed tidal shore areas that provided a protected habitat for four species of clams (Groesbeck *et al.*, 2014). The clam gardens are viewed by scientists as a special example of learning TEK in a *nonintuitive ecological area* (Jackley *et al.*, 2016).

Other Native knowledge seems grounded in alternative realities that preclude normal observation according to Western epistemology but are clearly possible in Native epistemologies (Stoffle *et al.*, 2020). Examples include knowledge of alternate dimensions, movement through portals, and travel to stars and planets. These are ancient knowledge domains commonly shared among contemporary Native Americans. At Chaco National Historic Park in New Mexico, for example, there is a prominent butte called Fajada, and along one edge are the remnants of structures that housed star watchers. At eye level on each house's roof are peckings (grounded depressions) on the sandstone wall of Fajada Butte that represent planets or stars. Above these rooms is the most elaborate carved social calendar in North America called the Sun Dagger (Sofaer, 1982; Sofaer *et al.*, 1979; Sofaer and Sinclair, 1982; Sofaer, 2021). According to representatives of the Acoma, Hopi, Santa Ana, and Zia pueblos and the Navajo Nation who participated in our ethnographic study, Fajada Butte was a university for training star watchers and having conferences of experts (Stoffle *et al.*, 1994). From this location, they both studied and went to the solar system. Other star watching and travel portals that are marked by star and planet patterns ground into bedrock have been identified by pueblo representatives at Tonto National Monument, Arizona (Stoffle *et al.*, 2009), and Hovenweep National Monument, Utah (Stoffle *et al.*, 2019).

4.4 Paiute Science: Ethnobotany

Recent studies document that Native peoples occupied the western United States for at least 37,000 years before the present (Rowe *et al.*, 2022). During this time, they used and coadapted with native plants and came to understand their potential use for medicine, food, tools, and construction. The use of tobacco seeds has been documented 12,300 years ago (Duke *et al.*, 2021; Gamillo, 2021). In this analysis, we build on new science and add findings to a diachronic ethnobotanical analysis of native plant use in an ancient Native homeland in southern Utah, currently known as the region surrounding Canyonlands National Park or CANY (Stoffle *et al.*, 2017).

Archaic Period: 1200 BC to AD 0

During the Archaic period, American Indian interactions with native plants were key to their livelihoods. Through time, probably throughout the Paleo-Indian and Archaic periods, American Indian people continued to learn about plants and their ecosystems, and how to use them in more efficient ways. Over more than 12,000 years, spanning the Pleistocene to Holocene transition, the climate of the western United States underwent some major ecological changes. This directly impacted the ability of plants and people to coadapt. We know from the archaeology of dry caves that, during this long period, American Indian people used dozens of wild plants over thousands of years.

Hockett's (2010: 221–2) data from Bonneville Estates Rockshelter, Utah, document that there were major changes in the occupational intensity of this area, which began about 10,900 BC, largely due to the onset of a much drier climate. Data from this study support the nutritional ecology model, which couples climate change in the early Holocene with dietary diversity – especially an increasing reliance on seeds – and increases in population numbers and residential stability. Rhode and Louderback (2010) analyzed the wild plants observed in this cave and concluded that American Indian people at this time (about 10,000 BC), and for a thousand years after, had a diet that was heavily dependent on small seeds. Table 4.1 lists twenty plants that were identified from this cave and that were being utilized by the American Indian people. According to Rhode and Louderback (2010), the sample is biased toward the more resilient seeds; although it is likely that roots, tubers, berries, and greens from these and other plants would have been used, they were not well preserved. This seed-gathering component of their diet was eventually strengthened by the use of pickleweed (*Salicornia utahensis*), which became a common plant due to increased aridity within the area. Rhode and Louderback concluded that broad-scale adaptation to resources such as desert grass seeds and cacti may already have had quite a long history, despite their use only being documented from the start of the Holocene.

Table 4.1 *Paleo-ethnobotanical resources (cultural component plant remains) from Bonneville Estates Cave, Utah*

Scientific name	Common name
Achnatherum hymenoides	Ricegrass seed
Artemisia ssp.	Sagebrush seed
Asteraceae	Sunflower (*Helianthus*) seed
Atriplex argentea	Saltbush fruit
Brassicaceae	Mustard family seed
Cactaceae	Cactus family seed, spine, cluster
Carex ssp.	Sedge seed
Chenopodiaceae	Goosefoot seed
Cleome serrulata	Beeweed (Rocky Mountain bee plant) seed
Fabaceae	Pea family seed
Forsellesia spinecens	Greasebush seed
Leymus cf. *cinereus*	Wild rye seed
Liliaceae	Lily family bulb
Mentezlia cf. *albicaulis* and *M. laevigata*	Blazingstar seed
Poaceae	Grass family seed
Rosa cf. *woodsii*	Wild rose seed
Salicornia utahensis	Pickleweed
Scirpus ssp.	Bulrush seed
Sporobolus ssp.	Dropseed sandgrass seed
Symphoricarpos cf. *longiflorus*	Snowberry seed

At some point in the process of learning about plants, use practices shifted from the passive gathering of seeds to a more active and knowledgeable engagement called horticulture. Horticulture, as we use the term here, involves the conscious use, modification, and redistribution of natural plants to serve human needs. Horticulture is based on the TEK of plant species, management outcomes, and overall ecology. As such, it involves deliberate engagements such as pruning pinyon trees, burning Indian ricegrass, and carrying selected seeds to new habitats to increase their accessibility (Stoffle *et al.*, 2003). An intimate knowledge of plant genetics has been suggested as a major "cultural focus" of desert-dwelling American Indian people (Shipek, 1991; Anderson and Lake, 2016). These desert-dwelling people developed sustainable adaptive strategies that maximized the carrying capacity of the land (Stoffle *et al.*, 1982). Over tens of thousands of years of coadapting with plants, American Indian people developed complex understandings of both plants and their ecosystems (Stoffle *et al.*, 1990; Halmo *et al.*, 1993). This knowledge persists today, despite the ravages of European animals on the desert landscape ecosystem and human depopulation due to European diseases (Stoffle and Evans, 1976; Stoffle *et al.*, 1995, 1999).

Coadaptation also involves establishing and maintaining relationships between people and animals, including water-management animals such as beavers (Albert and Trimble, 2000; Hao, 2013; Maenhout, 2013). Before being decimated by early trappers, such as the French-Canadian Denis Julien who trapped the Canyonland National Park region in 1832 and 1844, beavers were common along the Colorado River (Williams, 2013: 100). They lived in riverbank dens and in its tributaries, where they stabilized the ecosystem with stream-wide dams and ponds. In recent years, the Pueblo of Zuni has reaffirmed their traditional relationship with beavers through their successful Ecological Wetlands Restoration project on their reservation (Albert and Trimble, 2000).

American Indian people in the western United States, therefore, have been doing more than just gathering and using plants and animals for the past 23,000 years. The distribution, genetics, and ecology of this region's plants have changed because these plants have been actively engaged by American Indian people (Nabhan *et al.*, 1981).

Irrigated Farming Period: AD 0 to AD 1350

The time and way of life that archaeologists define as the Archaic period slowly transformed into a period that we are terming here the irrigated farming period. Other types of farming that included fields watered by floodwaters, rainfall, and seepage persisted in some areas (Nickens, 1981: 93–4). The largest concentrations of people during this period, however, were supported by the larger carrying capacity of irrigated farming. There are a number of other terms that have been used to define the archaeological traditions in this period, but we are avoiding the use of those because they imply that artifacts present in these periods (often as few as four items) can indicate a connection or disconnection with contemporary cultural groups. Instead, in this chapter, the "irrigated farming period" refers to a livelihood and a time when irrigated farming dominated the landscape. Since the term is ethnically neutral, it does not imply a connection with specific contemporary American Indian peoples.

With the adoption of cultigens like corn (*Zea mays*), squash (*Cucurbita* ssp.), and beans (*Phaseolus vulgaris*) as early as 500 BC, farming was added to horticulture and together these practices become a permanent foundation for a new and more resilient livelihood involving the management and use of wild plants and cultivars. After corn, beans, and squash arrived north of the Colorado River, they certainly could have moved throughout Utah by AD 500. Cowboy Cave has an early corn date of 40 BC, but most corn dates are between then and AD 440 (Schroedl and Coulam, 1994: 13). Schroedl and Coulam's (1994: 24) reanalysis of Cowboy Cave seed data supports the interpretation that there was a general trend of increasing intensification of plant husbandry during what they call the Terminal Archaic period, which ended for the Cowboy Cave around AD 600 (Winter and Hogan, 1986). Given the apparent addition of corn to a well-established livelihood, Schroedl and Coulam (1994) concluded that corn was initially just another grass that was incorporated into an existing horticultural subsistence pattern.

As cultivars became prevalent in the livelihoods of the people, their population numbers and the distribution of their sedentary communities greatly expanded (Pierson, 1980). Lindsay and Loosle (2006: 17) note that storage features reflecting early farming activities in the Uinta Basin date back to around AD 1. Talbot and Wilde (1989) conclude that between AD 700 and AD 800, the frequency of farming sites in the Great Salt Lake area increased more than tenfold. This suggests that populations were growing, becoming more sedentary, or both as a consequence of increasing reliance on maize. During this period in the American Indian Crossing of the Colorado River (AICC), the central irrigated fields and settlements were primarily in the Moab Valley, but farming villages also occurred in Castle Valley and in other stream-fed valleys around the La Sal Mountains. Such farming villages also continued until the mid-1870s along streams fed by the Abajo Mountains, such as Indian Creek and Salt Creek.

Irrigated farming continued to flourish because of increased precipitation and warmer annual temperatures. By around AD 1150, the carrying capacity of the land was maximized due to a full set of cultigens being grown in optimized irrigation networks and broad-spectrum gathering of flourishing natural plants. To the north of the AICC in the Uinta Basin, Lindsay and Loosle's (2006: 10) study of the Johnson Rockshelter, dated AD 880, documented the presence of dent corn (a regionally unique variety), pumpkin or squash, and beans.

In CANY, Chandler (1990: 97) inventoried the plants used by American Indians at the extensive Bighorn Sheep Ruin (dated AD 1164–1282) on Salt Creek. She notes diverse plants, which were modified to make other objects including all of the following:

Basketry, cordage, sandals, cloth, quids, arrowshafts, worked wood, a painted squash rind pendant, and perforated corn shanks. The following parts of 13 taxa of plants were used by the site occupants in the manufacture of modified vegetal artifacts: dogbane (*Apocynum*), yucca, grass, and cotton fiber; squash pericarp; *Dicotyledoneae*, cottonwood, and willow wood, *Gymnospermae* resin, juniper bark; reed (*Phragmites*) and sedge cub, and various corn parts.

(Chandler, 1990: 97; see also analysis by Matthews, 1988)

The gathering of wild plants continued during the peak of irrigated agriculture according to coprolite analysis from Bighorn Sheep Ruin. Chandler (1990) documents that:

Cotton was locally grown and processed with looms into textiles and cordage. While the people living at Bighorn Sheep Ruin would have used some of these cotton products, cotton and items manufactured from it generally were prime trade goods and passed along the American Indian trail through the CANY to distant Indian villages. It is important to note, although cotton typically was grown far to the south, cotton growing this far north documents higher rainfall and a warmer climate in southeastern Utah.

The Bighorn Sheep Ruin study is especially important for this ethnographic analysis because the study documents that, even at the height of irrigated agriculture, natural plant resources continued to be important. Four of the wild plants recorded at Bighorn Sheep Ruin had been used by American Indian people more than 10,000 years ago (Table 4.2). All of the plants identified at Bighorn Sheep Ruin are used by contemporary American Indian peoples from the AICC.

To the north of CANY, the Uinta Basin irrigation farmers continued to participate in seasonal hunting and gathering, despite the demands of farming. The authors conclude that the extensive subterranean storage pits at the Johnson Rockshelter and elsewhere support Yoder's (2005) hypothesis that, between AD 550 and AD 950, the Uinta Basin farmers were semisedentary, perhaps seasonally leaving their primary farming residences after the maize harvest for hunting and other resource procurement trips.

Modern Native Americans: AD 1350 to Today

In the western United States, the peak of Native population density and environmental carrying capacity ended with the beginning of the Little Ice Age (about AD 1300 to AD 1400). For the next 500 years, or until about 1850, the climate was drier, less predictable, and cooler. People moved around seeking solutions, but all of North America experienced similar climate change conditions. Most populations shrank in place, social structure was simplified, and transhumant movements out and back from central village locations during the seasons became the norm. The diet included more wild and semidomesticated plants, but farming at permanent water continued.

Because patterns of both wild plant management and animal hunting continued during the rise and intensification of irrigated farming, the ancient TEK regarding the uses for, the conservation of, and the spiritual meanings essential to these ancient gathering activities persisted (see Table 4.2). So, despite the common academic perspective that Native farmers diminished their commitment to gathering and hunting when they shifted to farming, these activities continued and TEK was transferred over generations.

A contemporary study of Paiute and Shoshone plant use identified 364 unique plants used in southern Nevada (American Indian Writers Subgroup, 1996). All ethnographic studies conducted by the authors of this analysis over

Table 4.2 *Plant use over time*

Domestication status	Scientific name	Common name	Used 10,000 years ago?	Used 2,000 to 750 years ago?	Contemporary use?
Cultigens	*Cucurbita*	Squash		Yes	Yes
	Gossypium	Cotton		Yes	Yes
	Phaseolus ssp.	Beans		Yes	Yes
	Zea mays	Corn		Yes	Yes
Non-cultigens	*Achnatherum hymenoides*	Indian ricegrass	Yes	Yes	Yes
	Amaranthus ssp.	Pigweed		Yes	Yes
	Apocynum ssp.	Dogbane		Yes	Yes
	Artemisia ssp.	Sagebrush	Yes		Yes
	Asteraceae ssp.	Sunflower	Yes		Yes
	Carex ssp.	Sedge	Yes	Yes	Yes
	Chenopodium ssp.	Goosefoot	Yes	Yes	Yes
	Cleome ssp.	Beeplant	Yes		Yes
	Descurainia ssp.	Tansymustard	Yes	Yes	Yes
	Juniperus ssp.	Juniper		Yes	Yes
	Opuntia ssp.	Prickly pear	Yes	Yes	Yes
	Phragmites australis	Common reed		Yes	Yes
	Physalis ssp.	Groundcherry		Yes	Yes
	Poaceae	Grass	Yes	Yes	Yes
	Populus ssp.	Cottonwood		Yes	Yes
	Portulaca ssp.	Purslane		Yes	Yes
	Rhus trilobata	Three-leaf sumac		Yes	Yes
	Salix ssp.	Willow		Yes	Yes
	Suaeda ssp.	Seepweed		Yes	Yes
	Yucca ssp.	Yucca		Yes	Yes

the past forty years have documented extensive plant knowledge among Native Americans (Stoffle *et al.*, 1990, 2018; Halmo *et al.*, 1993). Our analysis documents that, while complete Paiute plant TEK is no longer concentrated in individuals as it must have been in the distant past, collectively understood Paiute plant TEK can be reconstructed with up to sixteen interviews (Stoffle *et al.*, 1999).

Paiute Science: Whiptail Lizard Eye Surgery

Another example of environmental knowledge possessed by Paiute people is the use of the tails of whiptail lizards in cataract eye surgery. Paiute medicine, like other forms of Native American medicine, is similar in many ways to Western medicine. Diagnosis and curing experimentation are found in both the Western world and Native communities. One type of Paiute medicinal practice that is the result of multiple generations of environmental learning is eye surgery using the tail of a certain species of whiptail lizard.

Paiutes were among many societies around the world that developed a form of couching or cataract eye surgery. Paiutes learned to use elements of their environment such as whiptail lizards and rattlesnake weed to develop a style of surgery uniquely theirs. A Southern Paiute elder explained how eye surgery was conducted:

When I was a little girl my grandma would tell me all sorts of stories about who we were as Southern Paiute people and how we are tied to the land. She often said that Paiute culture comes from the land, you know. Anyway, there was one story in particular I loved hearing about. It always fascinated me. It was about how certain medicine men would perform eye surgery using a tail of a whiptail lizard.

When a person like you or me got those clouds in their eyes … you know cataracts, they would need them removed so they can see. They would go see a medicine man. A ceremony was held for this. What he would do would be to catch a certain type of whiptail. See not the ones with the black strips on their sides because those are bad. They will run up your leg and come get you. Once they caught the right whiptail, they would say prayers and then snap the tail right off! It didn't hurt the lizard because his tail would grow back. Then the medicine man would go to the patient and use the jagged edge of the tail to cut out the cataract from the eye. The whole time they would be singing songs and saying prayers to ensure nothing would go wrong. You have to pray in everything you do, don't want to upset anything. Anyway, after the cataract was cut out, the eye was very bloody. So it needed to be covered up so the bleeding would stop and it wouldn't get infected. The medicine man would use something like rattlesnake weed which was a medicine plant for the eyes. They would place it on the eye, and it would help it heal.

(Van Vlack, 2007)

The species of whiptail lizard used in this form of eye surgery is known as the Great Basin whiptail lizard (*Aspidoscelis tigris*). This species has a slim body and is just over a foot long: four inches of body and about nine inches of tail. The lizard has smooth scales and its back has dark spots with light-colored strips. The Great Basin whiptail is found in a variety of ecosystems; the lizards prefer hot and dry open areas with sparse foliage, but they can be found living in chaparral, sagebrush, woodland, or riparian areas. The Great Basin whiptail lizard is found throughout the Great Basin and into New Mexico and west Texas. They tend to live in areas at sea level to about 7,000 feet above sea level (Behler and King, 1979).

There are many species of whiptail lizards found in traditional Southern Paiute territory, but the Great Basin whiptail lizard is the preferred choice. When asked about other species of whiptails, Southern Paiute people describe one particular species (desert grassland whiptail [*Cnemidophorus uniparens*]) as dangerous. This species is easily distinguishable from the Great Basin whiptail lizard because the desert grassland whiptail has noticeable black stripes down its side.

When a person developed cataracts, they could visit a medicine man so he could remove them. Only specially trained medicine men had the authority to perform such a dangerous and complex ceremony. Cataract surgery involved a special ceremony, medicinal plants, and a tail of the Great Basin whiptail lizards.

When the patient was diagnosed by the medicine man to have a cataract, the medicine man would perform a doctoring ceremony. First, he would watch the whiptail lizard and he would say prayers and explain to the lizard what was about to occur and ask it for help. If the lizard obliged, then the medicine man would snap off its tail. Quickly after doing this, the jagged end of the still wiggling tail was used to remove the cataract from the eye. During this process, songs were sung and prayers were said to ensure good thoughts for the patient and the lizard who sacrificed its tail. As a Kaibab Paiute elder (Van Vlack, 2007) said, "You have to pray in everything you do, don't want to upset anything." After the cataract was cut out, the eye was very bloody and prone to infection. The eye needed to be covered to make the bleeding stop and to prevent germs from entering the wound, and something would be needed to reduce inflammation. The medicine man made a poultice using the milky sap from the stems, leaves, and flowers of rattlesnake weed (the Paiute term, *tuvipaxghaiv*, means "necklace for the ground"). Rattlesnake weed is used as a wash to treat eye ailments including cataracts (Stoffle *et al.*, 1990). This poultice was placed on the eye and the medicine man said more prayers. After the ceremony, the tail of the lizard and other items used would be properly disposed of, usually by burning. The lizard, within some amount of time, would regenerate a tail and would continue living a normal, productive life.

During a traditional cultural property eligibility study of a ceremonial area on the Nevada Test Site, a Southern Paiute elder recounted a story of her grandmother performing this surgery on herself in the 1900s:

I remember my grandma actually doing this to herself when I was a young girl. She never let me watch because she said it was always too bloody and dangerous, but I always wanted to see. They used to do this all the time, not too long ago. Paiute medicine is a lot better than what we have today. Ours comes from the Earth . . . the other kind comes from a lab . . . you know it's hard to trust it.

(Van Vlack, 2007)

What makes this procedure unique is that the medicine man uses a whiptail lizard tail instead of other normal doctoring tools like obsidian and chert points. The jagged edge of the tail had to have been just sharp enough to remove the cataract but not sharp enough where it would destroy the entire eye. The end of the tail is also probably the right thickness to put inside the eye to cut out the cataract. A stone point could have been too large and sharp.

This procedure begs the question: How did the Paiutes know to use the broken tail of a Great Basin whiptail lizard to remove cataracts? The answer can be found in the idea that the longer people are in a place the more they learn, and the people develop deep spiritual understandings and relationships with their environment. Whiptail lizard eye surgery is a common practice among many Great Basin groups. Omer Stewart (1942) documented this practice for the Southern Paiutes and Utes. Contemporary ethnography also documents this activity for the Western Shoshones (Van Vlack, 2007). What is interesting to note is that Stewart (1942: 320) recorded that his Navajo informants did not practice this type of eye surgery, whereas the San Juan Paiutes, who share a boundary with the Navajo, did. This is further evidence that this type of knowledge is obtained through thousands of years of environmental learning and interaction in the same ecosystem.

Discussion

Whiptail lizard eye surgery is a type of optometric procedure known as "couching." Couching is a technique that dislocates the cataractous lens through the use of a sharp object or by blunt manipulation. This technique is found in many places around the world and, in some places (such as rural sub-Saharan Africa, Morocco, India, Nepal, and China), it is still the preferred form of cataract treatment. In Old World societies, "couching" can be traced back to 1700 BC and the Assyrian Code of King Hammurabi in the Middle East; in India, "couching" was first documented around 700 BC by the Hindu surgeon Susruta (Schrader, 2004). Around the world, highly specialized traditional doctors performed this surgery. In many societies, these doctors have been predominantly male, and the knowledge was passed on from fathers to sons. This form of surgery is so successful that many patients will regain full vision of the infected eye. During a recent study of traditional healers in Nigeria, scientists learned that, of the sixty-five eyes they examined that had been subjected to the eye surgery, 65 percent showed a corrected visual acuity between 20/20 and 20/40 (Schrader, 2004: 155).

4.5 Paiute Science: Portals, Other Dimensions, and the Solar System

It is beyond the ken of Western science that there are portals through which people, animals, and songs can pass to other dimensions of existence. Furthermore, in the epistemology that guides Western science, it is impossible for Native people to know about such portals and to have used them to move between dimensions and travel to the solar system. Yet, in Kanab Creek, as it flows toward the Grand Canyon and Colorado River, there is a space travel cave that has been used in recent times by a member of the Kaibab Paiute tribe. On the celling of this space travel cave are paintings of where the traveler went and what he saw while there. The cave is a portal used by specially prepared spiritual leaders facilitated by a drink from a piece of the eastern root of sacred datura. To Western scientists, such a description presents what we have called an *epistemological divide* in environmental communication. It is a point at which epistemologies clash, not permitting the other to be correct, that is, not veristic. We have written about this epistemological problem before (Stoffle *et al.*, 2016) and those discussions need not be repeated here. For this discussion of Paiute science, it is sufficient to note that what is impossible in one culture is very possible and even expected in another. During a Cry Funeral ceremony, a Paiute lead singer can travel to visit with the deceased at a place where the songs have taken him on his thousand-mile journey during the funeral. The singer brings back news to the funeral gathering that the deceased has arrived at a certain place on the Salt Song Trail and appreciates that his family and community are singing him to the afterlife. When the funeral is completed, the formerly deceased person reappears as an animal (usually a bird) in the morning in front of a family member to document that he is well. Over the next year, spiritual leaders will visit the afterlife to check on the condition of the deceased, who is now living with relatives. Soon, the deceased will be born as an animal to live out another life cycle in this dimension. Matter can be neither created nor destroyed in Paiute epistemology.

The following text is provided by Richard Arnold to illustrate Paiute connections to the solar system. Only a brief text is appropriate, given that the TEK regarding this knowledge is culturally restricted.

Stars Dancing (Shared by Richard Arnold for This Chapter)

Tung-wuh-vi-gai = to rattle

Woont tu ve un kunt = doctor who talks to the stars, moon, and sun

Ceremonial rattles are used for a variety of purposes to perpetuate change and restore balance in the world with the passing of loved ones. Salt and Silver Songs (old version tied to Pahrump, Ash Meadows, and Good Springs) describe our journey to the afterlife and interactions with the spirits, people, animals, insects, and the sky, along with the wind, stars, and planets (constellations).

Rattles have distinct sounds called *tung-wuh-vi-gai* created to mimic the rattlesnake (known as Grandfather). Often, rattles incorporate designs used to pay tribute to the natural elements, which are grounded in respect and with reverence to our relative(s) and the old ones. Often, rattles can contain a series of holes to enhance the sound but, moreover, to call upon the power of the stars to sustain balance in the world by listening to the voices of the stars.

Interactions with the stars require a special connection and systematic preparation, similar to what special Indian doctors called *Woont-tu-ve-un-kunt*. These doctors have the ability to communicate with the universe above that includes the power of the Sun, Moon, and stars, including those stars that live in the Milky Way, travel across the night sky, or demonstrate different behaviors when the world is out of balance or requires restoration.

Specific prayers and a series of songs are connected to the winter season and relied upon to doctor people using direct communication and interaction with people needing assistance. Indian people are not permitted to tell these stories out of season or else a person or family member will get bitten by a snake or struck by lightning for disrespecting established protocols.

Indian doctors use their knowledge to conduct ceremonies, which establish a foundation for explaining who, what, and why the interactions or disconnection is occurring at different times or intervals. When singing, the doctor or singer(s) travel to the sky to interact with the old ones who take care of the stars and guide us during the present and after-life. During this time, we must introduce ourselves to the spirits and elements then describe the purpose of our visit or request. Thereafter, we are guided upon a spiritual journey that is vocalized through songs, or stories, and tributes to the person.

This practice was told to us during Creation when the world was new. We continue to practice these ceremonies as a means of expressing the voices of our worldly elements that keep us strong.

Why Are Paiute People Scientists?

All those who learn by observing nature, check their findings through time and with others, and incorporate these understandings into their lifeways so as to increase their resilience and quality of life are scientists. Of course, if people do not live in a natural environment or move frequently, they will know less about their environment. Perhaps they just know more about the built urban environment.

Gary Nabhan (1997: 2) begins *Cultures of Habitat* with conservation observations: (1) when human populations have stayed in the same place for the greatest duration, fewer plants and animals have become endangered species; (2) the more stable a community is, the better it can buffer native plants and animals from otherwise pervasive threats; and (3) naturally diverse regions are also culturally diverse. These observations have stood the test of time and research (see Pretty *et al.*, 2009).

North America before 1492 was, for 37,000 years, the home of people, plants, and animals. This massive ecosystem had its own perturbations, but these are to be contrasted with, not compared to, what occurred after the arrival of Europeans, their animals, and their diseases. In many areas, Native Americans, before European intrusion, permanently resided in their homelands for thousands, perhaps tens of thousands, of years.

Paiute people, and their cousins the Utes and Shoshone people, meet all of Nabhan's criteria for conservation. They were culturally diverse, reflecting in part where they lived; they had stable communities; and they had been in the region for thousands of years. Our research argues, based on the presence of sacred items that document religious continuity in the archaeology record, that Paiutes have lived in their homelands for more than 8,000 years (Stoffle *et al.*, 2020). Paiute people stipulate that they were created in these lands and have never lived elsewhere (Figure 4.3). Together with their cousins, they occupied most of the western United States, including California east of the Serra Nevada Mountains, all of the Great Basin, the Mojave Desert, and the Colorado Plateau to the Sea of Cortez.

Our environmental learning model, presented earlier, argues that, with time, the knowledge of specific fauna and flora will combine with integrated topographic understandings and recognition of ecological zones. Importantly, the ecological services of keystone species will be understood, and they will be celebrated and protected. Places that call down the rain, like the largest mountains, will become perceived as origin mountains. Species knowledge will become integrated with observations of trophic levels and biodiversity and biocomplexity interactions.

Most importantly, however, Paiutes learned from their environment because all elements of their homelands were perceived as living and sentient, with needs and feelings and with the agency to conduct their own affairs. In such a world, Paiutes are taught to watch and listen to everything, from the water to the air, in order to see what the world wishes of them. They are taught that, if you pay attention to the world's elements, they will guide your behavior. And, of course, Paiute people have been doing this for more than 37,000 years.

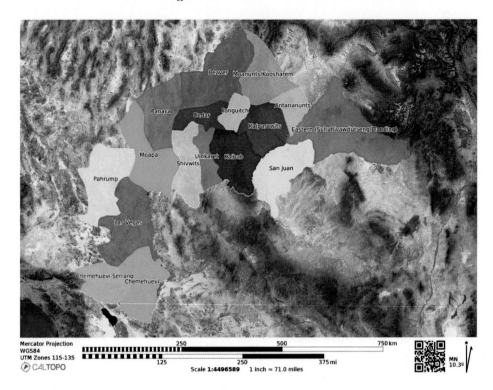

Figure 4.3 Traditional Southern Paiute territory.

References

Albert, S. and Trimble, T. (2000). Beavers are partners in riparian restoration on the Zuni Indian Reservation. *Ecological Restoration*, **18**(2), 87–92.

American Indian Writers Subgroup (1996). *American Indian Assessments: Final Environmental Impact Statement for the Nevada Test Site and Off-Site Locations in the State of Nevada*, Appendix G. Las Vegas, NV: US Department of Energy Nevada Operations Office.

Anderson, K. and Lake, F. (2016). Beauty, bounty, and biodiversity: the story of California Indians' relationship with edible native geophytes. *Fremontia*, **44**(3), 44–51.

Barker, J. (1997). Review: *Red Earth, White Lies*. *Studies in American Indian Literatures*, **9** (2), 84–7.

Behler, J. L. and King, F. W. (1979). *National Audubon Society Field Guide to North American Reptiles and Amphibians*. New York: Knopf.

Bennett, J. W. (1969). *Northern Plainsmen: Adaptive Strategy and Agrarian Life*. Arlington Heights, IL: AHM Publishing Corporation.

Berkes, F. (2018). *Sacred Ecology*, 4th ed. New York: Routledge.

Berkes, F. and Turner, N. (2006). Knowledge, learning and the evolution of conservation practice for social-ecological system resilience. *Human Ecology*, **34**(4), 479–94.

Chandler, S. (1990). Limited excavations at Bighorn Sheep Ruin (42SA1563) Canyonlands National Park, Utah. *Utah Archaeology*, **3**(1), 85–104.

Connell, J. (1978). Diversity in tropical rain forests and coral reefs. *Science*, **199**(4335), 1302–10.

Crowley, K. and Link, M. (1989). *The Sky Islands of Southeast Arizona*. Stillwater, MN: Voyageur Press.

Davidson-Hunt, I. J. (2006). Adaptive learning networks: developing resource management knowledge through social learning forums. *Human Ecology*, **34**(4), 593–614.

Deloria, Jr., V. (1995). *Red Earth, White Lies: Native Americans and the Myth of Scientific Fact*. New York: Scribner.

Deloria, Jr., V. (1998) Intellectual self-determination and sovereignty: looking at the windmills in our minds. *Wicazo Sa*, **13**(4), 25–31

Deloria, Jr., V. and Stoffle, R. (1998). *Native American Sacred Sites and the Department of Defense*. Washington, DC: United States Department of Defense.

Deur, D., Dick, A., Recalma-Clutesi, K., and Turner, N. (2015). Kwakwaka'wakw B clam gardens: motive and agency in traditional Northwest Coast mariculture. *Human Ecology*, **43**(2), 201–12.

Duke, D., Wohlgemuth, E., Adams, K., *et al.* (2021). Earliest evidence for human use of tobacco in the Pleistocene Americas. *Nature Human Behaviour*, **6**(2), 183–92.

Dwyer, P. (1985). Review of "Pigs for the Ancestors: Ritual in the Ecology of a New Guinea People" by Roy A. Rappaport, 1984. *Oceania: a Journal Devoted to the Study of the Native Peoples of Australia, New Guinea, and the Islands of the Pacific*, **56**, 151–4.

Gamillo, E. (2021). Humans' earliest evidence of tobacco use uncovered in Utah: the charred seeds suggest that people used tobacco over 12,000 years ago – much earlier than previously thought. *Smithsonian Magazine*, October 12. www.smithsonianmag .com/smart-news/humans-earliest-evidence-of-tobacco-use-uncovered-in-utah-180978850/.

Grayson, D. (1993). *Desert's Past: A Natural Prehistory of the Great Basin*. Washington, DC: Smithsonian Institution Press.

Groark, K. (1996). Ritual and therapeutic use of "hallucinogenic" harvester ants (*Pogonomyrmex*) in Native South-Central California. *Journal of Ethnobiology*, **16**(1), 1–30.

Groesbeck, A., Rowell, K., Lepofsky, D., and Salomon, A. (2014). Ancient clam gardens increased shellfish production: adaptive strategies from the past can inform food security today. *PLOS One*, **9**(3), e91235.

Halmo, D., Stoffle, R., and Evans, M. (1993). Paitu Nanasuagaindu Pahonupi (Three Sacred Valleys): cultural significance of Gosiute, Paiute, and Ute plants. *Human Organization*, **52**(2), 142–50.

Hao, S. (2013). Engineers of the San Pedro River. 50 *Earth Justice Blog*, June 21. https:// earthjustice.org/users/shirley-hao.

Hardin, G. (1968). Tragedy of the Commons: the population problem has no technical solution; it required a fundamental extension in morality. *Science*, **162**(3859), 1243–8.

Hernandez-Avila, I. and Varese, S. (1999). Indigenous intellectual sovereignties: a hemispheric convocation. *Wicazo Sa Review*, **14**(2), 77–91.

Hockett, B. (2010). Nutritional ecology of Late Pleistocene to Middle Holocene subsistence in the Great Basin. In K. Graf and D. Schmitt, eds., *Paleoindian or Paleoarchaic?: Great Basin Human Ecology at the Pleistocene–Holocene Transition*. Salt Lake City, UT: University of Utah Press, pp. 204–30.

Jackley, J., Gardner, L., Djunaedi, A. F., and Salomon, A. K. (2016). Ancient clam gardens, traditional management portfolios, and the resilience of coupled human-ocean systems. *Ecology and Society*, **21**(4), 20.

Lansing, S. (1991). *Priests and Programmers: Technologies of Power in the Engineered Landscape of Bali*. Princeton, NJ: Princeton University Press.

Lewis, H. (1982). *A Time for Burning*. Edmonton, AB: University of Alberta Press.

Lewis, H. (1989). Ecological and technological knowledge of fire: Aborigines versus park rangers in Northern Australia. *American Anthropologist*, **91**(4), 940–60.

Lewis, H. and Ferguson, T. (1999). Yards, corridors, and mosaics: how to burn a boreal forest. In R. Boyd, ed., *Indians, Fire, and the Land in the Pacific Northwest*. Corvallis, OR: Oregon State University Press, pp. 164–84.

Lin, F., Graham, L., Campbell, R., and Davies, P. (2007). Structural modeling of snow flea antifreeze protein. *Biophysical Journal*, **92**(5), 1717–23.

Lindsay, I. and Loosle, B. (2006). The Johnson Rockshelter (Site 42UN2580): an analysis of material remains and cultural contexts of a Fremont storage site. *Utah Archaeology*, **19**(1), 1–20.

Maenhout, J. (2013). Beaver ecology in Bridger Creek, a tributary to the John Day River. MSc Thesis, Oregon State University.

Matthews, M.H. (1988). Macrobotanical remains and worked vegetal artifacts from bighorn sheep ruin/42Sa1563. In S. M. Chandler, ed., *Archaeological synthesis of the Bighorn Sheep Ruin/42Sa1563, Needles District, Canyonlands National Park, Southeastern Utah*, Technical Report 51. Denver, CO: National Park Service, Rocky Mountain Regional Office, pp. A-1–A-74.

McCay, B. and Acheson, J. (1987). *The Question of the Commons: The Culture and Ecology of Communal Resources*. Tucson, AZ: University of Arizona Press.

Miller, A. M. and Davidson-Hunt, I. (2010). Fire, agency and scale in the creation of aboriginal cultural landscapes. *Human Ecology*, **38**(3), 401–14.

Mohawk, J. (1996). Review of "Red Earth, White Lies." *American Anthropologist*, **98**(3), 650–1.

Nabhan, G. (1997). *Cultures of Habitat: On Nature, Culture, and Story*. Washington, DC: Counterpoint.

Nabhan, G. (2018). *Mesquite: An Arboreal Love Affair*. White River Junction, VT: Chelsea Green Publishing.

Nabhan, G., Dobyns, H., Hevly, R., and Euler, R. (1981). Devil's claw domestication: evidence from southwestern Indian fields. *Journal of Ethnobiology*, **1**(1), 135–64.

Nickens, P. (1981). *Pueblo III Communities in Transition: Environment and Adaptation in Johnson Canyon*. Boulder, CO: Colorado Archaeological Society and University of Colorado.

Pierson, L. (1980). *A Cultural Resource Summary of the East Central Portion of Moab District*. Cultural Resource Series, No. 10. Moab, UT: U.S. Department of the Interior, Bureau of Land Management.

Pimm, S. (1991). *The Balance of Nature?: Ecological Issues in the Conservation of Species and Communities*. Chicago, IL: University of Chicago Press.

Pretty, J., Adams, B., Berkes, F., *et al.* (2009). The intersections of biological diversity and cultural diversity: towards integration. *Conservation & Society*, **7**(2), 100–12.

Rappaport, R. A. (1968). *Pigs for The Ancestors: Ritual Ecology of a New Guinea People*. New Haven, CT: Yale University Press.

Rappaport, R. A. (1984). *Pigs for The Ancestors: Ritual Ecology of a New Guinea People*, 2nd ed. Prospect Heights, IL: Waveland Press.

Rhode, D. and Louderback, L. (2010). Dietary plant use in the Bonneville Basin during the Terminal Pleistocene/Early Holocene transition. In K. Graf and D. Schmitt, eds., *Paleoindian or Paleoarchaic?: Great Basin Human Ecology at the Pleistocene Holocene Transition*. Salt Lake City, UT: University of Utah Press, pp. 231–50.

Rowe, T., Stafford, T., Fisher, D., *et al.* (2022). Human occupation of the North American Colorado Plateau ~37,000 years ago. *Frontiers in Ecology and Evolution*, **10**, 903795. https://doi.org/10.3389/fevo.2022.903795.

Sauer, C. (1925). *The Morphology of Landscape*. Berkeley, CA: University of California Press.

Sauer, C. (1956). The agency of man on the earth. In W. L. Thomas, Jr., ed., *Man's Role in Changing the Face of the Earth*. Chicago, IL: University of Chicago Press.

Schrader, W. E. (2004). Traditional cataract treatment and the healers perspective: dialogue with western science and technology in Nigeria, West Africa. *Annals of African Medicine*, **3**(3), 153–8.

Schroedl, A. and Coulam, N. (1994). Cowboy Cave revisited. *Utah Archaeology*, **7**(1), 1–34.

Shipek, F. (1991). *Delfina Cuero, Her Autobiography and Account of her Last Years and Her Ethnobotanic Contributions*. Menlo Park, CA: Ballena Press.

Sofaer, A. (1982). Lunar markings on Fajada Butte. In A. Aveni, ed., *Archaeoastronomy in the New World*. Cambridge, UK: Cambridge University Press, pp. 169–81.

Sofaer, A. (2021). Solstice project papers. https://solsticeproject.org/solstice-project-papers/.

Sofaer, A. and Sinclair, R. M. (1982). *Astronomical Marking Sites on Fajada Butte*. Chaco Canyon Center, National Park Service, New Mexico.

Sofaer, A., Zinser, V., and Sinclair, R. M. (1979). A unique solar marking construct. *Science*, **206**(4416), 283–91.

Stewart, O. C., 1942. *Culture Element Distributions: XVIII: Ute-Southern Paiute*. Berkeley, CA: University of California Press.

Stewart, O., Lewis, H., and Anderson, M. K. (2009). *Forgotten Fires: Native Americans and the Transient Wilderness*. Norman, OK: University of Oklahoma Press.

Stoffle, R. W. and Arnold, R. (2003). Confronting the angry rock: American Indians' situated risks from radioactivity. *Ethnos*, **68**(2), 230–48.

Stoffle, R., Arnold, R., and Bulletts, A. (2016). Talking with nature: Southern Paiute epistemology and the double hermeneutic with a living planet. In G. Tully and M. Ridges, ed., *Collaborative Heritage Management*. Piscataway, NJ: Gorgias Press, pp. 75–99.

Stoffle, R., Chmara-Huff, F., Van Vlack, K., and Toupal, R. (2004). *Puha Flows from It: The Cultural Landscape Study of the Spring Mountains*. Prepared for the United States Forest Service Humboldt Toiyabe National Forest Spring Mountains National Recreation Area. Tucson, AZ: Bureau of Applied Research in Anthropology, University of Arizona.

Stoffle, R. W. and Evans, M. J. (1976). Resource competition and population change: a Kaibab Paiute ethnohistorical case. *Ethnohistory*, **23**(2), 173–97.

Stoffle, R., Evans, M., Nieves Zedeño, M., Stoffle, B., and Kesel, C. (1994). *American Indians and Fajada Butte Ethnographic Overview and Assessment for Fajada Butte and Traditional (Ethnobotanical) Use Study for Chaco Culture National Historical Park, New Mexico*. Prepared for the National Park Service. Tucson, AZ: Bureau of Applied Research in Anthropology, University of Arizona.

Stoffle, R., Halmo, D., and Evans, M. (1999). Puchuxwavaat Uapi (to know about plants): traditional knowledge and the cultural significance of Southern Paiute plants. *Human Organization*, **58**(4), 416–29.

Stoffle, R. , Halmo, D., Evans, M., and Olmsted, J. (1990). Calculating the cultural significance of American Indian plants: Paiute and Shoshone ethnobotany at Yucca Mountain, Nevada. *American Anthropologist*, **92**(2), 416–32.

Stoffle, R. W., Jake, M. C., Bunte, P., and Evans, M. J. (1982). Southern Paiute peoples' SIA responses to energy proposals. In C. C. Geisler, D. Usner, R. Green, and P. West, eds., *Indian SIA: The Social Impact Assessment of Rapid Resource Development on Native Peoples*. Ann Arbor, MI: School of Natural Resources, University of Michigan, p. 107.

Stoffle, R. W., Jones, K. L., and Dobyns, H. F. (1995). Direct European immigrant transmission of Old-World pathogens to Numic Indians during the nineteenth century. *American Indian Quarterly*, **19**(2), 181–203.

Stoffle, R., Naranjo, A., Sittler, C., and Slivk, K. (2018). Grandfather tree: Ute horror at the killing of a heritage tree. *International Journal of Intangible Heritage*, **13**, 36–49.

Stoffle, R., Pickering, E., Sittler, C., *et al.* (2017). [CANY] *Ethnographic Overview and Assessment for Canyonlands National Park.* Tucson, AZ: School of Anthropology, University of Arizona.

Stoffle, R., Sittler, C., Albertie, M., *et al.* (2019). *Ethnographic Overview and Assessment for Hovenweep National Monument.* Tucson, AZ: School of Anthropology, University of Arizona.

Stoffle, R., Sittler, C., Van Vlack, K., Pickering, E., and Lim, H. (2020). Living Universe or GeoFacts: stone arches in Utah National Parks – epistemological divides in heritage environmental communication. *International Journal of Intangible Heritage*, **15**, 16–27.

Stoffle, R., Toupal, R., Van Vlack, K., *et al.* (2009). *Native American Ethnographic Study of Tonto National Monument.* Prepared for National Park Service. Tucson, AZ: Bureau of Applied Research in Anthropology, University of Arizona.

Stoffle, R. W., Toupal, R., and Zedeño, N. (2003). Landscape, nature, and culture: a diachronic model of human–nature adaptations. In H. Selin and A. Kalland, eds., *Nature Across Cultures*. Dordrecht, Netherlands: Kluwer Academic Publishers, pp. 97–114.

Talbot, R. K. and Wilde, J. D. (1989). Giving form to the formative: shifting settlement patterns in the eastern Great Basin and northern Colorado Plateau. *Utah Archaeology*, **2**(1), 3–18.

Thomas, W. L. (1956). *Man's Role in Changing the Face of the Earth.* Chicago, IL: University of Chicago Press.

Toupal, R. (2001). Landscape perceptions and natural resources management: finding the "social" in the "sciences." PhD Dissertation, School of Renewable Natural Resources, University of Arizona.

Treat, J. (1997). Review: *Red Earth, White Lies. Nova Religion: The Journal of Alternative and Emergent Religions*, **1**(1), 163–5.

Turner, N. J. (2003). The ethnobotany of edible seaweed (*Porphyra abbottae* Krishnamurthy and related species; Rohdophyta: Bangiales) and its use by First Nations on the Pacific Coast of Canada. *Canadian Journal of Botany*, **81**(4), 283–93.

Turner, N. J. (2016). We give them seaweed: social economic exchange and resilience in the Northwestern North America. *Indian Journal of Traditional Knowledge*, **15**(1), 5–15.

Turner, N. J. and Berkes, F. (2006). Coming to understanding developing conservation through incremental learning. *Human Ecology*, **34**(4), 495–513.

Turner, N. J., Ignace, M. B., and Ignace, R. (2000). Traditional ecological knowledge and wisdom of Aboriginal Peoples in British Columbia. *Ecological Applications* **10**(5), 1275–87.

Turner, N. J., Lepofsky, D., and Deur, D. (2013). Plant management systems of British Columbia's First Peoples. *The British Columbian Quarterly*, **179**, 107–33.

Van Vlack, K. (2007). Traditional ecological knowledge and resilience of the Southern Paiute High Chief System. MA Thesis, Department of American Indian Studies, University of Arizona.

Vayda, A. (1976). *War in Ecological Perspective.* New York: Plenum Press.

Vayda, A., Walters, B., and Setyawati, I. (2004). Doing and knowing questions about studies of local knowledge. In A. Bicker, P. Sillitoe, and J. Pottier, eds., *Investigating Local Knowledge: New Directions, New Approaches*. London: Ashgate Publishing.

Williams, D. B. (2013). *A Naturalist's Guide to Canyon Country*, 2nd ed. Guilford, CT: Falcon Guides.

Winter, J. C. and Hogan, P. F. (1986). Plant husbandry in the Great Basin and adjacent northern Colorado Plateau. In C. J. Condie and D. D. Fowler, eds., *Anthropology of the Desert West: Essays in honor of Jesse D. Jennings*. Salt Lake City, UT: University of Utah Press, pp. 117–44.

Yoder, D. T. (2005). Fremont storage and mobility: changing forms through time. MA Thesis, Department of Anthropology, Brigham Young University.

5

"To Get More Harvest"

Natural Systems, Cultural Values, and Indigenous Resource Management in Northwestern North America

NANCY J. TURNER AND DOUGLAS DEUR[*]

People ... think that Nature just grows on its own. But our people felt, to get more harvest, and bigger berries, they did these things [pruning, burning]. Same thing a farmer does ...
(Dr. Daisy Sewid-Smith, personal communication to NT, 1996)

Mockorange. Something like that yew. If you trim it, it'll last a lot longer.
(Dr. Luschiim Arvid Charlie, personal communication to NT, 2011)

5.1 Introduction

For at least 15,000 years, and possibly much longer, Indigenous peoples have occupied the northwestern corner of North America. Over this span of time, these peoples have relied on hundreds of species of plants, animals, fungi, and algae – providing them with foods, materials, and medicines for their survival. This chapter focuses on the region extending from southern Alaska and the Yukon and south across British Columbia and the US northwestern states to central California and east to the Continental Divide. The geographic and biological diversity of this mountainous region is immense, encompassing a range of marine and coastal

[*] We extend deep gratitude to the many Native knowledge holders and researchers who have contributed to the research summarized in this chapter. Among them, we especially wish to thank Clan Chief Adam Dick (Kwaxsistalla), Dr. Daisy Sewid-Smith (Mayanilth), Kim Recalma-Clutesi (Ogwilogwa), Elsie Claxton, Earl Claxton, Louis Claxton, Nicholas XEMŦOLTW Claxton, Seliliye Belinda Claxton, John Elliott, Dr. Luschiim Arvid Charlie, Violet Williams, John Thomas, Dr. Mary Thomas, Ida Jones, Chief Charlie Jones (Queesto), Dr. Richard Atleo (Chief Umeek), Chief Earl Maquinna George, Lena Jumbo, Stanley Sam, Roy Haiyupis, Thomas Dawson, Flora Dawson, Christine Joseph (Whata), Don Assu, Guujaaw (Gidansda), Captain Gold, Diane Brown (Gwaganad), Barbara Wilson (Kii'iljuus), Florence Davidson, Emma Matthews, Chief Johnny Clifton, Helen Clifton, William White (Xelimulh/Kasalid), Marven Robinson, Leigh Joseph, Christopher Paul, Karen Evanoff, Justine James, Jr., Harvest Moon, Chris Morganroth, Patti Gobin, Ray Fryberg, Marilyn Hall, Dino Herrera, Mary Ann Wright, Orin Kirk, Neva Eggsman, Joe Hobbs, Lynn Schonchin, Debbie Riddle, Joe Scovell, Robert Kentta, David Harrelson, Patricia Whereat Phillips, Brenda Beckwith, Randy Bouchard, Dan Cardinall, Catherine Jacobsen, Dr. Reece Halter, Dr. Dana Leposky, Dr. Ken Lertzman, Dr. Leslie Main Johnson, Dr. Dorothy Kennedy, Melissa K. Grimes, Dr. Harriet Kuhnlein, Dr. Andrea Laforet, Dr. Trevor Lantz, Abe Lloyd, Dr. Sandra Peacock, Dr. Tricia Gates Brown, Jamie Hebert, Rochelle Bloom, Robert D. Turner, Dr. Wendy Wickwire, and Dr. Brenda Beckwith. Research for this paper was undertaken in part through a grant to NT from the Social Sciences and Humanities Research Council of Canada (#410–2000.1166 and others).

habitats; coniferous and hardwood forests; river valleys, lakes, rivers, marshes, and muskeg; subalpine meadows; and rocky mountaintops – their configurations shifting over time and across geographic space. Here, Indigenous peoples' survival has been contingent upon access to a range of habitats – each with characteristic species – within their traditional territories, even as conditions have fluctuated over time. Their livelihoods have also been sustained by the sharing of knowledge and goods with others through social, political, and economic arrangements that, among many other things, reduce the risks associated with environmental change and unpredictability.

However, the peoples of this region have not been passive harvesters of natural resources, simply reacting to particular environments and adapting to ecological change. Rather, over the millennia, they have been active participants in the ecosystems of their homelands, developing a wide range of strategies to maintain and enhance the biological resources on which they have depended for food, medicines, materials, and a range of other cultural, economic, and spiritual purposes. A growing body of evidence points to diverse and sophisticated practices that have enabled people to live – and, in many cases, thrive – sustainably, since time immemorial, throughout this region.

This chapter presents an overview of these practices – widely known collectively as traditional land and resource management (TLRM)[1] – some of which continue to be followed by Indigenous communities into the present day and have potential relevance to modern land-use and stewardship practices. The information we present here reflects a large and rapidly growing literature that documents the intersection of TLRM practices with Indigenous peoples' traditional ecological knowledge (TEK) or Indigenous and local knowledge (ILK) systems (e.g., Williams and Hunn, 1982; Ford and Martinez, 2000; Minnis and Elisens, 2000; Anderson, 2005; Deur and Turner, 2005; Turner and Berkes, 2006; Deur, 2009; Lepofsky, 2009; Turner, 2014; Berkes, 2017, 2021; Boyd, 2022). As these sources attest, Indigenous management systems incorporate a broad spectrum of strategies – tailored to specific natural habitats and guided by specific values and beliefs. Taken together, they suggest certain characteristic modes of resource management, sharing, and intensification that can be detected on the basis of archaeological, ethnographic, and biophysical research across the region, spanning millennia of human occupation.

Accordingly, in this chapter, we summarize information drawn from diverse sources: from ethnographic, historical, and ethnobotanical literatures; from the

---

[1] Although broadly applied in reference to Indigenous practices, this term is not without some controversy, as some suggest that "management" is an incompatible concept, rooted in the Cartesian dualities and industrial logic of the Western world. Yet, if the term is taken from its derivation from Latin, *manus* (meaning "hand"), it fits the subtleties of many of the Indigenous approaches to maintaining and enhancing their lands and resources.

first-hand knowledge, observations, and experiences of traditional resource practitioners within Indigenous communities; from the analysis of historical change in culturally modified plant communities; and, in some cases, from experimental outcomes of replication of traditional management techniques. While certain land and resource management traditions continue within present-day Native communities and can be studied directly, many of these management practices have not been undertaken for generations, due largely to the disruptions of the colonial period; in these cases, Native oral traditions have provided evidence that can often be tested through archaeological and biophysical studies.

Here, we start with a description of northwestern North America, including the Indigenous peoples who have occupied these lands for millennia and the plant habitats on which they have relied. Following an overview of TEK systems that have been widespread among Indigenous communities of the Northwest, we summarize the features of various TLRM approaches and how traditional resource managers have applied these techniques in specific contexts. We stress the overarching influences of people's worldviews and the relationality with the species on which they rely – which not only permeates Indigenous cultural and cosmological systems, but also has been shown to leave specific and tangible imprints upon culturally modified habitats. We discuss the relevance of traditional management values and practices in contemporary times, in the end addressing how TLRM can be recognized and applied in contemporary land management, with Indigenous knowledge contributing potentially valuable guidance amidst today's crisis of global biodiversity loss and climate change.

5.2 Indigenous Peoples and Environments of Northwestern North America: Ecocultural Diversity within Complex Natural Landscapes

The Indigenous peoples of the study region are remarkably diverse in language and culture. They comprise over fifty distinct language or major-dialect groups within several recognized language families, including two stand-alone languages without affiliated language groups (Haida and Ktunaxa). While peoples across the region shared certain attributes – a reliance on Pacific salmon and other anadromous fish for sustenance, for example, and certain cultural and spiritual practices – the technologies and social organization of Native communities has varied markedly across the region. Major areas of cultural similarity include the peoples of the western Subarctic, the Northwest Coast, and the Interior Plateau culture areas, with some groups representing cultural transition to other areas, such as the Great Basin and California cultural regions (Helm and Sturtevant, 1982; Suttles, 1990; Walker, 1998; Turner, 2014). In each region, one finds numerous factors that shape the ways these peoples interact with their environments, from geographic and ecological variability to diverse cultural and historical influences.

Among these influences, environmental variability looms large. The climate in the study region includes the cold, long winters and shorter bright, warm summers of the Subarctic; the relatively mild conditions of the Pacific Coast, where maritime influences ensure that winters are rainy and summers are generally cool; and the much drier continental climate of the Interior Plateau and south into the Great Basin area, with notable extremes of temperature – hot, dry summers and cold, usually snowy, winters. Within each region, climate varies over short distances too, due to factors such as elevation, proximity to water, and the rain shadow effect of mountains. Here, mountain peaks rise from sea level to over 6,000 meters' elevation (approximately 20,000 feet), amplifying localized diversity of plant and animal life. Much of the region is forested. This includes the boreal forest, or taiga, and mixed deciduous–evergreen subboreal forests of the northern areas; the low- and high-elevation coniferous forests of the coastal areas (the wettest climates of North America, with some places receiving over seven meters of annual rainfall); the dry pine (*Pinus* spp.) and Douglas fir (*Pseudotsuga menziesii*) forests of the interior rain-shadowed valleys; and the interior wet forests in the wet belt toward the Rocky Mountains, with different species at upper altitudes. The milder, Mediterranean-type climate in certain coastal areas and in the interior valleys ranging from northern California to southeastern Vancouver Island supports oak (*Quercus* spp., *Lithocarpus densiflorus*) woodlands and prairies, and east of the coast and the Cascade Range are areas of cold desert, with bunchgrass (*Pseudoroegneria spicata*) and sagebrush (*Artemisia* spp.) and many places experiencing precipitation well below one meter per year. In the higher mountains throughout are subalpine parkland with patches of scrubby, slow-growing trees and treeless alpine areas with year-round snow and glaciers at the altitude extremes. Within each of these broad-scale vegetation zones are numerous habitats and ecosystems at different successional stages, adding to the overall biodiversity (Franklin and Dyrness, 1973; Schoonmaker *et al.*, 1997).

Approximately 2,500–3,000 species of native vascular plants exist within the region, many sharing common patterns of distribution and ecological requirements. The area's bryophytes, algae, fungi, and lichens are similarly diverse. In all, over 500 of these plant species are known to have specific cultural roles and names in one or more of the region's Indigenous languages. Integral to Indigenous TEK systems are many aspects of these plants, their habitats, and their unique biogeographies: their distribution, phenology, modes of regeneration, propagation and dissemination, associations with animals, and other features.

Furthermore, since time immemorial, the lifeways of Indigenous peoples have been synced with plants' predictable growth patterns and locales. For generations, the localized and seasonal availability of culturally preferred species has influenced people's travels through their home territories over the course of growing seasons.

So regular and patterned were these travels that they became cyclical "seasonal rounds," in which whole communities or families moved (and some continue to move) from place to place, harvesting seasonally available resources within their traditional territories and capitalizing on the habitat diversity afforded by the complex terrain. Specific geographic and temporal sequences have persisted over time, with harvest sites and seasonal encampments enduring year after year, generation after generation, together contributing to the overall natural resource diversity, food security, and resiliency of Native communities. Cues in the landscape – the appearance of salmonberry (*Rubus spectabilis*) blossoms, a particular bird, or a species of insect, and certain other species – have often served as an indicator that certain species were available at a distance, reflecting the traditional phenological knowledge of Native people accumulated over generations of observing these correlations (Lantz and Turner, 2003).

Traditional management has not only enhanced the quantities and culturally valued qualities of these plants, but also often served to place staple plants in predictable concentrations close to human communities – in many instances reducing the labor and scheduling conflicts associated with more diffuse natural distributions of these species. Indeed, such priorities seem to have facilitated certain agricultural or nearly agricultural developments among Indigenous peoples in North America, and may undergird the rise of agriculture more generally around the world (Deur, 2000; Smith, 2005).

5.3 Indigenous and Local Knowledge Systems

All Indigenous and local communities, namely those that are long-term residents within particular locations and environments, form and retain significant knowledge about their homeplaces and about the habitats and species that share their homelands. Such knowledge often becomes highly nuanced and sophisticated, especially when it applies to the key species on which communities rely for food, medicine, materials, and other purposes – and even when it is not recorded in written form. This knowledge, alongside related practices and worldviews, is passed down through generations verbally and otherwise and comprises what have been known as TEK systems. In recent literature, researchers have referred to these systems of knowledge by other terms, most notably ILK systems (Ford and Martinez, 2000; Turner and Berkes, 2006; Berkes, 2017, 2021; see also Chapter 1). Encompassed within ILK systems are certain forms of knowledge and practice, as well as mechanisms for gathering and transferring environmental information, including:

- **practical knowledge** – in-depth knowledge about the species living within a group's territory; their life cycles and seasonality; their potential use as foods, materials, and medicines; effective ways of harvesting, processing, storing, and using them; their relationships with other species and habitats; and similar information;

- **modes of knowledge transmission** – culturally mediated ways of passing on this knowledge, such as through experiential learning, imitation, storytelling, ceremonies, and direct instruction and specific modes of training;
- **modes of cultural mediation** – social, political, and economic institutions that help to oversee and mediate use, sharing, responsibility and relationships with these species, including ownership and proprietorship over places and culturally important species, the division of labor, the specialization of roles, planning and decision-making, and the intergenerational transmission of roles relating to land use;
- **perspectives and beliefs** – worldviews and values relating to lands, waters, other species, and human–environment interactions, including concepts of interspecific respect, reciprocity, and kincentricity, based on spiritual beliefs and fostered through ceremonial recognition, often with tangible effects upon patterns of land and resource management (Sewid-Smith *et al.*, 1998; Turner *et al.*, 2000; Turner and Berkes, 2006; Kimmerer, 2013; Turner, 2014; Berkes, 2017; Deur *et al.*, 2020a, b).

In some ways, ILK systems share important traits with the formal systems of scientific knowledge developed in Western and other academic traditions. These ILK systems retain information regarding temporal and geographic patterns in natural phenomena, cause–effect relationships, trends in resource abundance, and other matters of enduring interest to the natural sciences. Yet, there are key differences as well. Practitioners of Western science strive to maintain objectivity, to base their theories on accumulated lines of evidence, and to draw conclusions from systematic, unbiased experimental testing. These formal academic and scientific approaches generally reject key aspects of ILK systems: spiritual foundations for knowledge, the acknowledgment of phenomena that are not readily replicable, or knowledge that has been transmitted intergenerationally by "word of mouth" between people without direct empirical experience – that is, anything that cannot be proven through scientific methods or falsified through the systematic analysis of empirical data (Popper, 1959; Aikenhead and Mitchell, 2010; Snively and Williams, 2016; Berkes, 2017).

A key element of most ILK systems involves the mechanisms that people have developed for maintaining and enhancing – both qualitatively and quantitatively – the plants and animals on which they depend. This aspect of ILK addresses certain urgent needs of traditional societies: How do we use the plants and animals in ways that do not cause depletion or deterioration in their populations through overharvesting or inappropriate use? How do we enhance key resources to meet the needs of growing populations? How do we bring essential natural resources closer to the people who rely upon them? It is within the context of these urgent questions that we see diverse systems of Indigenous stewardship – TLRM systems – take on critically important roles in sustaining people's livelihoods and well-being across generations.

5.4 Traditional Management of Plants and Plant Habitats: The Human Role in Natural Processes

Traditional land and resource management is a concept that encompasses many different mechanisms for enhancing the quality, quantity, and predictability of culturally preferred natural resources. For our purposes, TLRM can be defined as:

the conscious accumulation, application and adaptation of any combination of techniques and methods drawn from Traditional Ecological Knowledge systems, mediated by particular beliefs and worldviews, that sustain or enhance the availability, abundance, productivity, diversity and/or quality of a plant or animal population or of an entire resource area or habitat over a period of years or generations. (Turner, 2014: 238)

Many people have recognized and written about aspects of TLRM in North America and elsewhere in the world – with a substantial proportion of recent work focusing on the Indigenous peoples of northwestern North America, California, the Desert Southwest, and Mesoamerica (e.g., Posey and Balée, 1989; Blackburn and Anderson, 1993; Anderson, 1996, 2005, 2009; Ford and Martinez, 2000; Minnis and Elisens, 2000; Davidson-Hunt, 2003; Hunn *et al.*, 2003; Deur and Turner, 2005; Turner and Peacock, 2005; Turner *et al.*, 2005, 2009; Menzies, 2006a, b; Turner and Berkes, 2006; Lepofsky and Lertzman, 2008; Thornton, 2008; Deur, 2009; Lepofsky, 2009; Johnson and Hunn, 2010; Fowler and Lepofsky, 2011; Trosper and Parrotta, 2012; Kimmerer, 2013; Lightfoot and Lopez, 2013; Lightfoot *et al.*, 2013a, b; Berkes, 2017, 2021; Boyd, 2022).

Indigenous management strategies tend to share certain nearly universal characteristics, such as the use of fire to enhance the availability of culturally preferred plant and animal species – a practice found widely in Indigenous societies across North America and worldwide. Yet these management strategies are also remarkably specialized and fitted to localized circumstances; local specialization, and the resulting diversity of management practices within complex environments, reflects generations of close observation and practical experience in situ within the traditional territories of Indigenous peoples. Generally, TLRM works with natural processes that are observable in their outcomes, namely by enhancing the reproductive and regenerative capacity of plants; maintaining and increasing plant populations through dissemination by seed, root, or other propagules; increasing the size and productivity of a particular plant resource and/or the duration of time that it is available; creating optimal habitat conditions of substrate, sunlight, moisture, and/or nutrients for given species; preventing herbivory or insect damage; and reducing interspecies and even intraspecies competition. Often, management strategies address many of these goals simultaneously.

Indigenous management operates along a continuum of geographic scales, from individual plants (such as enhancing the output of a single productive berry bush), through particular habitats (such as traditionally burned prairies), to vast multifaceted landscapes (such as entire regions characterized by anthropogenic vegetation, for instance the oak savannahs of the Puget Lowlands and Willamette Valleys of the Pacific Northwest) (Peacock and Turner, 2000). Indigenous management also operates across a continuum of timescales – ranging from one-time interventions, through those applied during a single season, to practices carried out annually across decades, centuries, or perhaps even millennia (Shebitz *et al.*, 2009; Weiser and Lepofsky, 2009).

Parallel strategies also exist in the traditional management of animal resources, including fish, shellfish, and game, which are manifested in such practices as traditional prohibitions of overharvesting, the enhancement of animals' food sources or habitat conditions, or even the transplantation of animal species to places of relative scarcity (Hunn *et al.*, 2003; Lepofsky and Caldwell, 2013; Deur *et al.*, 2015; Thornton, 2015). In some cases, these animal enhancement strategies are directly connected and intertwined with plant management. For example, in many parts of northwestern North America, landscape burning has enhanced the regrowth of grasses and other herbs, providing fresh forage for deer or elk, while also reducing insect pests, increasing available nutrients, and stimulating the germination and regrowth of some plant species such as camas (Camassia spp.) and beargrass (Xerophyllum tenax; Figure 5.1). Similarly, plant management strategies are often embedded within management systems for fish, shellfish, and other maritime resources. For example, in this region, people often burned or otherwise enhanced berry patches near salmon fishing stations so that families could harvest resources in tandem during fishing seasons. They also constructed "clam gardens" to concentrate clam production in the vicinity of traditionally managed plant communities. Some Native communities also report dispersing fish waste on culturally preferred species such as stinging nettles (*Urtica dioca* – used, among other things, in the manufacture of fishing nets) and berry bushes (*Vaccinium* spp., *Rubus* spp., and others) to enhance productivity as an ancillary part of the fish harvest.

Another key element of TLRM systems of Indigenous peoples of western North America is that, especially in the central and northern parts of the region, most plant resources are perennial, with a capacity not only for reproduction from seeds or spores, but also for vegetative regeneration through the meristematic tissues in their roots, shoots, bark, and buds – sometimes referred to as "meristem banks" (John Zasada,

Figure 5.1 Camas (*Camassia quamash*), a species whose productivity has been enhanced by traditional harvesters through regular burning and selective harvesting. (Photograph courtesy of N. Turner.)

personal communication, 2000). These meristematic tissues allow the ready propagation of many species by partial transplantation of roots, shoots, and other materials – a practice observed widely across the region (Turner *et al.*, 2013; Turner, 2021).

While much TLRM has been transmitted to Indigenous resource managers as cultural knowledge emanating before recoverable time, this knowledge has clearly been elaborated and fine-tuned by millennia of observations within particular environments. The accounts of Indigenous peoples, as reported in the TLRM literature cited here, suggest that their appreciation of natural processes has been enhanced by observing the effects of animals in herbivory and dissemination and of naturally occurring fires, floods, and other disturbances on plant growth and productivity. Such phenomena have demonstrated relationships of cause and effect in natural systems and have illuminated the potentials for human management, for example:

• beavers demonstrate how certain trees such as willows (*Salix* spp.) (Figure 5.2) and shrubs such as hazelnut (*Corylus cornuta*) can regrow robustly after being cut

Figure 5.2 Willow (*Salix* spp.) in the Columbia River valley; the lead branch has been cut off by a beaver and the tree is responding by producing several long, slender branches just below the cut. Human harvesters have also pruned willow to produce the long straight shoots used in basketry and other traditional crafts. (Photograph courtesy of N. Turner.)

back, with long straight shoots perfect for use in baskets and other crafts – human managers learned to prune back these plants as well;

- black bears eating berries such as red huckleberries (*Vaccinium parvifolium*) and soapberries (*Shepherdia canadensis*) tend to drop a few berries in the process, with new berry bushes sprouting readily from the scattered fruits – intentional human scattering of fruits was integral to both material and ceremonial practices meant to enhance berry output in some societies;

- grizzlies reveal how digging for edible underground plant parts such as the bulblets of northern riceroot lily (*Fritillaria camschatcensis*) helps to create niches for the plants, fragmenting and scattering propagules left behind so that they regenerate more quickly and broadly (Figure 5.3) – such distribution of riceroot bulblets was practiced by human harvesters along the British Columbia coast;

- natural fires create open spaces in the forest canopy for the growth of important berry species, such as black mountain huckleberry (*Vaccinium membranaceum*) (Figure 5.4) or blackcap (*Rubus leucodermis*) – this phenomenon was replicated by human managers;

- prograding sediments in estuaries or rock emerging from beneath glaciers were soon colonized by successional vegetation of high food value, such as wild strawberry (*Fragaria* spp.) and Pacific silverweed (*Potentilla egedii*) or nesting sites for culturally significant birds – these places were later burned or otherwise cleared by Indigenous harvesters to sustain the productivity of those species, which both released nutrients and sustained insolation while preventing the

Figure 5.3 Northern riceroot (*Fritillaria camschatcensis*) scattered bulblets at the bottom of a hole dug by a grizzly bear. Each of the small bulblets and the central bulb are capable of growing a new plant, prompting traditional harvesters to only take a portion and leave the rest to grow for later harvest. (Photograph courtesy of N. Turner.)

Figure 5.4 Black mountain huckleberries (*Vaccinium membranaceum*) and Cascade bilberries (*V. deliciousum*) – these species are enhanced by periodic burning of their montane habitat by traditional harvesters. (Photograph courtesy of N. Turner.)

development of later successional stages that might engulf these resources in dense coastal forest.

The response of each species to different biological stimuli and ecological processes has depended on their particular growth forms and habitats. By bearing witness to natural choreographies, analyzing patterns of cause and effect in natural systems, and transmitting this knowledge across generations, Indigenous peoples have been able to operationalize this knowledge in ways that have sustained and enhanced a diverse range of culturally preferred species.

5.5 Worldview, Sustainability, and Resource Harvests: Recurring Cultural Motifs Guiding Northwestern Traditional Land and Resource Management

Plants – and the habitats that sustain them – hold special meaning and significance within the cultures of Indigenous peoples in northwestern North America. While there is considerable variation in the teachings, songs, stories, and other cultural practices relating to plants among these peoples, certain themes appear almost universally in this region and reflect patterns seen well beyond the region in other Indigenous societies as well.

Here, anthropogenic habitats were places of enduring investment by a community – places where the people not only exerted labor but also sustained ongoing reciprocal relationships between human and plant communities. If the people took good care of the plant communities, Elders commonly report, then the plant communities reciprocated by taking good care of the people. Related to this sense of connection and obligation, especially in the northern part of our study area, anthropogenic plant communities were "owned" in various ways by communities and community leaders such as clan chiefs. Yet, in this setting, Indigenous peoples construed ownership through a distinctive lens, with hereditary chiefs serving as ongoing caretakers of resource sites sustained intergenerationally. These resource "owners" therefore had clear obligations: Those who owned Pacific crabapple

Figure 5.5 Pacific crabapple (*Malus fusca*) – a species whose prime trees or groves were often owned by families or clans among Northwest Coast peoples. (Photograph courtesy of N. Turner.)

(*Malus fusca*) trees (Figure 5.5) or gardens of edible Pacific silverweed and spring-bank clover (*Trifolium wormskioldii*) in the intertidal zone, for example, would receive a prime portion of each harvest, but they were, in return, expected to share these food resources with the village and village allies, on behalf of the family or clan. They were also obligated to ensure that the resource would be available to future generations of chiefs for the same purpose (Boas, 1921; Drucker, 1951; Turner *et al.*, 1981, 2005; Turner and Kuhnlein, 1983; Blackman, 1990; Johnson, 1994; D. Sewid-Smith, personal communication to NT, 1997; Thornton, 1999; Deur *et al.*, 2020b). These practices have been supported by a host of social and economic traditions that manifest these values in myriad ways: sharing plants and plant knowledge, devising technologies and economic mechanisms to support plant use and distribution, and more (Box 5.1).

This sense of interspecific and intergenerational obligation cannot be fully understood without acknowledging that these societies assert the fundamental sacredness and sentience of all life-forms on which they depend (Sewid-Smith *et al.*, 1998;

Box 5.1

Selected Social and Economic Practices Relating to Plant Habitat Management and Harvest

Plant Community Ownership/Proprietorship. Individuals or cultural groups hold rights (usually inherited) to use particular resources or harvesting areas: camas patches (Beckwith, 2004; Suttles, 2005), highbush cranberry and other berry patches and crabapple stands (Turner *et al.*, 2005), and estuarine root gardens (Deur, 2005).

Plant Community Monitoring. Groups or individuals have the responsibility to watch over certain resources and harvesting areas, monitoring for predation, competing users, resource conditions, etc. Examples include edible red laver seaweed (Turner, 2003), black tree lichen (Crawford, 2016), and various berry species, edible cambium, and fiber plants (e.g., Turner *et al.*, 2005).

Socially Determined Plant Conservation. Such conservation occurs through ceremonial promotion or protection of particular places, species, and populations. Examples include sword fern fronds (Turner and Hebda, 2012), devil's club stalks for medicine (Turner and Thompson, 2006), and edible seaweed (Turner, 2003; Turner and Clifton, 2006).

Teamwork and Division of Labor in Plant Harvest and Management. Different task groups are formed within a community specializing in different aspects of harvesting and processing plant resources. This is widely practiced by Northwest Indigenous Nations (e.g., Turner, 2005)

Distributed Seasonal Access to Plant Resource Areas. This especially manifests as "seasonal rounds" and is widely practiced by Northwest Indigenous Nations. Examples

---

Box 5.1 (cont.)

include spring harvesting of edible red laver seaweed (Turner, 2003; Turner and Clifton, 2006) and montane harvesting in summer (Turner *et al.*, 2011).

Trade and Exchange of Plant Products. Plant materials are exchanged through intertribal or kin-based trade networks, with trading of surplus production. This is widely practiced by Northwest Native Nations. Examples include camas bulbs, wapato tubers, seaweed, and crabapples (e.g., Turner and Loewen, 1998).

Plant Feasting and Sharing. Feasting and sharing of plant foods takes place within communities, with elites and leaders taking on primary roles; this serves as a way of distributing plant resources and ensuring community investment and labor in plant production. It is widely practiced by Northwest Native Nations (e.g., Turner and Hebda, 2012).

Plant Knowledge Transmission. Knowledge and experiences relating to plant resource management and conservation are passed on through participatory and experiential learning, stories, ceremonies, art, discourse, and focused instruction. This is widely practiced by Northwest Native Nations (e.g., Turner and Berkes, 2006)

Plant Access Technologies. Technologies are used that allow improved access to particular plant resources, improved plant transport, and, in turn, plant resource intensification, such as trails, camp shelters, and specialized freight canoes. This is widely practiced by Northwest Native Nations (e.g., Lepofsky and Lertzman, 2008).

Plant Harvest Technology Innovations. Tools are developed to facilitate harvest, and habitat modification and fire technologies are used; in addition, improvements are made to tools and approaches for processing and storing food and other plant materials. Examples include improved berry "combs" to run through bushes and efficiently remove berries, digging sticks, baskets, mats, drying racks, smoking, and pit-cooking. These are widely practiced by Northwest Native Nations (e.g., Lepofsky and Lertzman, 2008).

Integrated Multi-resource Plant Management. Combined management strategies include the effects and outcomes of two or more management strategies, applied to two or more species or entire habitats over time and space. Examples include estuarine root gardens, eelgrass beds, cedar groves, and the berry gardens of central coastal peoples (Deur and Turner, 2005).

Brown *et al.*, 2009; Atleo, 2011; Charlie and Turner, 2021; Courchene *et al.*, 2021). Oral traditions and protocols assert that human communities must extend concepts of kinship, respect, and reciprocity toward non-human life-forms – a point reflected in Indigenous stories, ceremonies, art, dance, teachings, and harvesting protocols. In Indigenous oral traditions, many plants – for example western redcedar (*Thuja plicata*) and yellow cedar (*Callitropsis nootkatensis*), western larch or "tamarack" (*Larix occidentalis*), red alder (*Alnus rubra*), trembling aspen (*Populus tremuloides*), blueberries (*Vaccinium* spp.), spiny wood fern (*Dryopteris expansa*), desert parsley (*Lomatium macrocarpum*), skunk cabbage (*Lysichiton americanus*), stinging nettle, giant horsetail (*Equisetum telmateia*), devil's club (*Oplopanax horridus*), and tree fungus (*Ganoderma applanatum* or related species) – have direct associations with original supernatural figures having human characteristics. According to some narratives, even tree pitch was formerly a human – Pitch Man. Other plants are specifically known for the legacy of their creation and transformation, sometimes through profound sacrifice on the part of an ancestor (Turner, 2014). Plants are widely described as being aware of human actions and as sometimes choosing to respond favorably to "respect," while responding unfavorably to "disrespect." Given the sense of kinship with plants and the oral traditions ascribing sentience and intentionality to plants, it is no wonder they have been treated with a level of respect and gratitude uncommon in industrialized societies and that the lives of plants are held in reverence.

Many Indigenous terms and phrases convey the ethical underpinnings of plant-harvesting strategies. For example, the Skidegate Haida phrase *Gina 'waadluxan gud ad kwaagid* means "everything depends on everything else," with the accompanying explanation (Marine Planning Partnership Initiative, 2015): "Our way of life teaches respect for all life. We live between the undersea and sky worlds that we share with other creatures and supernatural beings. Our responsibilities to the sea and land are guided by ancestral values." In addition, *Yahguudang* or *Yakguudang* (meaning "respect") has the following explanation: "Respect for each other and all living things is rooted in our culture. We take only what we need, we give thanks, and we acknowledge those who behave accordingly" (Marine Planning Partnership Initiative, 2015). Nearly identical phrases and concepts have been reported by Indigenous communities throughout the region (Sewid-Smith *et al.*, 1998; Atleo, 2004, 2011; Deur, 2009).

These beliefs and values also guide the practices associated with the harvest, management, and use of culturally significant species (Nelson, 1983; Anderson, 1996; Hunn *et al.*, 2003; Turner, 2005, 2014; Berkes, 2017; Deur *et al.*, 2020a, b; Turner and Mathews, 2020). In fact, plants and animals are widely regarded as sentient life-forms and as generous relatives of humankind who provide their bodies willingly, provided that humans show proper respect. While refracted through a distinctive and anthropomorphic lens, it seems clear that this cosmological position was informed by cause–effect relationships observed, for example, during times of overexploitation – when forms of "disrespect" came back to haunt subsequent resource harvesters in future times – and these values had tangible outcomes in the natural resource management traditions of Indigenous peoples. Accordingly, being able to care for plants has been central to Indigenous peoples' underlying worldviews and to their enduring identities today. Maintaining the root-digging grounds, berry patches, cedar stands, seaweed-picking rocks, and other culturally tended habitats and landscapes is not just practical; ensuring that these species survive and thrive has profound spiritual meaning and helps Indigenous people to fulfill cultural obligations prescribed in foundational oral traditions of creation.

The sentience and intentionality of plants is also manifested in the First Fruits, First Roots, and other ceremonies that accompany the harvesting of many species in parts of the region. In these traditions, people show respect to the plants in exchange for the plants' sacrifice; if this is done correctly, traditional Indigenous resource managers report, "the plants will come back better." Ethnographer Charles Hill-Tout (in Maud, 1978: 116) discussed the meaning of these ceremonies in the Salishan context:

> Concerning the meaning and object of these ceremonies I have been led to the opinion by my studies of the Salish and other tribes that they were always propitiatory in intent. They were intended to placate the spirits of the fish, or the plant, or the fruit, as the case may be, in order that a plentiful supply of the same might be vouchsafed to them . . . if these ceremonies were not properly and reverently carried out there was danger of giving offense to the spirits of the objects and being deprived of them. . . . For it must be remembered that, in the mind of the [Indigenous person], the salmon, or the deer, or the berry, or the root, was not merely a fish or an animal or a fruit, in our sense of these things, but something more.

Ceremonial and medicinal plants are harvested with particular care and respect in this light, reflecting their perceived potency and importance to human welfare – the seeds of *Lomatium* species, blades of sweetgrass (*Hierochloe hirta* subsp. *arctica*), boughs of western redcedar (*Thuja plicata*) or western hemlock (*Tsuga heterophylla*), sword fern (*Polystichum munitum*) fronds for ceremonial use, and many others all receive such special treatment.

Arguably, these same values are manifested in anthropogenic plant communities of many kinds. Selective harvest, the transplantation of species, scattering berries or other propagules densely in harvest areas, and the use of fire to enhance plants were all seen not just as mechanical ways of enhancing output, but as manifestations of "respect" shown to the plants, which the plants reciprocated by coming back more abundantly. Killing outright or completely replacing one habitat with a monoculture were commonly seen as perilous acts from a cosmological perspective, never mind the ecological consequences. Plant habitat enhancement was typically an incremental process, favoring native species already living in an area rather than supplanting unwanted species outright – likely a reflection of this worldview.

To accentuate this point, such values and practices can be seen even in the selective harvest of plant parts, that is, leaving the larger plant alive as a sign of such respect. This is especially visible in the case of "culturally modified trees," in which bark or even entire staves or boards would be taken from a tree, with thanks, apologies, or other signs of respect – leaving the larger tree intact. This is seen, for example, in the harvest of bark or wood from standing western redcedars, which have special significance in oral tradition, belief, and ceremony among many Indigenous communities of the maritime Northwest. As Curtis (1915: 11) reported for the Kwakwaka'wakw:

A standing tree from which boards have been split is called *keto'q* ("begged from"), and it is said that, since trees are believed to have sentient life, the ancients before obtaining boards in this way would look upward to the tree and say: "We have come to beg a piece of you today. Please! We hope you will let us have a piece of you." The same request was made of a yew tree [*Taxus brevifolia*] before cutting off a piece for making tools.

Also speaking of Kwakwaka'wakw tradition, anthropologist Franz Boas (1921: 616–17) likewise explained the absolute importance of keeping a cedar tree alive when one harvests bark from it:

They do not take all of the cedarbark, for the people of the olden times said that if they should peel off all the cedar-bark . . . the young cedar would die, and then another cedar-tree nearby would curse the bark-peeler so that he would also die. Therefore, the barkpeelers never take all of the bark off a young tree.

This aligns with the teachings of traditionally trained Kwakwaka'wakw cultural leaders of recent times, such as Clan Chief Adam Dick (Kwaxsistalla) and Dr. Daisy Sewid-Smith (Mayanilth), who attest to the need to show respect to the cedar, to "praise" the tree, and to do no harm to the tree if the bark, wood, or other materials are to be recruited to support the spiritual and material needs of humanity (Turner, 2014: 513–14). Trees harvested in this way can still be found widely in the Northwest forests, that is, scarred trees still living centuries after the harvest, which

Figure 5.6 Culturally modified western redcedar (*Thuja plicata*) tree showing where bark had been harvested several decades previously. (Photograph courtesy of N. Turner.)

are signposts not only of traditional resource harvest practice, but of the underlying values that shaped those practices over deep time (Turner *et al.*, 2009) (Figure 5.6).

5.6 Techniques and Methods Applied to Plants and Related Resources in the Traditional Land and Resource Management Systems of Northwestern North America

Here, we seek to illuminate the remarkable range and diversity of TLRM strategies by focusing on a number of examples from across northwestern North America. Knowing that the forms of TLRM suggested by the available literature are so numerous – well beyond what could be summarized in a single chapter, or indeed a single book – we chose our examples advisedly. The examples we highlight here run the gamut geographically, from the stewardship of individual plants to the management of vegetation at landscape scales. These examples also operate at a range of temporal scales, from individual instances of plant enhancement to practices carried out seasonally from the beginning of recoverable time in certain Indigenous societies.

Plants known to have been sustained and enhanced by such methods include food plants (root vegetables or geophytes, green vegetables, berries and other fruits, nuts such as hazelnut and acorns, seeds, seaweeds, and mushrooms), species used for tea, tree species whose inner bark and cambium tissues were eaten, all types of plant materials (bark and wood pieces, fibrous bark strips, sheets of outer bark, fibrous stems and leaves of herbaceous perennials, materials for glues and dyes, fire starter, tinder, and fuel), and medicinal plants valued for their bark, roots, or greens. In fact, it can be argued that virtually all plants harvested and used by Indigenous peoples were managed in some way. Each of these management strategies increases the productivity and/or localized abundance of specific plant resources and particular habitat types, enhancing the quantity, quality, and/or geographic concentrations of plants used for food, medicines, materials, and other purposes. These practices are numerous and widespread. Taken together, they tentatively suggest that humans have had significant effects upon the natural habitats of northwestern North America and have increased the region's overall biodiversity. Each management method applied practical strategies within the context of environmental possibility, with management techniques appropriate to the habitats and biological requirements of each culturally preferred species. Similarly, each management method has been situated within overarching socially guided protocols and spiritual beliefs, all learned from an early age and embedded in language, stories, and ceremonies. In all of these ways, management techniques have reflected the deeper cultural and ecological contexts of their genesis.

In Box 5.2, we summarize information drawn from multiple published sources, including those mentioned in Section 5.5, as well as a range of ethnobotanical sources from our own research with Indigenous plant specialists in British Columbia (Turner, 2005; 2014[2]), including published and unpublished ethnobotanies of coastal peoples (Coast Salish, Nuu-chah-nulth, Kwakwaka'wakw, Haida, Nuxalk, Heiltsuk, Haisla, Ts'msyen, Nisga'a) and interior peoples (Nlaka'pamux, Syilx [Okanagan], Ktunaxa, Stl'atl'imx, Secwepemc, Tsilhqot'in, Dakelh, Gitxsan), as well as various ethnobotanical sources from Washington, Oregon, Alaska, northern California, and elsewhere in the region (e.g., Chestnut, 1902; Boas, 1921; Gill, 1984; Hunn and Selam, 1990; Blackburn and Anderson, 1993; Thornton, 1999; Peacock and Turner, 2000; Hunn *et al.*, 2003; Anderson, 2005; Deur and Turner, 2005; Shebitz *et al.*, 2009; Weiser and Lepofsky, 2009; Ignace *et al.*, 2016, 2017; Phillips, 2016; Deur *et al.*, 2020; Charlie and Turner, 2021; Deur, 2022). We provide selected sources for each practice referenced.

The Box 5.2 organizes the TLRM strategies used for various groups of plants based on growth form and habit, with examples. We also provide an estimated total count of species with similar features that either are documented as being managed by one or more Indigenous groups in these ways or are reasonably assumed (based on available literature sources and reported similarities/parallels of use) to have been managed in similar ways by Indigenous resource managers.

5.7 Discussion: A Spectrum of Practices

As can be seen from the many examples in Box 5.2, "getting more harvest" is not just a single, straightforward strategy; many plant species have been maintained and enhanced through an entire range of techniques and approaches, intersecting at different scales of time, geographic space, and ecological space, and mediated within a range of social, cultural, and economic practices and conventions. In all, allowing for a slight duplication across the different categories of plants, approximately 500 different culturally important plant species have been documented as being managed or cared for to some extent in northwestern North America, often in multiple and intersecting ways, by Indigenous peoples whose homelands overlap with the plants' natural ranges.

---

[2] The plants listed and included in this section are drawn in large part from a survey of plant species with cultural applications that are named in three or more languages in a set of fifty Indigenous languages and major dialects in northwestern North America in a supplementary appendix of Turner (2014), as well as data on Indigenous peoples' plant use in Oregon and northern California from the references cited.

Box 5.2

Examples of Traditional Land and Resource Management Strategies, by Plant Type

1. **Culturally Important Annual Plants: Greens, Seeds, and Grains**
   Main Uses/Applications: mostly edible seeds and grains (parched for pinole), greens for food, medicine, and smoking or chewing (e.g., *Nicotiana* spp.).
   Major Habitat Types: coastal prairies, valley grasslands, chaparral, and oak savannas; north California to Haida Gwaii; tobaccos grown in enclosed plots and cleared patches.
   Management Strategies: harvesting pinole seeds with seed beater scatters seed for regrowth; some seeds saved and sown, often near dwellings; patches burned over in fall after harvesting; tobacco patches enclosed, soil fertilized with wood ash; ownership of seed gathering grounds; plants thinned for greens or leaves harvested selectively from multiple plants.
   Examples of Species in This Category: *Amsinckia lycopsoides* (tarweed fiddleneck), *Bromus marginatus* (mountain brome), *Calandrinia ciliata* (fringed redmaids), *Claytonia perfoliata* (miner's lettuce), *Madia sativa* (coast tarweed), *Nicotiana attenuata* (wild tobacco), *Nicotiana quadrivalvis* (wild tobacco), and *Trifolium ciliolatum* (foothill clover).
   Total Number of These Species Documented in the study Region: ~40.
   Selected References: Chestnut (1902), Turner and Taylor (1972), Anderson (2005), Lightfoot *et al*. (2013a, b), Phillips (2016).

2. **Culturally Important Herbaceous Perennial Plants Used for Seeds/Fruits**
   Main Uses/Applications: seeds harvested for food or for decoration.
   Major Habitat Types: coastal prairies, valley grasslands, chaparral, and oak savannas; north California to Vancouver Island.
   Management Strategies: harvesting scatters seed for regrowth; some seeds saved and sown, often near dwellings; patches burned over in fall after harvesting; methods to enhance the spatial concentration of *wokas*/pond lily in predictable locations proximate to settlement sites; *wokas* harvested in a zig-zag pattern to leave some for the next harvest; also marsh burned at the edges to get rid of competing plants; *wokas* scattered in these recently cleared areas ritually as part of a **first food ceremony** similar to that used for huckleberries; not as dense or productive; strawberries fertilized, cere-monially managed, and runners selectively harvested.
   Examples of Species in This Category: *Balsamorhiza deltoidea* (deltoid balsamroot); *Balsamorhiza sagittata* (balsamroot or spring sunflower), *Fragaria chiloensis* (seaside strawberry), *Lithospermum ruderale* (stoneseed [beads]), *Nuphar lutea* ssp. *polysepala* (**wokas**) (yellow pond-lily), *Wyethia longicaulis* (Humboldt mulesears), and *Typha latifolia* (cattail [stuffing]).

Box 5.2 (cont.)

Total Number of These Species Documented in the Study Region: ~10.
Selected References: Turner et al. (1981, 2021), Thornton (1999), Moss
(2005), Deur (2009)

3. **Culturally Important Geophytes (Herbaceous Perennial Plants that Die Back to Subterranean Storage Organs: Bulbs, Corms, Taproots, Rhizomes, Tubers)**
Main Uses/Applications: mainly food; some harvested for medicine and some for materials for weaving and dyeing.
Major Habitat Types: tidal marshes, freshwater marshes, and lake edges; coastal prairies, grasslands, and subalpine meadows.
Management Strategies: soil enhancement with digging sticks by tilling and loosening the soil; bulblets and smaller parts left in the ground to regrow or intentionally replanted; ground weeded during harvest to remove excess grasses, etc.; harvesting often during time of seed release; selective harvesting (e.g., only vegetative plants [leaving flowering plants to produce seed] or medium-sized plants [leaving older "mother plants" and younger plants to continue growing]); patches owned and monitored; harvesting overseen by experienced people; first roots ceremony held for some species – no one to harvest roots for themselves before this; ceremonial scattering of seeds for some; patches and meadows often burned over to maintain open prairies and provide nutrients; some beds built up as terraces with rock or wood.
Examples of Geophyte Species: foods: *Allium cernuum* (nodding onion), *Balsamorhiza sagittata* (arrowleaf balsamroot), *Camassia quamash* (common camas), *Dichelostemma capitatum* (blue dicks), *Erythronium grandiflorum* (yellow glacier lily), *Fritillaria camschatcensis* (northern riceroot), *Lewisia rediviva* (bitterroot), *Lomatium cous* (cous or biscuitroot), *Potentilla egedii* (Pacific silverweed), *Perideridia oregana* (Oregon yampah), *Sagittaria latifolia* (wapato), *Zostera marina* (eelgrass), *Pteridium aquilinum* (bracken fern); **medicines:** *Eurybia conspicua* (showy aster), *Ligusticum canbyi* (Canby's lovage), *Valeriana sitchensis* (mountain valerian), *Veratrum viride* (false hellebore); **materials:** *Carex barbarae* (white root), *Schoenoplectus acutus* (tule).
Total Number of These Species Documented in the Study Region: ~90.
Selected References: Norton (1979a, b), Anderson and Rowney (1998), Peacock (1998), Beckwith (2004), Stevens (2004a, b), Deur (2005, 2009), Turner (2006), Lloyd (2011), Joseph (2012), Cullis-Suzuki *et al.* (2015), Hoffmann *et al.* (2016), Lyons *et al.* (2018).

4. **Culturally Important Greens: Herbaceous Perennial Plants (Shoots, Stems, or Leaves)**
Main Uses/Applications: food and tea, fibrous materials for basketry and cordage, medicine, and spiritual uses.

Box 5.2 (cont.)

Major Habitat Types: forest edges, coastal areas, wetlands, and village sites.

Management Strategies: leaves and stems harvested selectively from
multiple plants, in spring (for edible greens) or late summer or fall (for fibrous
plants), allowing regeneration from the roots or rhizomes; in some cases only
vegetative plants are harvested (e.g., *Triglochin*); patches owned and monitored;
patches burned over periodically; harvesting and burning overseen by experienced people; roots/rhizomes of prime plants transplanted in new places, such as
around village sites; ceremonial management.

Examples of Herbaceous Perennial Shoot, Stem, or Leaf Species: foods:
Epilobium angustifolium (fireweed), *Heracleum maximum* (common cow-parsnip),
Lomatium nudicaule ("wild celery"), *Opuntia fragilis* (brittle pricklypear cactus),
Rumex aquaticus var. *fenestratus* (western dock), *Triglochin maritima* (seaside
arrowgrass), *Equisetum telmateia* (giant horsetail); **medicines:** *Achillea millefolium*
(common yarrow), *Artemisia tilesii* (Tilesius' wormwood or "caribou leaves"),
Goodyera oblongifolia (western rattlesnake plantain), *Mentha arvensis* (field mint),
Plantago major (broadleaved plantain); **materials:** *Apocynum cannabinum* (Indian-
hemp), *Carex obnupta* (slough sedge), *Iris missouriensis* (western blue flag),
Leymus mollis (American dunegrass), *Schoenoplectus acutus* (hardstem bulrush),
Typha latifolia (cattail), *Urtica dioica* (stinging nettle), *Xerophyllum tenax* (common
beargrass); **ceremonial:** *Polystichum munitum* (western swordfern).

Total Number of These Species Documented in the Study Region: ~105.

Selected References: Lewis (1993), Kimmerer and Lake (2001), Stevens
(2004a, b), Moss (2005), Shebitz *et al.* (2009).

5. **Culturally Important Berries and Other Fleshy Fruits: Perennial, Woody
   Species (for Wood/Bark See Section 7 of This Box)**
   Main Uses/Applications: mostly food, some as nutraceuticals.
   Major Habitat Types: forests and forest edges, early successional clearings,
   and edges of wetlands.
   Management Strategies: harvested for food seasonally without damage to
   the plants; sometimes handfuls scattered, both incidentally in picking (e.g.,
   soapberries, huckleberry/*Vaccinium* species) and in a ceremonial "giving back";
   first fruits ceremony held for some fruits like huckleberries (no fruit allowed to be
   picked before this took place); needs of bears and birds considered and some fruit
   left for them; often different varieties recognized and named; bushes pruned,
   usually during or after harvesting; bushes/plants fertilized with fish remains, ashes
   from fire, etc.; patches burned over periodically; habitats sometimes created by
   felling trees, providing stumps for prime substrate (e.g., salal, red huckleberry); roots
   dug up and transplanted to new locales; patches owned and monitored.

Box 5.2 (cont.)

Examples of Woody Berry/Fruit Species: *Amelanchier alnifolia*
(saskatoonberry), *Crataegus douglasii* (black hawthorn), *Gaultheria shallon* (salal),
Malus fusca (Pacific crabapple), *Prunus virginiana* (choke cherry), *Ribes divarica-
tum* (coastal black gooseberry), *Rubus idaeus* (wild raspberry), *Rubus spectabilis*
(salmonberry), *Rubus ursinus* (trailing blackberry), *Sambucus nigra* ssp. *cerulea*
(blue elderberry), *Shepherdia canadensis* (soapberry), *Vaccinium membranaceum*
(black mountain huckleberry), *Vaccinium parvifolium* (red huckleberry), *Viburnum
edule* (highbush cranberry).
Total Number of These Species Documented in the Study Region: ~55
Selected References: Chestnut (1902), Thornton (1999), Mack and
McClure (2002), McDonald (2005), Deur (2009), Trusler and Johnson (2008),
Routson *et al.* (2012), Turner *et al.* (2013).

6. **Perennial Woody Nuts/Large Seed Species (for Wood/Bark See Section 7 of This
   Box)**
   Main Uses/Applications: nuts/acorns; mostly for food, some for medicine and
   materials.
   Major Habitat Type: open woodlands, savannahs, valley bottoms, in well-
   drained soils; whitebark pine grows in subalpine forests.
   Management Strategies: areas regularly burned to maintain oak savannah;
   seeds/acorns scattered during harvesting; sometimes seeds/saplings planted to
   distant locales; hazelnuts pruned and coppiced; thickets maintained by regular
   landscape burns; stands monitored.
   Examples of Species in This Category: *Chrysolepis chrysophylla* var.
   chrysophylla (golden chinkapin), *Corylus cornuta* (beaked hazelnut),
   Lithocarpus densiflorus (tanoak), *Pinus albicaulis* (whitebark pine), *Pinus lam-
   bertiana* (sugar pine), *Quercus kelloggii* (California black oak).
   Total Number of These Species Documented in the Study Region: ~12.
   Selected References: Kimmerer and Lake (2001), Anderson (2005),
   Lepofsky and Lertzman (2008), Kimmerer (2013), Lightfoot *et al.*
   (2013a, b), Phillips (2016), Armstrong *et al.* (2018).

7. **Culturally Important Woody Species (Trees/Shrubs as Sources of Wood, Bark,
   Roots, and Boughs for Food, fiber, Sheets, Containers, and Fuel)**
   Main Uses/Applications: pitch/resin for chewing and for medicine; edible
   inner bark; bark sheets and fibers for weaving, mats, clothing, and containers;
   bark for medicine, dyes, and decorative overlay; wood for implements and fuel;
   roots for basketry; branches for rope and basketry; boughs for medicine, incense,
   bedding, and harvesting herring eggs.

   Major Habitat Type: old growth and successional forests, woodlands,
   riparian and marshy areas, and savannahs; from sea level to subalpine.

Box 5.2 (cont.)

Management Strategies: pitch harvested by cutting into the bark of living
trees from cuts or bark blisters; some pitch trees maintained for generations;
edible inner bark and bark for decoration and medicine, removed by strips or
vertical patches without girdling the tree; birch bark and cherry bark for basketry
removed from the outer layer of trees or branches, without damaging the growing
cambium layer; cedar roofing and bark strips for weaving hats, mats, and clothing
harvested in patches or strips; roots harvested selectively from living trees;
branches and strips of wood from standing trunk cut selectively from living
trees and shrubs, for tools, snowshoes, paddles, digging sticks, wedges,
bows, arrows, spear shafts, and many other implements (e.g., Pacific yew,
crabapple, maples, saskatoon bushes, hazel, and oceanspray); wood for
construction (planks split from living trees – cedar; withes for smokehouse
and sweathouse frames – red-osier dogwood); some pruned, burned,
coppiced, and/or thinned to get many long withes or whips of the right type;
all harvested from living trees and shrubs in quantities that allow the plants
to re-sprout or heal over through growth of meristematic tissues: "meristem
bank"; some species transplanted.

Examples of Species in This Category: *Abies lasiocarpa* (subalpine fir:
boughs, pitch blisters), *Cupressus nootkatensis* (yellow-cedar: inner bark, wood),
Juniperus communis (common juniper: boughs), *Juniperus scopulorum* (Rocky
Mountain juniper: wood, boughs), *Picea sitchensis* (Sitka spruce: pitch, roots),
Pinus ponderosa (ponderosa pine; pitch, wood), *Pseudotsuga menziesii*
(Douglas fir: pitch, bark slabs, boughs), *Taxus brevifolia* (Pacific yew: wood,
bark), *Thuja plicata* (western redcedar: outer and inner bark, wood, boughs,
branches, roots), *Alnus rubra* (red alder: wood, bark for dye, medicine),
Amelanchier alnifolia (saskatoonberry: withes, branches, bark), *Artemisia
tridentata* (big sagebrush: branches), *Betula papyrifera* (paper birch: outer
bark), *Cercis canadensis* var. *texensis* (California redbud: withes), *Cornus
sericea* (red-osier dogwood or "red willow": withes, bark), *Elaeagnus
commutata* (silverberry: bark), *Fraxinus latifolia* (Oregon ash: wood),
Mahonia aquifolium (tall Oregon-grape: stems, bark), *Malus fusca* (Pacific
crabapple: wood, bark), *Oplopanax horridus* (devil's-club: stems, bark),
Philadelphus lewisii (mockorange: withes, branches, bark), *Prunus emargi-
nata* (bitter cherry: bark, sticks), *Rhamnus purshiana* (cascara: bark, wood),
Rhododendron groenlandicum (Labrador-tea: branches), *Salix exigua*
(sandbar willow or rope willow), *Salix lucida* ssp. *lasiandra* (Pacific
willow), *Symphoricarpos albus* (common snowberry or waxberry).

Total Number of These Species Documented in the Study Region: ~140.

Box 5.2 (cont.)

Selected References: Anderson (1991, 1999), Eldridge (1997), Anderson and Rowney (1998), Sewid-Smith *et al.* (1998), Turner (2006), Turner *et al.* (2009), Earnshaw (2017a, b), Mathews (2017), Armstrong *et al.* (2018).

8. **Culturally Important Mushrooms, Puffballs, and Fungi**
   Main Uses/Applications: some species eaten and some used for tinder and/or medicine.
   Major Habitat Type: woods, on and under trees and rotten logs; meadows.
   Management Strategies: soil covered up after harvesting (mushrooms); partial harvesting of tree fungi; use of fire to create an ideal environment, especially for pyrophilous mushrooms that grown better after fires dispersal of spores with harvesting; various fungi will regrow with partial harvesting.
   Examples of Species in This Category: *Tricholoma magnivelare* (pine mushroom), *Inonotus obliquus* (cinder conk fungus), *Morchella* spp. (morel mushrooms), *Lycoperdon* spp. (puffballs).
   Total Number of These Species Documented in the Study Region: ~10–20 pyrophilous mushrooms.
   Selected References: Turner *et al.* (1986), Johnson (1992), Anderson and Lake (2013), Turner and Cuerrier (2022).

9. **Culturally Important Seaweeds, Lichens, and Mosses**
   Main Uses/Applications: seaweeds eaten, used to harvest herring eggs; black tree lichen eaten after cooking; sphagnum moss harvested for diapers and dressings.
   Major Habitat Type: marine algae in intertidal zone; tree lichens on pine and fir; sphagnum mosses on the ground in peat bogs; other mosses on the ground.
   Management Strategies: partial harvesting, leaving growing tissues behind; seasonal harvesting.
   Examples of Species in This Category: *Macrocystis integrifolia* (giant kelp), *Postelsia palmaeformis* (sea-palm), *Pyropia abbottiae* (red laver seaweed), *Bryoria fremontii* (black tree lichen), *Sphagnum* spp. (peat moss).
   Total Number of These Species Documented in the Study Region: ~20.
   Selected References: Anderson (2005), Turner and Clifton (2006), Crawford (2016).

Across these groups of plants, common management strategies can be seen: rotational and partial or selective harvesting; scattering or replanting propagules; pruning/coppicing bushes or trees; seasonal fallow; weeding/removing competing species; burning; clearing; transplanting; tilling; cultivating; aerating; structural changes to create or extend habitats; creating patches and stands; managing water levels, air and soil moisture, sunlight, and nutrients; diversification of harvesting; and, perhaps most of all, social and ceremonial conventions that guide human relationships with other species and the land. Figure 5.7 shows schematically the interplay among techniques, activities, and considerations applied at different scales of time and space, although it is impossible to capture the overall complexity of interactions and practices in a single figure.

It is likely that these practices and approaches have been developed and enhanced over time and often in situ; however, it is equally likely that many are ancient, extending back to "time immemorial," shared across generations and cultural and geographic space, and later modified according to specific ecological, biological, and cultural situations (Turner, 2014; Hoffmann *et al.*, 2016). The combined results have been well-defined "anthropogenic habitats" with specific characteristics including an abundance of one or more culturally preferred species: berry patches, camas prairies, tidal marsh root gardens, wapato (*Sagittaria* spp.) beds, subalpine meadow gardens,

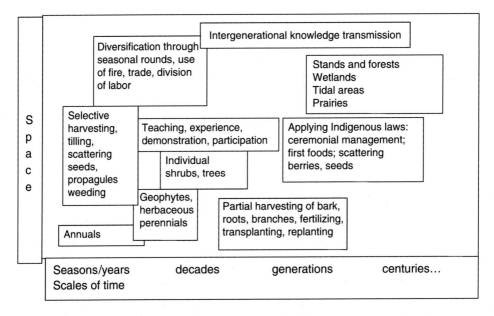

Figure 5.7 Schematic diagram showing the interplay among the region's traditional management techniques, activities, and considerations, applied at different scales of time and space.

managed cedar stands, Pacific yew (*Taxus brevifolia*) stands, tobacco (*Nicotiana* spp.) gardens, basket "grass" patches, and many others. We also see other culturally modified habitats containing not one or two target species, but multiple species of cultural importance enhanced together – mosaics of habitat and multispecies complexes of particular plant resources that have been called "orchard gardens," "wetland gardens," and "berry gardens," for example (Kimmerer and Lake, 2001; Deur and Turner, 2005; Turner, 2005, 2014; Turner and Peacock, 2005; Turner *et al.*, 2005; Weiser and Lepofsky, 2009; Hoffmann *et al.*, 2016; Armstrong, 2017; Lepofsky and Armstrong, 2018; Lyons *et al.*, 2018).

These anthropogenic habitats and the cultural practices underlying them have now been relatively well documented, despite decades of inattention by earlier generations of academic researchers. The lack of academic attention by earlier researchers reflected European observers' unfamiliarity with many traditional management methods. Indeed, colonists arriving in this region often assumed that its habitat variegation and species richness were the product solely of natural forces, rather than the interplay between natural and cultural processes over time (Deur and Turner, 2005). Such views are ubiquitous in colonial writings regarding the region. Before becoming the first Governor of British Columbia, to cite one example, then Hudson's Bay Company employee James Douglas (1842: 7) surveyed the site of his future capital in Victoria, noting, "we saw several Acres of Clover growing with a Luxuriance and Compactness more resembling the close Sward of a well-managed Lea than the Produce of an uncultivated Waste." Often, the text is ambiguous as to whether Douglas and those of his time recognized that these were indeed "well-managed leas" rather than "uncultivated waste," but Douglas and his peers certainly had motives to discount Indigenous stewardship of lands targeted for dispossession and reoccupation. Within northwestern North America, traditionally managed habitats were often the first to be targeted for colonial reoccupation – these well-managed leas being, for example, some of the largest available clearings for agricultural resettlement in otherwise rugged and densely forested terrain. In the wake of European encroachment into North America, European styles of agriculture and forestry soon replaced many Indigenous practices (King, 2004; Turner and Turner, 2008; Joseph and Turner, 2020; Turner, 2020a). In the years that followed, grazing by domesticated livestock, the introduction of invasive species, pollution, and development further eroded anthropogenic plant communities. Cultural knowledge about these places among Indigenous people also eroded markedly in the wake of riveting technological, economic, and demographic changes after colonization; land dispossession by European newcomers; the strategic destruction of Indigenous peoples' substance resources in times of conflict; enforced bans on burning, bark harvesting, and other Indigenous management strategies; and the loss of knowledge caused by removing

Indigenous children from home centers of learning and placing them into residential schools, where cultural practices were denigrated and forbidden (Turner *et al.*, 2008; Turner, 2020b; Dick *et al.*, 2022) and indigenous language.

Despite these trends, many Indigenous individuals have remembered the original practices, due to the robustness of oral tradition and the continuation of certain practices on the margins of settler societies. Their accounts have helped facilitate the recognition of anthropogenic habitats in recent decades, often in combination with the biological, archaeological, and phytogeographical evidence of those practices through multidisciplinary documentation efforts. Similarly, the landscapes of northwestern North America abound in detectable remnant and relict anthropogenic plant communities. Abandoned estuarine root grounds are still detectable based on species composition and archaeological features, for example, and widespread prairies have slowly become overgrown with shrub and forest in the absence of regular fire management, but there are still remnants of camas and other prairie species. These landscapes not only serve as telling aspects of the larger "landscape archaeology" – that is, the practices and even the values of ancestral people still inscribed clearly on the land – they also serve as the final bastions for certain species that have become less common in the region and as venues for potential future habitat restoration.

Increasingly, these practices also provide inspiration or direct guidance for environmental restoration that also sustains cultural institutions – what sometimes has been termed "ecocultural restoration" (Anderson and Barbour, 2003; Senos *et al.*, 2006; Shebitz *et al.*, 2009; Gomes, 2012; Joseph, 2012). A key direction in modern research of anthropogenic habitats addresses the ways in which TLRM can be recognized and applied in Indigenous-led land management within the context of today's global crisis in biocultural diversity loss and climate change (Maffi and Woodley, 2012; Kimmerer, 2013; Berkes, 2017; Lepofsky and Lertzman, 2019). Increasing food security and food sovereignty for Indigenous peoples is also a major reason for reinstating TLRM regimes, with improved access to traditional lands and waters, as well as to the plant and animal species sustained by generations of intentional management (Kuhnlein *et al.*, 2013).

5.8 Conclusions

Indigenous peoples' TLRM systems are not only culturally significant to Native peoples and their enduring relationships with the landscape. These systems of management and care have been highly impactful on the plant communities and habitats of northwestern North America and, indeed, of many other parts of the world. Traditional burning alone has reshaped vegetation communities across the region, creating oak savannahs, camas prairies, huckleberry patches, and other features that are still seen – often in relict form – today. In many places, these anthropogenic plant habitats still define the biota and habitat conditions of entire bioregions within the larger Northwest, arguably contributing to the region's overall biodiversity. The Indigenous peoples of this region regularly utilized some 500 species of plants and, in some manner, traditional management techniques were applied to virtually all of these species. These plants have been intentionally seeded or transplanted, managed through selective harvest, and tended through weeding, pruning, coppicing, soil tillage and amendment, or other techniques. Moreover, Indigenous communities have "owned" or otherwise invested in specific plant harvest sites, underscoring the labor and also the sense of interspecific connection with these places.

These management systems have helped to sustain Indigenous communities, providing key foods, medicines, and materials and supporting a wide range of other cultural, social, economic, and spiritual needs. Reflecting the reciprocal logic of Indigenous cosmology, Indigenous communities have cared for culturally preferred plants for countless generations and, in exchange, these plants have sustained Indigenous communities. The sense of reciprocal obligation between humans and the species they consume emerges from cornerstone elements of traditional cosmologies shared by Indigenous peoples across this region. This sense of reciprocal obligation arguably shaped harvest and management practices and the biophysical manifestation of these practices in the natural world – as seen in the bark removal scars on ancient western redcedar trees and in the incremental expansion of culturally significant native plants such as camas in situ (rather than completely removing and replacing native vegetation with introduced plant species).

In the complex environments of northwestern North America, which have significant topographical and botanical variegation, Indigenous communities had access to a wide range of plant species and habitats, adding to the diversity and wealth of plant products. This also allowed for the application of different management techniques in different environments. To be sure, Indigenous peoples' understanding of environmental cause and effect within such natural settings has been key to the success and endurance of these management practices. Regardless of whether we refer to this knowledge as TEK or ILK, it is clear that these systems of understanding were highly sophisticated. They reflected inherited cultural knowledge spanning generations of experience, while integrating the outcomes of ongoing, direct observations of environmental phenomena, which allowed Indigenous societies to refine, adapt, and tailor such cultural knowledge to local circumstances. Information regarding propagation techniques, appropriate places and times for burning, localized phenology, weather and climate patterns, tides and currents all formed part of this body of knowledge, which has been transmitted and refined across time. Such knowledge has been passed on to successive generations, each interpreting and refining the information in light of their own times and observations of natural phenomena. While not originally a written system of knowledge, Indigenous protocols placed high value on detailed recollection of central oral traditions, songs, stories, and other mechanisms of cultural transmission.

Not only did TLRM systems represent attempts by Indigenous communities to "get more harvest" – that is, to enhance the quantity, quality, and localized availability of plant resources – these management systems have also manifested the cornerstone values of Indigenous societies. They reflect a fundamental sense of kinship and reciprocity between humans and other species, manifested through efforts to harvest without inflicting lasting harm on culturally preferred plants and plant habitats, as well as efforts to show respect through material and spiritual intervention on behalf of those plants consumed by human harvesters. In this sense, the traces we see of traditional plant management on modern landscapes, such as culturally modified trees or anthropogenic berry patches, are not merely manifestations of Indigenous ecological knowledge and practice, they are signposts of traditional values and cosmology, inscribed clearly on the land. These signposts are, in some manner, intelligible among those who have engaged Indigenous knowledge holders to understand both the environmental and the cultural imperatives that have shaped the management strategies of their ancestors over time.

In addition, among some Indigenous communities, these remnant anthropogenic habitats have also become important signposts of another kind, namely proof of their ancestors' historical presence on certain lands that have been lost due to dispossession and alienation of tribal lands over the last two centuries. Anthropogenic plant communities were often the first places to be reoccupied by arriving settlers and quickly fell out of the control of Native communities. Indigenous peoples were displaced not only from plants and plant harvest areas, but also, too often, from the knowledge of plant management sites and practices through generations of enforced acculturation, missionization, and forced entry into residential schools (Dick *et al.*, 2022; papers in Turner, 2020b). Culturally modified plant communities, especially those in which one can see clear traces of ancestral knowledge, practice, and cosmology on the land, have been important touchstones in the endurance and revival of Indigenous identities into the present day. In a growing number of cases, such places are becoming the focal point of legal efforts to sustain or reintroduce Indigenous stewardship, through such mechanisms as the modern treaty process in Canada, the designation of "traditional cultural properties" receiving special protections on US public lands, and "co-management" strategies involving both Native and non-Native resource managers working in collaboration (Pinkerton, 2019). By restoring access to anthropogenic plant habitats and by bringing Indigenous knowledge holders back to the land, all parties can realize a number of positive outcomes. In some contexts, the return of human management has allowed relict plant communities to be rehabilitated and allowed the restoration of botanical diversity that had been lost after generations of livestock grazing, industrial forestry, and other development (Dick *et al.*, 2022). Simultaneously, as Indigenous peoples regain access to anthropogenic plant communities, they have been able to sustain or even revitalize entire domains of traditional knowledge, for example food and medical knowledge, knowledge of craft traditions such as basketry, and cultural knowledge of the environmental processes regulating plant output. Just as these plant communities were formed at the intersection of natural and cultural systems, their restoration yields clear benefit in both natural and cultural domains today (Anderson and Barbour, 2003; Senos *et al.*, 2006; Gomes, 2012; Joseph, 2012). With time and persistence, in spite of global crises of climate and biodiversity, these efforts may yet achieve the goals of the traditional managers of long ago: sustaining plant communities and sustaining human communities, in all their rich interdependencies, now and for the benefit of future generations.

As Indigenous knowledge holders remind us, the need to acknowledge these traditions, to show respect to the plant world, and to sustain imperiled plant communities seems as urgent as ever. Quoting our friend and mentor, Kwakwaka'wakw cultural leader, Dr. Daisy Sewid-Smith (Mayanilth), the plants people cared for over the generations have long provided "the house that sheltered us, the foods we ate, the medicines we had . . . . And that is the reason why we respected nature as we did . . . . When you remove something, you have to put something back to make that plant, animal, fish, live again" (personal communication to NT, 1998). By continuing to give back to the plants, Indigenous peoples continue to honor ancient bargains and to ensure the livelihood of Native peoples and native plant communities far into the imaginable future.

References

Aikenhead, G. and Mitchell, H. (2010). *Bridging Culture: Indigenous and Scientific Ways of Knowing Nature*. London: Pearson.

Anderson, E. N. (1996). *Ecologies of the Heart: Emotion, Belief and the Environment*. New York: Oxford University Press.

Anderson, M. K. (1991). California Indian horticulture: management and use of redbud by the southern Sierra Miwok. *Journal of Ethnobiology*, **11**(1), 145–57.

Anderson, M. K. (1999). The fire, pruning, and coppice management of temperate ecosystems for basketry material by California Indian tribes. *Human Ecology*, **27**(1), 79–113.

Anderson, M. K. (2005). *Tending the Wild: Native American Knowledge and Management of California's Natural Resources*. Berkeley, CA: University of California Press.

Anderson, M. K. (2009). *The Ozette Prairies of Olympic National Park: Their Former Indigenous Uses and Management*. Port Angeles, WA: Olympic National Park.

Anderson, M. K. and Barbour, M. G. (2003). Simulated Indigenous management: a new model for ecological restoration in national parks. *Ecological Restoration*, **21**, 269–77.

Anderson, M. K. and Lake, F. K. (2013). California Indian ethnomycology and associated forest management. *Journal of Ethnobiology*, **33**(1), 33–85.

Anderson, M. K. and Rowney, D. L. (1998). California geophytes: ecology, ethnobotany, and conservation. *Fremontia*, **26**(1), 12–18.

Armstrong, C. G. (2017). Historical ecology of cultural landscapes in the Pacific Northwest. PhD dissertation, Simon Fraser University.

Armstrong, C. G., Dixon, M., and Turner, N. J. (2018). Management and traditional production of beaked hazelnut (k'áp'xw-az', Corylus cornuta; Betulaceae) in British Columbia. *Human Ecology*, **46**, 547–59.

Atleo, E. R. (Chief Umeek). (2004). *Tsawalk: A Nuu-chah-nulth Worldview*. Vancouver, BC: UBC Press.

Atleo. E. R. (Chief Umeek). (2011). *Principles of Tsawalk: An Indigenous Approach to Global Crisis*. Vancouver, BC: UBC Press.

Beckwith, B. R. (2004). The queen root of this clime: ethnoecological investigations of blue camas (Camassia quamash, C. leichtlinii; Liliaceae) landscapes on Southern Vancouver Island, British Columbia. PhD dissertation, University of Victoria.

Berkes, F. (2017). *Sacred Ecology*, 4th ed. London: Routledge.

Berkes, F. (2021). *Advanced Introduction to Community-Based Conservation*. Cheltenham, UK, and Northampton, MA: Edward Elgar.

Blackburn, T. C. and Anderson, M. K. (1993). *Before the Wilderness*. Menlo Park, CA: Ballena.

Blackman, M. B. (1990). Haida traditional culture. In W. P. Suttles, ed., *Handbook of North American Indians:* Northwest Coast, vol. 7. Washington, DC: Smithsonian Institution, pp. 240–60.

Boas, F. (1921). *Ethnology of the Kwakiutl*. Washington, DC: Smithsonian Institution.

Boyd, R., ed. (2022). *Indians, Fire and the Land in the Pacific Northwest*. 2nd ed. Corvallis, OR: Oregon State University Press.

Brown, F., Brown, K., Wilson, B., Waterfall, P., and Cranmer Webster, G. (2009). *Staying the Course, Staying Alive: Coastal First Nations Fundamental Truths*. Victoria, BC: Biodiversity BC.

Charlie, L. A. and Turner, N. J. (2021). *Luschiim's Plants: A Hul'q'umi'num' (Cowichan) Ethnobotany*. Madeira Park, BC: Harbour Publishing.

Chestnut, V. K. (1902). Plants used by the Indians of Mendocino County, California, Contributions from the U.S. *National Herbarium*, **7**, 295–408.

Courchene, D., Whitecloud, K., and Stonechild, B. (2021). Connecting spiritually with the land and each other, *Reconciling Ways of Knowing*. www.waysofknowingforum.ca/dialogue7.

Crawford, S. (2016). The ethnolichenology of *Wila (Bryoria fremontii)*: an important edible lichen of Secwepemc Country and neighboring territories. In M. B. Ignace, N. J. Turner, and S. L. Peacock, eds., *Secwepemc People and Plants: Research Papers in Shuswap Ethnobotany*. Tacoma, WA: Society of Ethnobiology, pp. 295–335.

Cullis-Suzuki, S., Wyllie-Echeverria, S., Dick, A. K., and Turner, N. J. (2015). Tending the meadows of the sea: a disturbance experiment based on traditional Indigenous harvesting of Zostera marina L. (Zosteraceae) the southern region of Canada's west coast. *Aquatic Botany*, **127**, 26–34.

Curtis, E. S. (1915). *The North American Indian: The Kwakiutl*, vol. 10. Evanston, IL: Northwestern University Library.

Davidson-Hunt, I. J. (2003). Indigenous lands management, cultural landscapes and Anishinaabe people of Shoal Lake, Northwestern Ontario. *Environments*, **31**(1), 21–41.

Deur, D. (2000). A domesticated landscape: Native American plant cultivation on the Northwest Coast of North America. PhD dissertation, Louisiana State University.

Deur, D. (2005). Tending the garden, making the soil: Northwest Coast estuarine gardens as engineered environments. In D. Deur and N. J. Turner, eds., *Keeping It Living: Traditions of Plant Use and Cultivation on the Northwest Coast of North America.* Vancouver, BC: UBC Press, pp. 296–330.

Deur, D. (2009). "A caretaker responsibility": revisiting Klamath and Modoc traditions of plant community management. *Journal of Ethnobiology*, **29**, 296–322.

Deur, D. (2022). *Gifted Earth: The Ethnobotany of Quinault and Neighboring Tribes.* Corvallis, OR: Oregon State University Press.

Deur, D., Dick, A., Recalma-Clutesi, K., and Turner, N. J. (2015). Kwakwaka'wakw "clam gardens": motive and agency in traditional Northwest Coast mariculture. *Human Ecology*, **43**, 201–12.

Deur, D., Evanoff, K., and Hebert, J. (2020). "Their markers as they go": modified trees as waypoints in the Dena'ina cultural landscape, Alaska. *Human Ecology*, **48**, 317–33.

Deur, D., Recalma-Clutesi, K., and Dick, A. (2020a). "When God put daylight on Earth we had one voice": Kwakwaka'wakw perspectives on sustainability and the rights of nature. In C. LaFollette and C. Maser, eds., *Sustainability and the Rights of Nature in Practise.* Boca Raton, FL: CRC Press, pp. 89–111.

Deur, D., Recalma-Clutesi, K., and Dick, A. (2020b). Balance on every ledger: Kwakwaka'wakw resource values and traditional ecological management. In T. Thornton and S. Bhagwat, eds., *The Routledge Handbook of Indigenous Environmental Knowledge.* London: Routledge, pp. 126–35.

Deur, D. and Turner, N. J. (2005). *Keeping It Living: Traditions of Plant Use and Cultivation on the Northwest Coast of North America.* Seattle, WA: University of Washington Press.

Dick, A., Sewid-Smith, D., Recalma-Clutesi, K., Deur, D., and Turner, N. J. (2022). "From the beginning of time": the colonial reconfiguration of native habitats and Indigenous resource practices on the British Columbia coast. *Facets*, **7**(1), 543–70.

Douglas, J. (1842). Letter to John McLoughlin, Hudsons Bay Company, July 12, 1842. In *Correspondence between the Chairman of the Hudson's Bay Company and the Secretary of State for the Colonies, relative to the Colonization of Vancouver's Island*, vol. 9, session 1849. London: Great Britain Parliamentary Sessional Papers: House of Lords, Accounts and Papers, pp. 5–7.

Drucker, P. (1951). *The Northern and Central Nootkan Tribes.* Washington, DC: Smithsonian Institution.

Earnshaw, J., ed. (2017a). Culturally modified trees, Part I. *The Midden*, **47**(2), 1–33.

Earnshaw, J., ed. (2017b). Culturally modified trees, Part II. *The Midden*, **47**(3), 1–53.

Eldridge, M. (1997). *The Significance and Management of Culturally Modified Trees.* Final report prepared for Vancouver Forest Region and CMT Standards Steering Committee. www.for.gov.bc.ca/ftp/archaeology/external/!publish/web/culturally_mo dified_trees_significance_management.pdf (accessed June 9, 2013).

Ford, J. and Martinez, D. eds. (2000). Traditional ecological knowledge, ecosystem science and environmental management. *Ecological Applications, Special Issue*, **10**(5), 1249–50.

Fowler, C. S. and Lepofsky, D. (2011). Traditional resource and environmental management. In E. N. Anderson, et al., eds., *Ethnobiology.* Hoboken, NJ: Wiley-Blackwell, pp. 285–304.

Franklin, J. and Dyrness, C. T. (1973). *Natural Vegetation of Oregon and Washington*, Technical Report PNW-8. Washington, DC: US Department of Agriculture General.

Gill, S. (1984). *Ethnobotany of the Makah People, Olympic Peninsula, Washington*. Neah Bay, WA: Makah Language Program.

Gomes, T. (2012). Restoring Tl'chés: an ethnoecological restoration study in Chatham Islands, British Columbia, Canada. MA thesis, University of Victoria.

Helm, J., and Sturtevant, W. C., eds. (1982). *Handbook of North American Indians: Subarctic*, vol. 6. Washington, DC: Smithsonian Institution.

Hoffmann, T., Lyons, N., Miller, D., et al. (2016). Engineered feature used to enhance gardening at a 3800-year-old site on the Pacific Northwest coast. *Science Advances*, **2**, e1601282.

Hunn, E. S., Johnson, D. R., Russell, P. N., and Thornton, T. F. (2003). Huna Tlingit traditional environmental knowledge, conservation, and the management of a "wilderness" park. *Current Anthropology*, **44**, S79–103.

Hunn, E. S. and Selam, J. (1990). *Nch'i-Wana, "The Big River": Mid-Columbia Indians and Their Land*. Seattle, WA: University of Washington Press.

Ignace, M., Ignace, R. E., and Turner, N. J. (2017). *Re Styecwmenúl'ecws-kucw/* How we look(ed) after our land. In M. Ignace and R. E. Ignace, eds., *Secwépemc People, Land, and Laws. Yeri7 re Stsq'ey's-kucw*. Montreal, QC: McGill-Queen's University Press, pp. 145–219.

Ignace, M. B., Turner, N. J., and Peacock, S. L. (2016). Secwepemc People and Plants: Research Papers in Shuswap Ethnobotany. Society of Ethnobiology, Contributions in Ethnobiology. Kamloops, BC: Shuswap Nation Tribal Council.

Johnson, L. M. G. (1992). Use of Cinder Conk (Inonotus obliquus) by the Gitksan of northwestern British Columbia, Canada. *Journal of Ethnobiology*, **12**(1), 153–56.

Johnson, L. M. G. (1994). Wet'suwet'en ethnobotany: traditional plant uses. *Journal of Ethnobiology*, **14**(2), 185–210.

Johnson, L. M. and Hunn, E. S. (2010). Landscape ethnoecology: reflections. In L. M. Johnson and E. S. Hunn, eds., *Landscape Ethnoecology: Concepts of Biotic and Physical Space*. New York: Berghahn Books, pp. 279–97.

Joseph, L. S. (2012). Finding our roots: ethnoecological restoration of Lhásem (Fritillaria camschatcensis (L.) Ker-Gawl), an iconic plant food in the Squamish River Estuary, British Columbia. MSc thesis, University of Victoria.

Joseph, L. S. and Turner, N. J. (2020). "The old foods are the new foods!": reviving Indigenous foods in northwestern North America. *Frontiers in Sustainable Food Systems, Special Issue*, **4**, 596237.

Kimmerer, R. W. (2013). *Braiding Sweetgrass. Indigenous Wisdom, Scientific Knowledge and the Teachings of Plants*. Minneapolis, MN: Milkweed Editions.

Kimmerer, R. W. and Lake, F. K. (2001). Maintaining the mosaic: the role of Indigenous burning in land management. *Journal of Forestry*, **99**, 36–41.

King, L. (2004). Competing knowledge systems in the management of fish and forests in the Pacific Northwest. *International Environmental Agreements: Politics, Law and Economics*, **4**, 161–77.

Kuhnlein, H. V., Erasmus, B., Spigelski, D., and Burlingame, B. (2013). *Indigenous Peoples' Food Systems & Well-being: Interventions & Policies for Healthy Communities*. Rome, Italy: Food and Agricultural Organization of the United Nations.

Lantz, T. and Turner, N. J. (2003). Traditional phenological knowledge of aboriginal peoples in British Columbia. *Journal of Ethnobiology*, **23**(2), 1–6.

Lepofsky, D. (2009). Traditional resource management: past, present and future. *Journal of Ethnobiology, Special Issue*, **29**(2), 184–212.

Lepofsky, D. and Armstrong, C. G. (2018). Foraging new ground: documenting ancient resource and environmental management in Canadian archaeology. *Canadian Journal of Archaeology/Journal Canadien d'Archéologie*, **42**, 57–73.

Lepofsky, D. and Caldwell, M. (2013). Indigenous marine resource management on the northwest coast of North America. *Ecological Processes*, **2**, 1–12.

Lepofsky, D. and Lertzman, K. P. (2008). Documenting ancient plant management in the northwest of North America. *Botany*, **86**, 129–45.

Lepofsky, D. and Lertzman, K. P. (2019). Through the lens of the land: reflections from archaeology, ethnoecology, and environmental science on collaborations with First Nations, 1970s to the present. *BC Studies*, **200**, 141–60.

Lewis, H. (1993). Patterns of Indian burning in California: ecology and ethnohistory. Original 1973. In T. C. Blackburn and M. K. Anderson, eds., *Before the Wilderness: Environmental Management by Native Californians*. Menlo Park, CA: Ballena, pp. 55–116.

Lightfoot, K. G., Cuthrell, R. Q., Striplen, C. J., and Hylkema, M. G. (2013a). Rethinking the study of landscape management practices among hunter-gatherers in North America. *American Antiquity*, **78**, 285–301.

Lightfoot, K. G., Cuthrell, R. Q., Boone, C. M., et al. (2013b). Anthropogenic burning on the central California coast in late Holocene and early historical times: findings, implications, and future directions. *California Archaeology*, **5**(2), 371–90.

Lightfoot, K. G. and Lopez, V. (2013). The study of Indigenous management practices in California: an introduction. *California Archaeology*, **5**(2), 209–19.

Lloyd, T. A. (2011). Cultivating the Taki'lakw, the ethnoecology of Tleksem, Pacific Silverweed …: lessons from Clan Chief Kwaxsistalla of the Dzawada7enuxw Kwakwaka'wakw of Kingcome Inlet. MSc thesis, University of Victoria.

Lyons, N., Hoffmann, T., Miller, D., et al. (2018). Katzie & the Wapato: an archaeological love story. *Archaeologies: Journal of the World Archaeological Congress*, **14**, 7–29.

Mack, C. A. and McClure, R. H. (2002). Vaccinium processing in the Washington Cascades. *Journal of Ethnobiology*, **22**(1), 35–60.

Maffi, L. and Woodley, E. (2012). *Biocultural Diversity Conservation: A Global Sourcebook*. Abingdon, UK: Routledge.

Marine Planning Partnership Initiative (2015). *Haida Gwaii Marine Plan*. Haida Nation & Province of British Columbia. www.haidanation.ca/wp-content/uploads/2017/05/MarinePlan_HaidaGwaii_WebVer_21042015-opt-1.pdf (accessed February 6, 2021).

Mathews, D. (2017). Savannahs of living fuel: the coast Salish management of Douglas-fir bark fuel. Society of Ethnobiology conference presentation, May, 11 2017, 15:15. https://ethnobiology.org/conference/abstracts/40/savannahs-living-fuel-coast-salish-management-douglas-fir-bark-fuel.

Maud, R. (1978). *The Salish People: The Local Contribution of Charles Hill-Tout:* The Mainland Halkomelem, vol. 3. Vancouver, BC: Talonbooks.

McDonald, J. (2005). Cultivating in the Northwest: early accounts of Tsimshian horticulture. In D. Deur and N. J. Turner, eds., *Keeping It Living: Traditions of Plant Use and Cultivation on the Northwest Coast of North America*. Vancouver, BC: UBC Press, pp. 240–73.

Menzies, C. R. (2006a). Ecological knowledge, subsistence, and livelihood practices: the case of the pine mushroom harvest in northwestern British Columbia. In C. R. Menzie, ed., *Traditional Ecological Knowledge and Natural Resource Management*. Lincoln, NB: University of Nebraska Press, pp. 99–116.

Menzies, C. R. (2006b). *Traditional Ecological Knowledge and Natural Resource Management*. Lincoln, NB: University of Nebraska Press.

Minnis, P. E. and Elisens, W. J. (2000). *Biodiversity and Native America*. Norman, OK: University of Oklahoma Press.

Moss, M. L. (2005). Tlingit horticulture: an Indigenous or introduced development? In D. Deur and N. J. Turner, eds., *Keeping It Living: Traditions of Plant Use and Cultivation on the Northwest Coast of North America*. Vancouver, BC: UBC Press, pp. 274–95.

Nelson, R. K. (1983). *Make Prayers to the Raven: A Koyukon View of the Northern Forest*. Chicago, IL: University of Chicago Press.

Norton, H. H. (1979a). The association between anthropogenic prairies and important food plants in western Washington. *Northwest Anthropological Research Notes*, **13**(2), 175–200.

Norton, H. H. (1979b). Evidence for bracken fern as a food for aboriginal peoples of western Washington. *Economic Botany*, **33**(4), 384–96.

Peacock, S. L. (1998). Putting down roots: the emergence of wild plant food production on the Canadian Plateau. PhD dissertation, University of Victoria.

Peacock, S. L. and Turner, N. J. (2000). "Just like a garden": traditional plant resource management and biodiversity conservation on the British Columbia plateau. In P. E. Minnis and W. J. Elisens, eds., *Biodiversity and Native North America*. Norman, OK: University of Oklahoma Press, pp. 133–79.

Phillips, P. W. (2016). *Ethnobotany of the Coos, Lower Umpqua, and Siuslaw Indians*. Corvallis, OR: Oregon State University Press.

Pinkerton, E. (2019). Benefits of collaboration between Indigenous and non-Indigenous communities through community forests in British Columbia. *Canadian Journal of Forest Research*, **49**(4), 387–94.

Popper, K. (1959). *The Logic of Scientific Discovery*. Abingdon, UK: Routledge.

Posey, D. A. and Balée, W. (1989). *Resource Management in Amazonia: Advances in Economic Botany*, vol. 7. New York: New York Botanical Garden.

Routson, K. J., Volk, G. M., Richards, C. M., et al. (2012). Genetic variation and distribution of Pacific crabapple. *Journal of the American Society of Horticultural Science*, **137** (5), 325–32.

Schoonmaker, P. K., Von Hagen, B., and Wolf, E. C. (1997). *The Rainforests of Home: Profile of a North American Bioregion*. Washington, DC: Island Press.

Senos, R., Lake, F., Turner, N. J., and Martinez, D. (2006). Traditional ecological knowledge and restoration practice in the Pacific Northwest. In D. Apostol, ed., *Encyclopedia for Restoration of Pacific Northwest Ecosystems*. Washington, DC: Island Press, pp. 393–426.

Sewid-Smith, D. M., Dick, A. K., and Turner, N. J. (1998). The sacred cedar tree of the Kwakwaka'wakw people. In M. Bol, ed., *Stars Above, Earth Below: Native Americans and Nature*. Pittsburgh, PA: Carnegie Museum of Natural History, pp. 189–209.

Shebitz, D. J., Reichard, S. H., and Dunwiddy, P. W. (2009). Ecological and cultural significance of burning beargrass habitat on the Olympic Peninsula, Washington. *Ecological Restoration* **27**(3), 306–19.

Smith, B. D. (2005). Low-level food production and the Northwest Coast. In D. Deur and N. J. Turner, eds., *Keeping It Living: Traditions of Plant Use and Cultivation on the Northwest Coast of North America*. Seattle, WA: University of Washington Press, pp. 37–66.

Snively, G. and Williams, W. L. (2016). *Knowing Home: Braiding Indigenous Science with Western Science*, book 1. Victoria, BC: University of Victoria Pressbooks.

Stevens, M. L. (2004a). White root (Carex barbarae). *Fremontia*, **32**(4), 3–6.

Stevens, M. L. (2004b). Ethnoecology of selected California wetland plants. *Fremontia*, **32** (4), 7–15.

Suttles, W. P. (1990). *Handbook of North American Indians: Northwest Coast*, vol. 7. Washington DC: Smithsonian Institution.

Suttles, W. P. (2005). Coast Salish resource management: incipient agriculture? In D. Deur and N. J. Turner, eds., *Keeping It Living: Traditions of Plant Use and Cultivation on the Northwest Coast of North America*. Vancouver, BC: UBC Press, pp. 181–93.

Thornton, T. F. (1999). Tleikw aaní, the "berried" landscape: the structure of Tlingit edible fruit resources at Glacier Bay, Alaska. *Journal of Ethnobiology*, **19**(1), 27–48.

Thornton, T. F. (2008). *Being and Place among the Tlingit*. Seattle, WA: University of Washington Press.

Thornton, T. F. (2015). The ideology and practice of Pacific herring cultivation among the Tlingit and Haida. *Human Ecology*, **43**(2), 213–23.

Trosper, R. L. and Parrotta, J. A. (2012). Traditional Forest-Related Knowledge. *Sustaining Communities, Ecosystems and Biocultural Diversity*. Dordrecht, Switzerland: Springer.

Trusler, S. and Johnson, L. M. (2008). "Berry patch" as a kind of place: the ethnoecology of black huckleberry in northwestern Canada. *Human Ecology*, **36**, 553–68.

Turner, N. J. (2003). The ethnobotany of "edible seaweed" (Porphyra abbottiae Krishnamurthy and related species; Rhodophyta: Bangiales) and its use by First Nations on the Pacific coast of Canada. *Canadian Journal of Botany*, **81** (2), 283–93.

Turner, N. J. (2005). *The Earth's Blanket. Traditional Teachings for Sustainable Living*. Vancouver, BC: Douglas & McIntyre.

Turner, N. J. (2006). From the roots: Indigenous root vegetables of British Columbia, their management and conservation. In Z. F. Ertug, ed., *Conference Proceedings of the Fourth International Congress of Ethnobotany (ICEB 2005), Ethnobotany: At the Junction of the Continents and Disciplines*. Istanbul, Turkey: Yeditepe University Yayinlari, pp. 57–64.

Turner, N. J. (2014). *Ancient Pathways, Ancestral Knowledge: Ethnobotany and Ecological Wisdom of Indigenous Peoples of Northwestern North America*, 2 vols. McGill-Queen's Native and Northern Series Number 74. Montreal, QC: McGill-Queen's University Press.

Turner, N. J. (2020a). From "taking" to "tending": learning about Indigenous land and resource management on the Pacific Northwest coast of North America. *ICES Journal of Marine Science*, **77**(7–8), 2472–82.

Turner, N. J. (2020b). *Plants, People, and Places: the Roles of Ethnobotany and Ethnoecology in Indigenous Peoples' Land Rights in Canada and Beyond*. Montreal, QC: McGill-Queen's University Press.

Turner, N. J. (2021). *Plants of Haida Gwaii. Xaadaa Gwaay guud gina k'aws (Skidegate), Xaadaa Gwaayee guu giin k'aws (Massett)*, 3rd ed. Madeira Park, BC: Harbour Publishing.

Turner, N. J., Ari, Y., Berkes, F., et al. (2009). Cultural management of living trees: an international perspective. *Journal of Ethnobiology*, **29**(2), 237–70.

Turner, N. J., Armstrong, C. G., and Lepofsky, D. L. (2021). Adopting a root: documenting ecological and cultural signatures of plant translocations in northwestern North America. *American Anthropologist*, **123**(4), 879–97.

Turner, N. J. and Berkes, F. (2006). Coming to understanding: developing conservation through incremental learning. *Human Ecology*, **34**(4), 495–513.

Turner, N. J., Bouchard, R., and Kennedy, D. I. D. (1981). *Ethnobotany of the Okanagan-Colville Indians of British Columbia and Washington*. Victoria, BC: British Columbia Provincial Museum.

Turner, N. J. and Clifton, H. (2006). "The forest and the seaweed": Gitga'at seaweed, traditional ecological knowledge and community survival. In C. Menzies, ed., *Traditional Ecological Knowledge and Natural Resource Management*. Lincoln, NB: University of Nebraska, pp. 65–86.

Turner, N. J. and Cuerrier, A. (2022). "Frog's umbrella" and "ghost's face powder": the cultural roles of mushrooms and other fungi for Canadian Indigenous peoples. *Botany* **100**(2), 183–205.

Turner, N. J., Deur, D., and Lepofsky, D. (2013). Plant management systems of British Columbia's First Peoples. *BC Studies: The British Columbian Quarterly*, **179**, 107–33.

Turner, N. J., Deur, D., and Mellott, C. R. (2011). "Up on the mountain": ethnobotanical importance of montane ecosystems in Pacific coastal North America. *Journal of Ethnobiology*, **31**(1), 4–43.

Turner, N. J., Gregory, R., Brooks, C., Failing, L. and Satterfield, T. (2008). From invisibility to transparency: identifying the implications (of invisible losses to First Nations communities). *Ecology and Society*, **13**(2), 7.

Turner, N. J. and Hebda, R. J. (2012). *Saanich Ethnobotany: Culturally Important Plants of the WSÁNEĆ (Saanich) People of Southern Vancouver Island*. Victoria, BC: Royal BC Museum.

Turner, N. J., Ignace, M. B., and Ignace, R. E. (2000). Traditional ecological knowledge and wisdom of aboriginal peoples in British Columbia. *Ecological Applications*, **10**(10), 181–93.

Turner, N. J. and Kuhnlein, H. V. (1983). Camas (Camassia spp.) and riceroot (Fritillaria spp.): two Liliaceous "root" foods of the Northwest Coast Indians. *Ecology of Food and Nutrition*, **13**, 199–219.

Turner, N. J., Kuhnlein, H. V., and Egger, K. N. (1986). The cottonwood mushroom (Tricholoma populinum Lange): a food resource of the interior Salish peoples of British Columbia. *Canadian Journal of Botany*, **65**, 921–27

Turner, N. J. and Loewen, D. C. (1998). The original "free trade": exchange of botanical products and associated plant knowledge in northwestern North America. *Anthropologica*, **XL**, 49–70.

Turner, N. J. and Mathews, D. (2020). Serving nature: completing the ecosystem services circle. In H. Bai, D. Chang, and C. Scott, eds., *A Book of Ecological Virtues: Living Well in the Anthropocene*. Regina, SK: University of Regina Press, 3–29.

Turner, N. J. and Peacock, S. (2005). "Solving the perennial paradox": ethnobotanical evidence for plant resource management on the Northwest Coast. In D. Deur and N. J. Turner, eds., *Keeping It Living: Traditions of Plant Use and Cultivation on the Northwest Coast of North America*. Seattle, WA: University of Washington Press, pp. 101–50.

Turner, N. J., Smith, R. Y., and Jones, J. T. (2005). "A fine line between two nations": ownership patterns for plant resources among Northwest Coast Indigenous peoples – implications for plant conservation and management. In D. Deur and N. J. Turner, eds., *Keeping It Living: Traditions of Plant Use and Cultivation on the Northwest Coast of North America*. Seattle, WA: University of Washington Press, pp. 151–80.

Turner, N. J. and Taylor, R. L. (1972). A review of the Northwest Coast tobacco mystery. *Syesis*, **5**, 249–57.

Turner, N. J. and Thompson, J. C. (2006). *Plants of the Gitga'at People. 'Nwana'a lax Yuup*. Hartley Bay, BC: Gitga'at Nation and Coasts Under Stress Research Project.

Turner, N. J. and Turner, K. L. (2008). "Where our women used to get the food": cumulative effects and loss of ethnobotanical knowledge and practice. *Botany*, **86**(1), 103–15.

Walker, D. E., Jr. (1998). *Handbook of North American Indians:* Plateau, vol. 12. Washington DC: Smithsonian Institution.

Weiser, A. and Lepofsky, D. (2009). Ancient land use and management of Ebey's Prairie, Whidbey Island, Washington. *Journal of Ethnobiology*, **29**(2), 161–66.

Williams, N. M. and Hunn, E. N. (1982). *Resource Managers: North American and Australian Hunter-Gatherers*. Boulder, CO: Westview.

6

Hunting and Trapping in the Americas: The Assessment and Projection of Harvest on Wildlife Populations

TAAL LEVI, CARLOS A. PERES, AND GLENN H. SHEPARD

6.1 Introduction

Humans have been hunting wildlife in the Americas since their arrival in the New World at least 13,000 years ago. While extant game species sustained this harvest for millennia, the Pleistocene megafauna of the Americas appears to have been driven extinct by a combination of Indigenous hunting and climate change (Alroy, 2001; Surovell et al., 2016). Hunting technologies used by Indigenous peoples of North and South America prior to European conquest included spears, arrows, clubs, blowguns firing poison-tipped darts, traps, and nets (Roosevelt, 1987; Scheinsohn, 2003; Castro et al., 2021). Indigenous hunting strategies, shamanic beliefs, and cosmologies throughout the Americas would appear to promote sustainable hunting, whether through ritual avoidance (Hickerson, 1965; Milton, 1997; Read et al., 2010), hunting zone rotation (Hames, 1980; Albert and Le Tourneau, 2007), notions of reciprocity between humans and animals (Fausto, 2007; Hill, 2011), or widespread beliefs that irresponsible hunting may be punished by illness or misfortune (Shepard, 2014; Fernández-Llamazares and Virtanen, 2020). Nevertheless, modeling data suggest that, even prior to European conquest, certain large, semisedentary Indigenous populations present in some parts of Amazonia may have driven harvest-sensitive species such as large-bodied monkeys and tapirs to local extinction (Shepard, Jr., et al., 2012). Moreover, numerous quantitative studies have shown that the introduction of firearms quickly overwhelms traditional safeguards, leading to local faunal scarcity even among small populations of seminomadic hunters (Alvard and Kaplan, 1991; Alvard et al., 1997; Chacon, 2001). Nonetheless, there is still robust evidence that Indigenous peoples' lands are a crucially important refuge for global biodiversity, including critically endangered non-human primate species (Estrada et al., 2022).

As human populations grow, modernize, and convert wildlands into anthropogenic landscapes, overharvesting in the remaining wildlands is a potent additive driver of wildlife population decline and extirpation. A key goal of contemporary wildlife conservation is to understand the conditions necessary for wildlife harvest to be sustainable and to design ethical policy frameworks to incentivize sustainability (Dobson *et al.*, 2019; Ingram *et al.*, 2021). This is critical to wildlife conservation because most protected areas in the neotropics contain people, and nearly all newly gazetted reserve acreage constitutes parks specifically designated to be managed by local communities (de Marques *et al.*, 2016). For example, the system of "sustainable development reserves" in Brazil, which was legally mandated in 1989, has grown to ~220 million hectares, compared with only ~50 million hectares in strictly protected areas (Peres, 2011). The wildlife in these reserves is critical for the food security of forest dwellers, Indigenous or otherwise, which can create tension between wildlife conservation and livelihoods. It is therefore a research priority to understand hunting, predict its long-term outcomes, and design politically viable management measures that will protect populations of vulnerable species while maintaining a protein source for humans.

Many approaches have been applied to assess the impact of hunting by determining whether or not it can be defined as "sustainable," but, while the concept of "sustainability" is frequently used, it is rarely precisely quantified. A proper understanding of sustainability requires specification of exactly what is being sustained, for whose benefit, the spatial extent under consideration, the duration over which the resource is being sustained, and how the impact of a particular use compares to realistic forms of alternative uses. For example, some species are particularly harvest sensitive and often succumb to depletion induced by hunters, while other prime game species may persist indefinitely under moderate hunting pressure. Prey profiles harvested by Amazonian hunters rapidly shift from low-fecundity harvest-sensitive species to high-fecundity harvest-insensitive species as local human population density increases (Peres, 2011). This is a consequence of the life history of each hunted species, which dictates age at first birth, fecundity, survivorship, and its "catchability," which is a measure of the effectiveness of hunting effort. The depletion of harvest-sensitive species is undesirable from the perspective of conservation or wildlife management, but local food security may be maintained by harvest-tolerant wildlife as long as protein acquisition rates remain high. This is an analogous problem to mixed stock fisheries, in which sustainable management of the target stock can incidentally induce a less productive stock to go extinct. Of course, it is desirable to pursue "win–win" scenarios in which harvest-sensitive taxa persist for the benefit of both wildlife conservation and food security, but there are often trade-offs. Most notably, under some conditions, access to firearms may increase food security by providing easier access to animal protein for harvest-tolerant taxa, despite causing local extirpation of harvest-sensitive ones.

Considerations of space and time are some of the largest challenges when assessing the impact or "sustainability" of hunting. Hunting studies often evaluate whether a species is expected to persist in the immediate vicinity of a community in which most "central-place" hunting takes place, but the landscape context and alternative land uses matter. This is particularly the case within large Indigenous reserves for which a small cost to wildlife conservation near communities is offset by the defense of a physically demarcated large reserve by a motivated stakeholder group. Indigenous reserves and other protected areas have been a key force in preventing deforestation from progressing north and west from the south-south-eastern fringes of the Amazon, where 80 percent of all Amazon deforestation has occurred (Schwartzman *et al.*, 2000). For example, the Kayapó control an enormous 10.6 million hectares in the Brazilian state of Pará (Zimmerman *et al.*, 2001), which they formally police through overflights and other means to prevent the intrusion of loggers and miners (Schwartzman and Zimmerman, 2005). Even when this is not the case, the presence of Indigenous stakeholders justifies the continued existence of protected areas that might otherwise be degazetted. Such large landholdings have come under threat of seizure in Brazil under the slogan "too much land for too few Indians," but their large size is critical for both long-term cultural integrity and wildlife conservation (Begotti and Peres, 2020). Areas of wildlife depletion are concentrated near communities because most hunting takes place within a day's walk, and distant areas serve a role as both de facto protected areas and wildlife source populations that continuously replenish hunted areas near communities. In addition to spatial context, the time horizon being assessed is critical because human populations grow and spread, introducing a potential conservation threat from within if hunting is not managed. In general, this internal threat is dwarfed by the immediate extrinsic threat of degazetting, downsizing, or downgrading the reserve to legally permit land conversion and development, such as mining and forestry. Whether or not the internal threat itself materializes is uncertain, given the challenges associated with forecasting cultural, technological, and policy changes many decades into the future. Nevertheless, there is a need for methods to project the impact of hunting in space and time under growing and spreading human populations under various management regimes for both long-term wildlife conservation and food security.

Cooperative wildlife management faces numerous legal, technical, and governance challenges (Yu *et al.*, 2010; Antunes *et al.*, 2019; de Mattos Vieira *et al.*, 2019). Although not universally adopted, accepted, or applicable, there are nonetheless numerous examples of participatory initiatives going back four decades that have met with varying degrees of success, and these case histories provide a crucial backdrop for developing ongoing game management strategies (Bodmer *et al.*, 1994; Townsend, 1997; Ohl-Schacherer *et al.*, 2007; Read *et al.*, 2010; de Mattos

Vieira *et al.*, 2015). A number of cultural elements have promoted, and continue to promote, the sustainability of wildlife harvest in the neotropics. The most important of these is hunting technology. Firearms have now been adopted by most communities, even in remote regions, because they initially increase catch per unit effort. However, their adoption can locally deplete some vertebrates, eventually resulting in lower catch per unit effort, as these species are only encountered far from communities. The reduced rate of kills per encounter allows bow hunting to maintain wildlife populations near communities while also retaining higher catch per unit effort in the long run. This result was not initially intuitive because early work found similar kill rates of large primates in bow-hunting and gun-hunting communities, leading to the suggestion that switching to firearms had limited impact on wildlife (Alvard, 1995). However, wildlife surveys revealed that spider monkeys were nearly absent within a day's walk from the gun-hunting communities (Mitchell and Raéz-Luna, 1991), resulting in a similar harvest by trading off a lower encounter rate with a higher rate of successful kills upon encounter as predicted by models (Levi *et al.*, 2009). Of course, bow hunting is itself a technological advancement that was adopted long after humans arrived in the Americas, spreading throughout South America as early as 3,500 years ago, and in most cases displacing other weapon systems such as atlatls and darts (Castro *et al.*, 2021). Across many weapon systems, species differ substantially in their vulnerability to overharvesting due to their own biology. As long as hunting effort is not too high, many species can remain harvest tolerant and can remain abundant, even in the immediate vicinity of communities. In some cases, food taboos and other cultural practices may prevent overharvesting of vulnerable species while allowing high levels of meat consumption from harvest-tolerant species (Ross *et al.*, 1978; Luzar *et al.*, 2012; Vieira and Shepard, 2017). This is particularly the case with large-bodied primates, as some communities have complete taboos against consuming them or partial taboos in which some species are avoided or consumed only seasonally. In other cases, a landscape that provides easy access to alternative sources of animal protein, such as freshwater fisheries, also protects harvest-sensitive prey species by reducing overall hunting pressure (Sampaio *et al.*, 2022). In general, the impact of hunting on wildlife is determined by a complex interaction between human population density, behavior, culture, the intrinsic population growth rate of each wildlife species, local habitat productivity, and the suite of species available to humans.

The key determinants of the impact of wildlife overharvesting are best clarified by careful modeling. In the past, hunting studies in the neotropics often attempted to define in a binary fashion whether exploitation of a given species within a catchment area is sustainable, but sustainability is not a yes-or-no question. Considering the intricacies of sustainability assessment is especially important when considering policy decisions. Here, we review how sustainability has been assessed, how new approaches rooted in the spatial and population ecology of humans and wildlife have improved upon sustainability assessments, and the requirements of future research to enable more accurate assessments of the impact of hunting on wildlife.

6.2 Sustainability Indices and Density-Dependent Population Growth

All assessments of wildlife harvest, and wildlife management in general, assume density-dependent population growth. The foundation of thinking about density-dependent population growth is the logistic growth equation, which is derived from the idea that per capita population growth declines with density as intraspecific competition for food and territory increases, leading to reduced birth rates and increased death rates. At its simplest, one assumes a linear decline of per capita population growth from the biologically constrained "maximum intrinsic growth rate" of that species, r, which occurs as density approaches zero and there is no intraspecific competition, to zero when the population is at carrying capacity, K (Figure 6.1). If the population exceeds carrying capacity, then per capita population growth is negative and the population declines. Per capita population growth is written as $\frac{1}{N}\frac{dN}{dt}$, where $\frac{dN}{dt}$ is a derivative representing the change in population size per unit time, which yields the following linear equation:

$$\frac{1}{N}\frac{dN}{dt} = r - \frac{r}{K}N$$

Absolute population growth (individuals per unit time) is just per capita population growth multiplied by the population, N, which produces the logistic growth equation familiar to all population ecologists:

$$\frac{dN}{dt} = rN\left(1 - \frac{N}{K}\right)$$

A key property of logistic population growth is that while, per capita population growth increases as density declines, absolute population growth is maximized at the intermediate density of $\frac{K}{2}$ and the amount of population growth at this value is called the maximum sustainable yield, *MSY* (see Figure 6.1). *MSY* is the maximum rate that a species can be harvested without causing the population to decline further, and its occurrence at intermediate densities is important because it is for this reason that wildlife can be resilient to hunting. As wildlife population densities are reduced, their per capita population growth rates increase, which can compensate for the harvest rate, provided it is not too great.

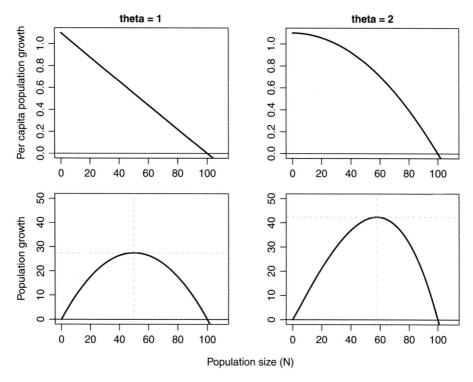

Figure 6.1 Per capita logistic population growth and total logistic population growth with $\theta = 1$, 2. Carrying capacity defines the x-intercept and here is $K = 100$. The maximum intrinsic growth rate defines the y-intercept of per capita logistic growth and the slope at $N = 0$ of logistic growth and here is $r = 1.07$. The vertical dotted lines represent the intermediate population size, N_{MSY}, associated with maximum sustainable yield, *MSY*, which is indicated by the horizontal dotted line. Both N_{MSY} and *MSY* are higher when $\theta > 1$.

In a management setting around the hunting of large mammals, the logistic growth equation is relatively conservative because it assumes that the per capita growth rate declines even while densities are low. In nature, density-dependent reductions in fecundity or increases in mortality may not materialize until densities are closer to K and competition for food and space become stronger, which results in higher population growth rates until density approaches carrying capacity. A generalization of logistic population growth introduces a third parameter, θ, that allows per capita population growth to decline nonlinearly, with values closer to $\theta = 2$ assumed to be more appropriate for longer lived mammals (Messier, 1994; Sinclair, 2003) (Figure 6.1):

$$\frac{dN}{dt} = rN\left(1 - \left(\tfrac{N}{K}\right)^{\theta}\right) \text{ and } \frac{1}{N}\frac{dN}{dt} = r\left(1 - \left(\tfrac{N}{K}\right)^{\theta}\right)$$

If $\theta > 1$, then population growth is maximized at population size above $\frac{K}{2}$, and the maximum sustainable yield is also higher.

Harvest can be added to logistic population growth in various forms, but it is pedagogically useful to start with a constant harvest rate, H:

$$\frac{dN}{dt} = rN\left(1 - \frac{N}{K}\right) - H$$

If the constant harvest rate is too high (above *MSY*; H_3 in Figure 6.2), then harvest rate is above the population growth rate, and the population growth rate, $\frac{dN}{dt}$, is negative for all population sizes and the population will go extinct. If the constant harvest rate is the same as *MSY* (H_2 in Figure 6.2), then theoretically the population will decline to approximately $\frac{K}{2}$ and stay there, but if it declines for any reason then it will go extinct because the harvest rate exceeds production. Harvesting at *MSY* is now typically seen as an upper limit to sound management under a precautionary principle. A lower harvest rate introduces two equilibrium population sizes, N_1 and N_2, where production and harvest are equal and the population is stable, $\frac{dN}{dt} = 0$. N_2 is referred to as a stable equilibrium because growth exceeds harvest for populations smaller than N_2 and the reverse is true for larger populations. By the same argument, N_1 is an unstable equilibrium. The population will grow if it is larger than N_1, but will fall to extinction if it is below N_1.

In practice, constant harvest rates like this do not happen because it becomes increasingly challenging to harvest a population as its population density declines. It is thus typical to consider harvest as a linear function of population size, with the

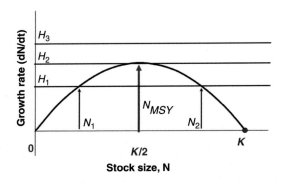

Figure 6.2 Three constant harvest rates overlaid on logistic population growth. The intersections are equilibria, with N_2 being stable and N_1 being unstable when the harvest rate is H_1. There is no equilibrium if the harvest is above *MSY*, as with H_3, and the population becomes extirpated. If the harvest just equals *MSY*, then there is a partially stable single equilibrium. In practice, harvest rates are rarely constant because hunting success rates depend on the abundance of the species.

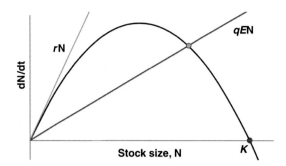

Figure 6.3 Logistic population growth and constant effort harvest with effort E and catchability q. As qE approaches r, which defines the initial slope, the equilibrium population declines toward 0.

slope defined by the effort, E, and the per capita effectiveness of that effort at harvesting animals or "catchability", q (Figure 6.3):

$$\frac{dN}{dt} = rN\left(1 - \frac{N}{K}\right) - qEN$$

Most hunting studies have distinguished between sustainable and unsustainable hunting using "sustainability indices" that typically take a snapshot of wildlife harvest and conclude whether it exceeds theoretical production or *MSY*. Robinson and Redford's (1991) model assumes that maximum population growth occurs at 60 percent of carrying capacity and that a different proportion, F, of that population can be harvested depending on whether the species is short or long lived. The scaling factor is somewhat arbitrarily defined as $F = 0.6$, 0.4, or 0.2 if the species' average lifespan is less than five years, greater than five but less than ten years, or greater than ten years, respectively (Table 6.1). One obvious drawback of this approach is misdiagnosing sustainability because harvest can be low, either because hunting is sustainable and there is a high-density equilibrium or because the stock is already depleted. That is, a species may be hunted in low numbers either because there is little hunting pressure or because the hunting pressure has been so severe that it has become locally rare and is now difficult to capture. Bodmer and colleagues (Bodmer, 1995; Bodmer *et al.*, 1997) developed a harvest model in which population size, female fecundity, and survivorship to average age of reproduction are input as parameters, which avoids the misdiagnosis problem (see Table 6.1). This approach requires a labor-intensive density census and previous data on fecundity and survivorship, although it is possible to use the F scaling factor from Robinson and Redford (1991) if such data are available. A simulation study showed that Bodmer's method was superior in preventing local extinction, whereas Robinson and Redford's method led to species extinction with almost all

Table 6.1 *Examples of static sustainability indices*

	Robinson and Redford	Bodmer
Equation used to calculate formula	$P = 0.6K(\lambda-1)*F$	$P = (0.5D)*(Y*g)*F$
Descriptions of the equation elements	λ, the intrinsic maximum rate of growth, is usually calculated with Cole's equation: $$1 = \lambda_c^{-1} + b\lambda_c^{-\alpha} - b\lambda_c^{-(\omega+1)}$$ α is the age at first reproduction b is the number of female offspring/adult females/time ω is the age at last reproduction Since this model assumes no mortality, F is a scaling factor based on life span designed to account for juvenile and adult mortality $F = 0.6, 0.4$, and 0.2 for species with life spans of less than five years, more than five and less than 10 years, and more than ten years, respectively	D is an estimate of the density in the catchment area based on census data. It is scaled by 0.5 assuming 50 percent of the population is female Y is the average number of young recorded per female harvested g is the average number of gestations per female per year Together, $Y*g$ is the average number of young produced per female per year F is either defined like in Robinson and Redford's model, or is the measured survivorship to reproductive age if those data are available

parameter combinations tested (Milner-Gulland and Akçakaya, 2001). Moreover, in the typical absence of reliable density estimates, the Robinson and Redford method tends to overestimate what could be defined as a sustainable harvest rate for Amazonian game species because most large-bodied Amazonian vertebrates occur at very low population densities (Peres, 2000).

When applying static sustainability indices, the population size and/or carrying capacity parameters come from an arbitrarily defined catchment area around a community. An unfortunate consequence of this is that the resulting sustainable level of production is very sensitive to the choice of hunting catchment. This occurs because area scales as the radius squared and so increasing the catchment area from a radius of 5 km (within which perhaps 60 percent of kills are made) to 8 km (within which perhaps 85 percent of kills are made) increases the catchment area by more than 2.5 times. Because the allowable harvest is directly proportional to the population size at carrying capacity, this 2.5-fold increase in the population allows for 2.5 times the harvest before being defined as unsustainable, despite the fact that there is no *a priori* reason to distinguish between a 5- and an 8-km radius. It is thus easy for a researcher to conclude that harvest is unsustainable by reducing the catchment area radius or sustainable by increasing it, which is an undesirable property of sustainability indices.

A second undesirable property of sustainability indices is the assumption that the entire catchment area has the same population density. Since these indices assume density-dependent growth, the maximum production estimate, and therefore the allowable offtake, is sensitive to the assumed density at carrying capacity. Population densities for hunted species increase with distance from a hunting community, from extirpation within some, often short, distance to carrying capacity far from the community (Peres and Lake, 2003; Levi *et al.*, 2009). A census may find an average population size of $\frac{K}{2}$, which, when applied to a static sustainability index, would lead to the conclusion that all individuals in the entire catchment area are reproducing near maximum production. The population might better be modeled as concentric annuli with different density-dependent responses, since annuli near the community and far from the community will not be producing at the maximum rate of growth due to population densities that are too low or too high, respectively (like in Figure 6.1).

A much more effective approach to sustainability assessment would be to bypass the idea of sustainability as a binary question and consider it instead as an exercise in projecting the impact of hunting in space and time to explore alternative scenarios. Being explicit about space and time allows for exploration of the impact of human population growth and spread, changes in regulations or behavior, and the role of source–sink dynamics in maintaining hunting as animals move from lightly hunted areas far from communities toward more frequently hunted areas near communities (Novaro *et al.*, 2000; Ohl-Schacherer *et al.*, 2007). Given the importance of unhunted wildlife refugia in maintaining source populations, it may be possible to use the location of communities as a landscape-scale proxy for sustainability. This approach has the advantage of being more readily monitorable than species offtake and more enforceable than typical wildlife management regimes such as offtake quotas in remote areas. Limiting the spread of communities is possible to incentivize via win–win conservation and development programs, namely by investing in infrastructure that people care about, which works as a centripetal social force that limits population spread and fissioning. Some possibilities include infrastructure investments in potable water, access to health care, decent schools, recreational facilities, small-scale animal husbandry, and fish farms.

6.3 Biodemographic Models

As an improvement on sustainability indices, Levi *et al.* (2009, 2011) developed a biodemographic model to project the impact of hunting in space and time by a potentially growing and spreading human population and a co-occurring animal population that is able to diffuse down its concentration gradient from lightly hunted sources to hunted sinks. The key observation underlying the biodemographic approach is that subsistence hunters are central-place foragers who concentrate their effort near human settlements (Lu and Winterhalder, 1997; Sirén *et al.*, 2004; Ohl-Schacherer *et al.*, 2007; Smith, 2008), which is the vast majority of cases. Given a certain amount of effort, the number of kills of a particular species will be a function of the desirability, vulnerability, and local abundance of that species. Levi and colleagues applied their model to assess the impact of subsistence hunting on large-bodied primates by Matsigenka people in Manu National Park within the Peruvian Amazon. Large-bodied primates are a key food resource for the Matsigenka and are also unusually susceptible to overhunting because they are vocal, diurnal, group-living, and have a very low maximum intrinsic population growth rate.

The approach of Levi and colleagues to assessing the impact of hunting was to overlay a 1 × 1 km grid over the landscape, with each square kilometer initialized with monkey population density at carrying capacity. Hunts were centered on human communities, and the distribution of hunt lengths followed a Rayleigh distribution (Figure 6.4), which is associated with the magnitude of distances when hunting effort follows a bivariate normal distribution centered on a community. The total amount of hunting effort incorporated the fraction of hunts at each distance, the number of cells at that distance, the number of hunts per hunter per year, and the population of hunters. The effectiveness of that effort is implemented with a parameter for the expected number of kills per encounter, which is influenced

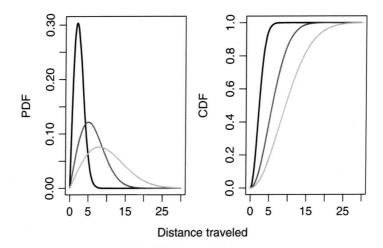

Figure 6.4 Examples of the probability density function (PDF) and cumulative distribution functions (CDF) of the Rayleigh distribution for $\sigma = 2, 5, 8$ given by black, dark-gray, and light-gray curves. The PDF defines the distribution of hunt lengths while the CDF defines the probability that a hunt ends prior to that distance.

Table 6.2 *Parameter values, their interpretation, and references*

Parameter	Definition	Value	References
K	Game species carrying capacity	25	Janson and Emmons (1990)
g	Maximum intrinsic population growth rate	*Ateles*: 0.07 *Lagothrix*: 0.12	Robinson and Redford (1991)
d	Monkeys killed per group encounter	bow: 0.1 gun: 0.9	Alvard and Kaplan (1991) Levi *et al.* (2009)
e	Encounter rate constant to convert game species density to group encounters per kilometer walked	0.02	Endo *et al.* (2010) Levi *et al.* (2009)
hphy	Hunts per hunter per year	40–80	Levi *et al.* (2009)
σ	Spatial spread of hunting effort	Manu: 5 Sarayacu: 7	Ohl-Schacherer *et al.* (2007) Sirén *et al.* (2004)
D	Diffusivity of monkeys	0.1	Levi *et al.* (2009)

by the hunting technology used. They parameterized the models with estimates obtained through a three-year field study in Manu National Park, Peru (Ohl-Schacherer *et al.*, 2007) (Table 6.2).

Analytical, Single-Settlement Mode

The simplest model is that of a single community in a landscape represented by a two-dimensional array (grid) of 1 km² bins, where $N_{x,y,t}$ represents the density of the hunted species in bin (x, y) at time t. Given a human population of size p, the population of the focal species in year $t + 1$ is a function of population growth R $(N_{x,y,t})$, offtake $O(N_{x,y,t}, p_t)$, which is a function of game species density $N_{x,y,t}$ and human population size at time t, p_t:

$$N_{x,y,t+1} = N_{x,y,t} + R\left(N_{x,y,t}\right) - O\left(N_{x,y,t}, p_t\right)$$

As before, a population grows following logistic growth but now implemented in discrete time and based on the local population size at location x, y and time t. In each bin, the total population production is:

$$R\left(N_{x,y,t}\right) = rN_{x,y,t}\left(1 - \left(\tfrac{N_{x,y,t}}{K}\right)^{\theta}\right)$$

Offtake in each gridcell is the product of the encounter rate, $E_{x,y,t}$; the kill rate, d, which depends on technology and hunter aptitude; and the hunting effort, $h_{x,y,t}$, which depends on how many hunters there are, how often they hunt, and how far they travel:

$$O\left(N_{x,y,t}, p_t\right) = \frac{encounters}{km\ walked} \times \frac{kills}{encounter} \times km\ walked = E_{x,y,t} \times d \times h_{x,y,t}$$

The kill rate is a constant dependent on the hunting technology employed, and the encounter rate is proportional to local population density $E_{x,y,t} = e_r \times N_{x,y,t}$, where e_r is empirically determined using "distance sampling," which relates encounter rates on line transects to population densities by accounting for the falloff of detection probability with perpendicular distance from the transect (Buckland *et al.*, 1993; Endo *et al.*, 2010). The hunting effort in each bin and year, $h_{x,y,t}$, is the most difficult to derive. A full derivation and justification for an approximation is given in Levi *et al.* (2011), but the equation for expected effort per hunt can be understood as the falloff of effort over space with every hunt imparting effort at the community itself (since all hunts start there) and less effort at increasingly greater distances. However, because we are operating in two dimensions, as distance increases that lower level of effort is distributed among more cells, which is approximated by the circumference +1 at that distance interval, to yield a falloff in effort for a gridcell s distance units away of $e^{-\frac{s^2}{2\sigma^2}} \frac{1}{2\pi s + 1}$. This has the desirable property of being equal to 1 at $s = 0$ (every hunt goes through the origin) and falling off to an effort of zero as distance increases. The actual hunting effort in cell x, y, at time t is the total number of hunts scaled by this effort distribution:

$$h_{x,y,t} = hphy \times p_t e^{-\frac{s^2}{2\sigma^2}} \frac{1}{2\pi s + 1},$$

where, p_t, is the number of hunters at time t, and *hphy* is the mean number of hunts per hunter per year. Plugging $h_{x,y,t}$ and $E_{x,y,t}$ into the offtake equation yields:

$$O\left(N_{x,y,t}, p_t\right) = E_{x,y,t} \times d \times h_{x,y,t} = e_r N_{x,y,t} \times d \times h_{x,y,t} = \left(d \times h_{x,y,t} \times e_r\right) \times N_{x,y,t}$$

where it can be seen that the harvest at any location is proportional to the local animal population size and dependent on effort, which itself depends on how many hunters there are, how often they hunt, how far they go, and the effectiveness of that effort at harvesting an animal.

Analytical Solutions. By determining when the growth and offtake terms are equal, we can solve for the equilibrium population size in bin (x, y) as a function of distance from a human settlement with a constant population size of hunters p to understand the predicted long-term impact of hunting by a single settlement. To do so, we set $N_{x,y,t+1} = N_{x,y,t} \equiv N_{x,y}$, which is equivalent to setting production, R, equal to offtake, O. The equilibrium population size of a game species s distance units from a community is given by:

$$N_s = max\left(0, K\left(1 - \frac{e_r \times d \times hphy \times p \times e^{-\frac{s^2}{2\sigma^2}}}{r(2\pi s+1)}\right)^{\frac{1}{\theta}}\right)$$

This equation is satisfying because an analytical solution allows us to easily see the influence of each component of the model of the equilibrial population size where harvest is balanced by population growth. Note that population p, hunts per hunter per year $hphy$, and kill rate d are multiplied together in one term such that any proportionate increase in one parameter, for example kill rate by switching to firearms, has the same expected impact as that proportionate increase in another parameter. One can use this same equation to solve for the distance interval where we expect a species to be extirpated by setting $N_{x,y} = 0$ and solving for s, which is referred to as the "extinction envelope." Thus, rather than asking "is hunting sustainable," this spatially explicit model instead produces a predicted distance interval over which a species will be extirpated as a function of human population size, hunting effort, and hunting effectiveness.

There are a few general lessons to be learned from the equilibrium solution. Notably, the expected rise in population density with distance is very steep, which arises because hunters access distant places less often and distant places have so much more area (imagine a 1-km thick annulus that is 10 km or 5 km from a community) (Figure 6.5). Because of this steep rise, the size of the extinction envelope effectively characterizes the impact of hunting. Access to hunting technology like firearms can greatly increase the kill rate, d, by an expected order of magnitude or more, which has a large impact on the size of the exclusion zone (Figure 6.5A). One could imagine that access to firearms could also reduce the frequency of hunting, $hphy$, which has the same proportionate impact but is unlikely to vary by an order of magnitude (Figure 6.5B). A two-fold change in $hphy$ does not qualitatively change the results, which is also reassuring since even two-fold uncertainty in the parameters in the product with $hphy$ do not change the qualitative picture. Another lesson from this equation is that the size of the extinction envelope does not depend on theta, which only influences the steepness of the increase from zero (Figure 6.5C), nor does it depend on carrying capacity, K, which influences the maximum density but not the shape of the increase with distance (Figure 6.5D). This is fortunate because both of these quantities are challenging to estimate. The model is sensitive to variation in the estimated maximum intrinsic population growth rate of the hunted wildlife, r, with values for spider monkeys ($r = 0.07$) producing much larger extinction envelopes than those for deer ($r = 0.4$) or collared peccary ($r = 1.25$) (Figure 6.5E). The average hunt length is $\sigma\sqrt{\frac{\pi}{2}}$ based on the mean of the Rayleigh distribution, which for values of σ of 4, 5, or 6 is associated with mean hunt

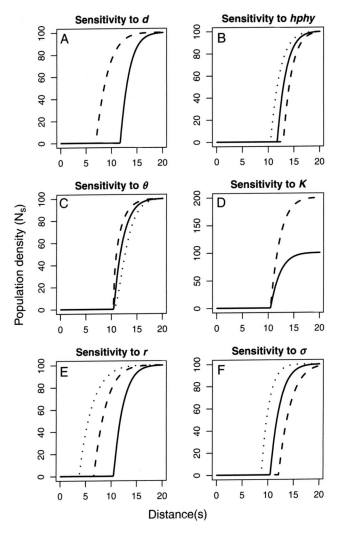

Figure 6.5 Sensitivity analysis of the equilibrium population density as a function of distance. (A) Kill rate, d, is varied from 0.1 (dashed), associated with bow hunting of large primates, to 1, more reflective of gun hunting, which results in a much larger extinction envelope when density is zero. (B) Because d and $hphy$ are multiplied together in the same term, a proportional change in either has the same impact. Here, $hphy$ is varied from 20 (dotted) to 40 (solid) to 80 (dashed) hunts per hunter per year, which increases the extinction envelope, but less than the order-of-magnitude increase in kill rate. (C) Varying θ from 0.5 (dotted) to 1 (solid) to 2 (dashed) does not change the extinction envelope but, instead, larger values of theta allow for a steeper recovery to carrying capacity with distance. (D) Varying K from 100 to 200 also does not influence the extinction envelope; it only influences the density that is recovered as distance from the community increases. (E) Varying r from 0.07 (solid) associated with spider monkeys to 0.4 (dashed) associated with brocket deer to 1.25 (dotted) associated with collared peccaries has a large effect on the size of the extinction envelope. This highlights how harvest-tolerant species can persist near communities even as harvest-sensitive ones are extirpated, but note that e_r also varies by species but has been held constant in this case. (F) Varying σ from 4 (dotted) to 5 (solid) to 6 (dashed) leads to a larger extinction envelope, highlighting the importance of understanding the spatial distribution of hunting effort.

$d = 0.1$ $d = 1$

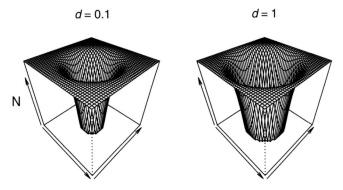

N

Figure 6.6 Visualization of the magnitude of depletion at equilibrium in three dimensions as the kill rate is varied from a value associated with bow hunting, $d = 0.1$, to gun hunting, $d = 1$.

distances of 5, 6.3, and 7.5 km walked, respectively. This relatively small variation in σ does impact the size of the exclusion zone and thus the impact of hunting, indicating that more studies should invest in estimating this parameter (Figure 6.5F).

Because the landscape is two-dimensional, the plots of density versus distance can be misleading because a small increase in an already large radius depletes a much larger area of wildlife than when the extinction envelope is initially small. The magnitude of the depletion can be visualized with a heat map or a three-dimensional perspective plot (Figure 6.6).

The relatively simple context of a single settlement is easily expanded into the context of many settlements of different population size, p_i, with potentially overlapping hunting zones. This requires proper accounting of hunting effort at any given spatial location given that each settlement i is located at $(x_{0,i}, y_{0,i})$, and the distance of location (x, y) to this settlement must now be specified as $\sqrt{(x_{0,i} - x)^2 + (y_{0,i} - y)^2}$ instead of left as s. Hunting effort at (x, y) summed over all contributing settlements becomes:

$$h_{x,y} = hphy \times \sum_{i=1}^{\text{settlements}} p_i e^{-\frac{(x_{0,i}-x)^2 + (y_{0,i}-y)^2}{2\sigma^2}} \frac{1}{2\pi \sqrt{(x_{0,i}-x)^2 + (y_{0,i}-y)^2} + 1}$$

This new expression for hunting effort can be used to solve for the equilibrium population at (x, y) in the context of multiple settlements by setting growth to offtake as before to obtain:

$$N_{x,y} = max\left(0, K^\theta\left(1 - \frac{e_r \times d \times hphy}{r}\sum_{i=1}^{settlement} p_i e^{-\frac{(x_{0,i}-x)^2+(y_{0,i}-y)^2}{2\sigma^2}}\frac{1}{2\pi\sqrt{(x_{0,i}-x)^2+(y_{0,i}-y)^2+1}}\right)\right)^{\frac{1}{\theta}}$$

There is no extinction envelope when multiple settlements are involved, but we can calculate a matrix of steady-state population density values if we specify the location and population of each settlement. The results can be summarized with cumulative distribution functions, which summarize the distribution of population depletion across the landscape as the proportion of cells below each population density. All cells are less than or equal to carrying capacity, but if the population is depleted at a landscape level, then a larger fraction of the landscape will have low population density.

The equilibrium catch per unit effort (CPUE) with multiple settlements can be considered either as the local CPUE for some subset of the total number of settlements or as the global CPUE, which is the total catch divided by the total effort.

Numerical, Multiple-Settlement Model with Source–Sink Dynamics

Analytical solutions are appealing because anyone can solve them by simply plugging numbers into an equation. However, analytical solutions have two limitations. First, they are used to determine the long-term impact of hunting rather than at any given point in time as communities grow and spread. Second, they cannot account for the movement of animals from far unhunted areas toward communities, which can locally maintain wildlife populations by source–sink dynamics. A numerical model that simulates one year at a time can be used to model wildlife populations at any point in a time series, which also makes it a useful validation tool because we can compare predicted and observed wildlife populations in any year after the establishment or removal of settlements and/or after changes in human population size or a management intervention.

The numerical simulation for the population in space and time now includes a migration term:

$$N_{x,y,t+1} = N_{x,y,t} + R(N_{x,y,t}) - O(N_{x,y,t}, p_t) + M(N_t)$$

Migration is modeled as a diffusion process as individuals move from higher to lower density bins, and the rate of movement is faster when the density difference between bins is higher. Migration is given by:

$$M(N_t) = D\nabla^2 N_t,$$

where D is the diffusivity constant (distance^2/time) that determines how fast animals move down their concentration gradient. The following equation is used to model diffusion in continuous space:

$$\nabla^2 N_t = \frac{\partial^2 N}{\partial x^2} + \frac{\partial^2 N}{\partial y^2}$$

This can be implemented numerically on a grid with the "five-point stencil" technique, which uses the values of the four nearest neighbors (up, down, left, right) to approximate derivatives on a grid. For bins 1-km across, and for a one-year time step, this becomes:

$$D\nabla^2 N_t \approx D\left(N_{x+1,y,t} + N_{x-1,y,t} + N_{x,y+1,t} + N_{x,y-1,t} - 4N_{x,y,t}\right)$$

One key consideration in the implementation of source–sink dynamics via diffusion is that the annual time-step may be too long to approximate the differential equation if the diffusivity is too high ($\gg 0.1$). The result can be standing waves rather than diffusion and can be remedied by breaking the year into multiple, shorter time intervals.

The analytical solutions provide equations that are relatively easy to use and suitable for many purposes, but this numerical model is the most flexible. As an example, consider a landscape with eleven settlements of different sizes (Figure 6.7). Each settlement can have a different population growth trajectory input via exponential growth or with actual values from the past and/or predicted values for the future. This contrasts with the single value for the size of population used with the analytical models. Furthermore, the progression of the impact of hunting can be visualized at any time (Figure 6.7A). The impact at a landscape scale can be summarized for all communities or the impact can be summarized locally around each community, using cumulative distribution functions. However, the summary depends critically on the area under consideration. For example, after fifteen years of hunting, nowhere has a population above 70 percent of K and already 43 percent of the area is below 10 percent of K when we consider a radius of 10 km around the largest community (Figure 6.7B). However, if we consider all land within 18 km of the community, then 38 percent is minimally impacted by hunting with a density over 90 percent of K and only 14 percent is near extirpation with a density below 10 percent of K.

6.4 Limitations and Advancements

Biodemographic models overcome many issues with sustainability indices by allowing for the projection of the impact of hunting in space and time to explore both present and future levels of wildlife depletion. However, these models have their own limitations that require continued development. These models assume that hunting effort is isotropically distributed around communities or households if each is modeled separately. This is an adequate approximation for the behavior of

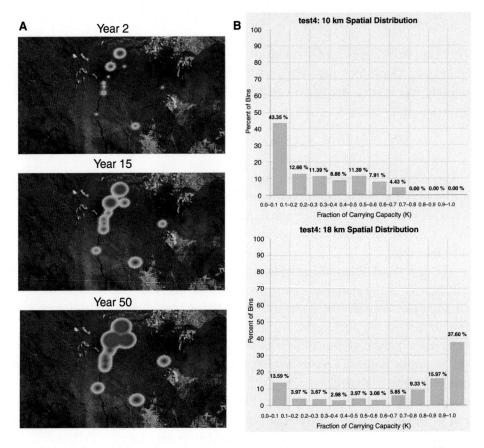

Figure 6.7 Example results from a numerical simulation of eleven communities with different sizes and population growth trajectories. Unlike the equilibrium solution, the numerical model allows us to inspect results at any point in time, allows populations to grow, allows new communities to form, and allows for changes in technology or behavior. (A) The results visualized at years 2, 15, and 50, with red indicating extirpation and yellow indicating intermediate densities. (B) The results can be summarized at a landscape scale or around a given community. Here, we show the proportion of the landscape around the largest community within a radius of 10 km and 18 km that fits within ten bins representing fractions of carrying capacity. Within 10 km, none of the landscape is unhunted and 43 percent is below 0.1 K. However, the summary changes substantially within a radius of 18 km, for which 38 percent of the area is above 0.9 K and only 14 percent is below 0.1 K. This highlights the sensitivity of sustainability assessments to the area being considered and why projections in space and time are more effective.

subsistence hunters in systems with limited road infrastructure and little access to motorboats. Such communities are still common in remote parts of the neotropics, but a large number of hunters now have a diverse economic portfolio with a mix of subsistence hunting and farming, on the one hand, and substantial market involvement via intermittent wage labor or sale and bartering of goods, on the other. These hunters often have access to roads or motorized boats that allow them to start hunts farther from population centers where wildlife-encounter rates are higher. That is, hunting is often far from, and isotropically distributed around, local communities. Some users of biodemographic models have addressed this issue by adding hunting effort along rivers with small "virtual" communities of declining size to emulate a fall in hunting effort away from communities, but this is not an elegant approach (Shaffer *et al.*, 2018 a, b). Hunting can also be anisotropic when communities are situated on the forest edge (like in mixed forest–savanna systems), when some vegetation types are more productive than others, or on the edge of land-use change.

The second limitation of biodemographic models is a lack of a multispecies context. For example, humans can be satiated by an abundant harvest-tolerant species. This can reduce harvest of the sensitive species because, on many hunts, the harvest-tolerant species is killed first and the hunter returns home rather than continues searching for the harvest-sensitive species. In the context of the biodemographic model, this could, for example, either reduce the hunt distance (σ) because many hunts are terminated early or reduce how frequently people hunt (*hphy*) because more hunts are successful. However, it is also possible that abundant stocks of harvest-tolerant species render hunting a worthwhile activity, thereby maintaining greater hunting effort than would be the case in their absence. This is the case of abundant prey species continuing to subsidize multispecies hunting efforts and therefore overexploitation of increasingly rare species, which is widespread in the fisheries industry (Tromeur and Doyen, 2019). A full multispecies model might have σ and *hphy* dynamically linked to the abundance of multiple wildlife species.

The application to multispecies systems would also benefit from consideration of diet breadth theory, which suggests that foragers specialize on high-profitability species when they are abundant and add new species into their diet as those species become less common (Levi *et al.*, 2011; Bugir *et al.*, 2021). To some degree, this does not present a problem to biodemographic models because we typically model large-bodied and energetically profitable species that are most vulnerable to hunting. However, as stated before, the availability of lower profitability species can maintain hunting efforts even as hunting would otherwise not be an energetically profitable activity relative to, say, fishing or agriculture (higher *hphy*), and can influence hunting distance as low-profit species are killed closer to home (Ohl-Schacherer *et al.*, 2007). Despite these limitations, in most applied settings the lack of a multispecies context is unlikely to qualitatively change model results for high-value species that are always in the dietary breadth as long as hunting effort does not respond substantially to changing prey community composition.

The most important improvement to the biodemographic modeling approach is incorporation of anisotropic distributions of hunter effort with a mix of linear travel along roads and streams followed by travel on foot in the forest. This can be readily incorporated into the existing model structure provided that excellent maps of elevation, drainage, and vegetation types are available. The most flexible models that include hunter decision-making in a multispecies setting and even with specific hunting grounds may require the use of individual based models. Levi *et al.* (2011) developed one such model in which hunters exhibit optimal behaviors subject to optimal foraging theory under the constraints of a daily hunt and the availability of ammunition as a function of whether, and how much, wild meat they have already obtained on the hunt.

6.5 Conclusion

Fewer humans live primarily subsistence lifestyles, as market involvement and wage labor continues to spread around the world, but wild meat is still key to the livelihoods of billions of people (Mwaura, 2022). Indigenous people in some regions disproportionately depend on wild meat, and both food security and wildlife conservation benefit from wildlife population assessments and population projections. This is particularly the case in the context of hunting technologies such as the widespread adoption of firearms or motorized vehicles that increase either the effectiveness or the spatial extent of hunting effort. However, while the biodemographic approach described here is a promising means to projecting wildlife populations in time and space for scenario planning and sustainability assessments; it does not itself influence wildlife populations. Cooperative implementation of direct wildlife management to replace unregulated hunting may be necessary in many places to maintain populations of harvested wildlife for continued harvest.

Implementing wildlife management is a huge challenge, particularly in low-governance environments (Yu *et al.*, 2010) and where hunting itself is stigmatized. Many countries have responded by making hunting illegal, which runs counter to the food security and sovereignty of Indigenous peoples. In contrast, the North American wildlife management model is specifically predicated on widespread hunting by people with sufficient resources to purchase licenses that fund management and top-down enforcement. Such approaches have not spread elsewhere; instead, the most promising alternative is community-based conservation. For example, communities along the Juruá River in the Brazilian Amazon have implemented fisheries management by capping the harvest of in situ stock counts at 20 percent. This has led to the recovery of giant arapaima fish (*Arapaima gigas*) for the benefit of both local food security and livelihoods because stock recovery trajectories result in large numbers of surplus arapaima that are increasingly sold at premium prices as sustainably harvested (Campos-Silva and Peres, 2016; Freitas *et al.*, 2020). There are promising avenues to extend this model to community-based management of river turtles (Norris *et al.*, 2019). This mode of conservation represents a departure from so-called fortress conservation that relies exclusively on strictly protected areas and represents a critical avenue for low-cost future conservation efforts because strictly protected areas without hunting cover a very small fraction of wildlands.

References

Albert, B. and Le Tourneau, F.-M. (2007). Ethnogeography and resource use among the Yanomami: Toward a model of "reticular space." *Current Anthropology*, **48**(4), 584–92.

Alroy, J. (2001). A multispecies overkill simulation of the end-Pleistocene megafaunal mass extinction. *Science*, **292**(5523), 1893–6.

Alvard, M. (1995). Shotguns and sustainable hunting in the neotropics. *Oryx*, **29**(1), 58–66.

Alvard, M. and Kaplan, H. (1991). *Procurement Technology and Prey Mortality among Indigenous Neotropical Hunters*. Boulder, CO: Westview.

Alvard, M. S., Robinson, J. G., Redford, K. H., and Kaplan, H. (1997). The sustainability of subsistence hunting in the neotropics. *Conservation Biology*, **11**(4), 977–82.

Antunes, A. P., Rebêlo, G. H., Pezzuti, J. C. B., *et al.* (2019). A conspiracy of silence: Subsistence hunting rights in the Brazilian Amazon. *Land Use Policy*, **84**, 1–11.

Begotti, R. A. and Peres, C. A. (2020). Rapidly escalating threats to the biodiversity and ethnocultural capital of Brazilian Indigenous Lands. *Land Use Policy*, **96**, 104694.

Bodmer, R. E. (1995). Managing Amazonian wildlife: biological correlates of game choice by detribalized hunters. *Ecological Applications*, **5**(4), 872–7.

Bodmer, R. E., Eisenberg, J. F., and Redford, K. H. (1997). Hunting and the likelihood of extinction of Amazonian mammals: Caza y Probabilidad de Extinción de Mamiferos Amazónicos. *Conservation Biology*, **11**(2), 460–6.

Bodmer, R. E., Fang, T. G., Moya, L., and Gill, R. (1994). Managing wildlife to conserve Amazonian forests: population biology and economic considerations of game hunting. *Biological Conservation*, **67**(1), 29–35.

Buckland, S. T., Anderson, D. R., Burnham, K. P., and Laake, J. L. (1993). *Distance Sampling: Estimating Abundance of Biological Populations*. London: Chapman and Hall.

Bugir, C. K., Peres, C. A., White, K. S., *et al.* (2021). Prey preferences of modern human hunter-gatherers. *Food Webs*, **26**, e00183.

Campos-Silva, J. V. and Peres, C. A. (2016). Community-based management induces rapid recovery of a high-value tropical freshwater fishery. *Scientific Reports*, **6**(1), 1–13.

Castro, S., Yebra, L., Cortegoso, V., *et al.* (2021). The introduction of the bow and arrow across South America's southern threshold between food-producing societies and hunter-gatherers. In J. B. Belardi, D. L. Bozzuto, P. M. Fernández, E. A. Moreno, and G. A. Neme (eds.), *Ancient Hunting Strategies in Southern South America*. Cham, Switzerland: Springer, pp. 137–58.

Chacon, R. J. (2001). *Testing the Energy Maximization and Time Minimization Hypotheses: The Effects of Shotgun Technology on Achuar Indian Hunting*. Santa Barbara, CA: University of California.

de Marques, A. A. B., Schneider, M., and Peres, C. A. (2016). Human population and socioeconomic modulators of conservation performance in 788 Amazonian and Atlantic Forest reserves. *PeerJ*, **4**, e2206.

de Mattos Vieira, M. A., de Castro, F., and Shepard, G. H. (2019). Who sets the rules? Institutional misfits and bricolage in hunting management in Brazil. *Human Ecology*, **47**(3), 369–80.

de Mattos Vieira, M. A. R., von Muhlen, E. M., and Shepard Jr, G. H. (2015). Participatory monitoring and management of subsistence hunting in the Piagaçu-Purus reserve, Brazil. *Conservation and Society*, **13**(3), 254–64.

Dobson, A. D., Milner-Gulland, E., Ingram, D. J., and Keane, A. (2019). A framework for assessing impacts of wild meat hunting practices in the tropics. *Human Ecology*, **47** (3), 449–64.

Endo, W., Peres, C. A., Salas, E., *et al.* (2010). Game vertebrate densities in hunted and nonhunted forest sites in Manu National Park, Peru. *Biotropica*, **42**(2), 251–61.

Estrada, A., Garber, P. A., Gouveia, S., *et al.* (2022). Global importance of Indigenous Peoples, their lands, and knowledge systems for saving the world's primates from extinction. *Science Advances*, **8**(31), eabn2927.

Fausto, C. (2007). Feasting on people: eating animals and humans in Amazonia. *Current Anthropology*, **48**(4), 497–530.

Fernández-Llamazares, Á. and Virtanen, P. K. (2020). Game masters and Amazonian Indigenous views on sustainability. *Current Opinion in Environmental Sustainability*, **43**, 21–27.

Freitas, C. T., Lopes, P. F., Campos-Silva, J. V., *et al.* (2020). Co-management of culturally important species: a tool to promote biodiversity conservation and human well-being. *People and Nature*, **2**(1), 61–81.

Hames, R. (1980). Game depletion and hunting zone rotation among the Ye'kwana and Yanomamo of Amazonas, Venezuela. In R. Hames, ed., *Working Papers on South American Indians*, No. 2. Bennington, Vermont: Bennington College, pp. 31–66.

Hickerson, H. (1965). The Virginia deer and intertribal buffer zones in the upper Mississippi Valley. In A. Leeds and A. Vayda, eds., *Man, Culture, and Animals: The Role of Animals in Human Ecological Adjustments*. Washington, DC: American Association for the Advancement of Science Publication 78, pp. 43–66.

Hill, E. (2011). Animals as agents: hunting ritual and relational ontologies in prehistoric Alaska and Chukotka. *Cambridge Archaeological Journal*, **21**(3), 407–26.

Ingram, D. J., Coad, L., Milner-Gulland, E., *et al.* (2021). Wild meat is still on the menu: progress in wild meat research, policy, and practice from 2002 to 2020. *Annual Review of Environment and Resources*, **46**, 221–54.

Janson, C. H. and Emmons, L. H. (1990). *Ecological Structure of the Nonflying Mammal Community at Cocha Cashu Biological Station, Manu National Park, Peru. Four Neotropical Rainforests*. New Haven, CT: Yale University Press, 314–38.

Levi, T., Lu, F., Yu, D. W., and Mangel, M. (2011). The behavior and diet breadth of central-place foragers: an application to human hunters and Neotropical game management. *Evolutionary Ecology Research*, **13**, 171–85.

Levi, T., Shepard, G. H., Ohl-Schacherer, J., Peres, C. A., and Yu, D. W. (2009). Modelling the long-term sustainability of indigenous hunting in Manu National Park, Peru: landscape-scale management implications for Amazonia. *Journal of Applied Ecology*, **46**(4), 804–14.

Lu, F. and Winterhalder, B. (1997). A forager-resource population ecology model and implications for indigenous conservation. *Conservation Biology*, **11**(6), 1354–64.

Luzar, J. B., Silvius, K. M. and Fragoso, J. (2012). Church affiliation and meat taboos in indigenous communities of Guyanese Amazonia. *Human Ecology*, **40**(6), 833–45.

Messier, F. (1994). Ungulate population models with predation: a case study with the North American moose. *Ecology*, **75**(2), 478–88.

Milner-Gulland, E. J. and Akçakaya, H. R. (2001). Sustainability indices for exploited populations. *Trends in Ecology & Evolution*, **16**(12), 686–92.

Milton, K. (1997). Real men don't eat deer. *Discover-New York*, **18**, 46–53.

Mitchell, C. L. and Raéz-Luna, E. F. (1991). *The Impact of Human Hunting on Primate and Game Bird Populations in the Manu Biosphere Reserve in Southeastern Peru.* New York: Wildlife Conservation International. New York Zoological Society New York, p33.

Mwaura, A. (2022). Intergovernmental Science-Policy Platform on Biodiversity and Ecosystem Services (IPBES) – media release. https://policycommons.net/artifacts/ 3680771/intergovernmental-science-policy-platform-on-biodiversity-and-ecosys tem-services-ipbes/4486613/ (accessed November 20, 2023).

Norris, D., Peres, C. A., Michalski, F., and Gibbs, J. P. (2019). Prospects for freshwater turtle population recovery are catalyzed by pan-Amazonian community-based management. *Biological Conservation*, **233**, 51–60.

Novaro, A. J., Redford, K. H., and Bodmer, R. E. (2000). Effect of hunting in source-sink systems in the neotropics. *Conservation Biology*, **14**(3), 713–721.

Ohl-Schacherer, J., Shepard, G. H., Kaplan, H., *et al.* (2007). The sustainability of subsistence hunting by Matsigenka native communities in Manu National Park, Peru. *Conservation Biology*, **21**(5), 1174–85.

Peres, C. A. (2000). Evaluating the impact and sustainability of subsistence hunting at multiple Amazonian forest sites. In J. G. Robinson and E. L. Bennett, eds., *Hunting for Sustainability in Tropical Forests*. New York: Columbia University Press, pp. 31–57.

Peres, C. A. (2011). Conservation in sustainable-use tropical forest reserves. *Conservation Biology*, **25**(6), 1124–9.

Peres, C. A. and Lake, I. R. (2003). Extent of nontimber resource extraction in tropical forests: Accessibility to game vertebrates by hunters in the Amazon basin. *Conservation Biology*, **17**(2), 521–35.

Read, J. M., Fragoso, J. M., Silvius, K. M., *et al.* (2010). Space, place, and hunting patterns among indigenous peoples of the Guyanese Rupununi region. *Journal of Latin American Geography*, **9**(3), 213–43.

Robinson, J. G. and Redford, K. H. (1991). Sustainable harvest of Neotropical wildlife. In J. G. Robinson and K. H. Redford, eds., *Neotropical Wildlife Use and Conservation*. Chicago: University of Chicago Press, pp. 415–29.

Roosevelt, A. C. (1987). Chiefdoms in the Amazon and Orinoco. *Chiefdoms in the Americas*, **1987**, 153–84.

Ross, E. B., Arnott, M. L., Basso, E. B., *et al.* (1978). Food taboos, diet, and hunting strategy: the adaptation to animals in amazon cultural ecology [and Comments and Reply]. *Current Anthropology*, **19**(1), 1–36.

Sampaio, R., Morato, R. G., Abrahams, M. I., Peres, C. A., and Chiarello, A. G. (2022). Physical geography trumps legal protection in driving the perceived sustainability of game hunting in Amazonian local communities. *Journal for Nature Conservation*, **67**, 126175.

Scheinsohn, V. (2003). Hunter-gatherer archaeology in South America. *Annual Review of Anthropology*, **32**, 339–61.

Schwartzman, S., Moreira, A., and Nepstad, D. (2000). Rethinking tropical forest conservation: Perils in parks. *Conservation Biology*, **14**(5), 1351–7.

Schwartzman, S. and Zimmerman, B. (2005). Conservation alliances with indigenous peoples of the Amazon. *Conservation Biology*, **19**(3), 721–7.

Shaffer, C. A., Milstein, M. S., Suse, P., *et al.* (2018). Integrating ethnography and hunting sustainability modeling for primate conservation in an indigenous reserve in Guyana. *International Journal of Primatology*, **39**(5), 945–68.

Shaffer, C., Yukuma, C., Marawanaru, E., and Suse, P. (2018). Assessing the sustainability of Waiwai subsistence hunting in Guyana by comparison of static indices and spatially explicit, biodemographic models. *Animal Conservation*, **21**(2), 148–58.

Shepard, G. H. (2014). Hunting in Amazonia. In H. Selin, ed., *Encyclopaedia of the History of Science, Technology, and Medicine in Non-Western Cultures*. Dordrecht, Netherlands: Springer, pp. 1–7.

Shepard, Jr., G. H., Levi, T., Neves, E. G., Peres, C. A., and Yu, D. W. (2012). Hunting in ancient and modern Amazonia: rethinking sustainability. *American Anthropologist*, **114**(4), 652–67.

Sinclair, A. R. (2003). Mammal population regulation, keystone processes and ecosystem dynamics. *Philosophical Transactions of the Royal Society of London. Series B: Biological Sciences*, **358**(1438), 1729–40.

Sirén, A., Hamback, P., and Machoa, E. (2004). Including spatial heterogeneity and animal dispersal when evaluating hunting: a model analysis and an empirical assessment in an Amazonian community. *Conservation Biology*, **18**(5), 1315–29.

Smith, D. A. (2008). The spatial patterns of indigenous wildlife use in western Panama: implications for conservation management. *Biological Conservation*, **141**, 925–37.

Surovell, T. A., Pelton, S. R., Anderson-Sprecher, R., and Myers, A. D. (2016). Test of Martin's overkill hypothesis using radiocarbon dates on extinct megafauna. *Proceedings of the National Academy of Sciences*, **113**(4), 886–91.

Townsend, W. (1997). La participación comunal en el manejo de vida silvestre en el oriente de Bolivia. In T. Fang, R. Bodmer, R. Aquino, and M. Valqui, eds., *Manejo de fauna silvestre en la Amazonía*. La Paz, Bolivia: UNAP University of Florida, UNDP/GEF, Instituto de Ecología, pp. 105–10.

Tromeur, E. and Doyen, L. (2019). Optimal harvesting policies threaten biodiversity in mixed fisheries. *Environmental Modeling & Assessment*, **24**(4), 387–403.

Vieira, M. and Shepard, G. (2017). A anta tem muita ciência: racionalidade ecológica e ritual da caça entre ribeirinhos amazônicos. In G. Marchand and F. F. Vander Velden, eds., *Olhares cruzados sobre as relações entre seres humanos e animais silvestres na Amazônia (Brasil, Guiana Francesa)*. Manaus, Brazil: Editora da Universidade Federal do Amazonas (EDUA), 17–31.

Yu, D. W., Levi, T., and Shepard, G. H. (2010). Conservation in low-governance environments. *Biotropica*, **42**, 569–71.

Zimmerman, B., Peres, C. A., Malcolm, J. R., and Turner, T. (2001). Conservation and development alliances with the Kayapó of south-eastern Amazonia, a tropical forest indigenous people. *Environmental Conservation*, **28**(1), 10–22.

7

On Fire and Water: The Intersection of Wetlands and Burning Strategies in Managing the Anthropogenic Plant Communities of Yosemite National Park

DOUGLAS DEUR AND ROCHELLE BLOOM[*]

7.1 Introduction

Indigenous Knowledge combined with analysis of wetland habitat species provides novel insights into the ingenuity of Sierran tribal peoples in undertaking habitat-specific tending practices. Tribal accounts and ethnographic literature have long provided evidence that traditional burning allowed residents of Yosemite Valley to enhance various culturally preferred plants and plant communities, such as black oak (Quercus kelloggii) groves or meadows of herbaceous plants such as sedges (Carex spp.), mints (Mentha spp.), milkweed (Asclepias spp.), and deer grass (Muhlenbergia rigens) (Anderson, 1988, 1990, 1993a, b; Clark, 1894; Reynolds, 1959). Some sources, especially Anderson (1988, 1990, 1993a, b), also note that burning exists alongside a range of other practices, such as pruning, coppicing, weeding, and reseeding of native plant species, which together have shaped the plant communities of Yosemite in myriad ways (Bibby, 1994; Deur, 2007).

In spite of this robust record of Yosemite plant management, we contend that this literature misses certain key aspects of traditional Native American land management and provides an insufficient picture of traditional practices on multiple counts. Among those deficiencies, we argue that these past sources overlook the elevated importance of wetland habitats in the traditional Native American use and management of Yosemite plant communities. We find that wetland plants are disproportionately significant within the plant-use traditions of Yosemite tribes, yet receive little attention within the literature addressing Yosemite (Deur, 2007; Bloom and Deur, 2022; Deur and Bloom, 2022). Moreover, Native interviewees and archival

[*] Tribal members who contributed to this work include Jay Johnson, Bill Tucker, Bill Leonard, Shirley Forga, Sandy Chapman, and Les James, all of the Southern Sierra Miwuk Nation (formerly the American Indian Council of Mariposa County). Without them, this work would have been impossible. We also wish to thank Tricia Gates Brown for editorial assistance. This work was partially supported by the National Park Service (NPS) through Cooperative Ecosystem Studies Unit (CESU) Task Agreement between the NPS and the University of Washington and Task Agreements between the NPS and Portland State University, issued under CESU Cooperative Agreement H8W07110001.

sources suggest that the hydrologic conditions within the valley historically inter-sected with traditional fire management in a number of important ways. Traditional Native American fire managers intentionally burned in a manner that used wetlands and flooded riparian areas as firebreaks, allowing small-scale burning from late spring until fall and the containment of any potentially large and catastrophic fires. This involved sequential small-scale burning of wetlands and wetland margins through the dry season as dropping water tables permitted. Tribal members attest that this nuanced interplay of fire and water – this intricate level of pyrodiversity within a relatively circumscribed area – enhanced a wide range of culturally significant species in both upland and wetland environments. This likely contrib-uted to the development of mosaics of culturally significant wetland margin habitats and, in turn, to the valley's overall biodiversity (Lewis and Ferguson, 1988; Jones and Tingley, 2021). While unreported in the existing literature pertain-ing to Yosemite, the practices we find at Yosemite have analogs and precedents. They bear certain resemblances to traditional burning practices of wetlands and wetland margins as reported in western and northeastern North America (e.g., Lewis, 1979, 1982; Anderson, 2009; Deur, 2009; Deur and Knowledge-Holders of the Quinault Indian Nation 2021), in the American South (Wharton *et al.*, 1982), and in tropical contexts, such as Aboriginal Australia (Whitehead *et al.*, 2003; Russell-Smith *et al.*, 2009; McGregor *et al.*, 2010). However, the use and traditional management of wetland plant habitats in Yosemite is unique in a number of respects and deserving of independent consideration; this is the principal objective of the narrative that follows.

The lack of wetland use and management practices at Yosemite mirrors the erosion of those habitats and the lack of Native American opportunities to access and manage such environments within what is today an internationally celebrated national park. Changes in wetland hydrology since the advent of park management have reduced floodplain connectivity and dropped water tables across much of the Yosemite Valley floor. This has negatively impacted many culturally significant habitats and reduced the potentials for traditional gathering and management – com-pounding the effects of other major interruptions such as park visitation and fire suppression. The displacement of Native peoples from the valley also contributed to the erosion of anthropogenic habitats – a phenomenon widely reported in US national parks (Bloom and Deur, 2020a; Deur and Bloom, 2020; Deur and James, Jr., 2020). The loss of wetland hydrology has amplifed the effects of fire suppression and the displacement of traditional managers, which together have allowed for the encroach-ment of a dominating conifer forest and have undermined the biological diversity and cultural values of the entire Yosemite region.

7.2 Methods

In an effort to understand traditional Native American plant and habitat management practices in Yosemite Valley, the authors undertook a series of research steps. Over an eighteen-year period, Deur has carried out intermittent ethnographic interviewing of and field visits with tribal members at Yosemite in relation to traditional land and resource use (Deur, 2007; Bloom and Deur, 2020a, b, 2022; Deur and Bloom, 2022). In addition, Deur has undertaken archival and literature reviews, expanding on earlier ethnographic studies by researchers like Bibby (1994) and Anderson (1988). While the objectives of each study varied over this period, Deur's interviews have consistently contained open-ended and inductive questions for tribal Elder, designed to identify the widest possible number of plants utilized in the living memory of tribal members. In turn, this corpus of largely unpublished reports and data has served to facilitate National Park Service (NPS) obligations to tribes, including the negotiation of possible future plant-gathering and land-management agreements between tribes and the NPS for lands within the park. These past ethnographic interviews have revealed a number of themes unreported or underreported in the existing published literature, including the traditional significance and management of wetland plant habitats. Reviewing this corpus of research materials, Deur has assembled the many references to wetland plant use and management in field notes and transcripts from his interviews with tribal members, in addition to data summarized in technical reports resulting from his work, although these are unavailable to the public due to their sensitivity, such as Deur (2007) and Bloom and Deur (2022). These original qualitative ethnographic data pertaining specifically to wetland plant habitat use and management are partially reported and summarized in the pages that follow.

Working in collaboration with Deur, Bloom mined the rich corpus of preexisting written historical and ethnographic material related to Yosemite Valley land and resource use to produce the Yosemite Ethnographic Database. This database consists of roughly 13,000 entries describing traditional Native American uses of Yosemite National Park lands and resources – primarily in Yosemite Valley (Bloom and Deur 2020a, b). This database incorporates most of the available ethnographic and historical literature – from the first written records of Yosemite in the 1850s to the ethnographic studies and tribal consultation notes of the present day – housed in park collections and other pertinent archival and library collections state- and nationwide. In compiling the database, a team of researchers and research assistants systematically reviewed written sources for references to lands and resources that were used, visited, or identified by tribal members as significant in Yosemite, as well as information relating to traditional management methods applied to the park's natural resources. This database incorporates

information derived from approximately 600 preexisting sources, including historical reports, early historical accounts written by visitors to Yosemite, ethnographies, ethno-ecological studies, oral histories, park publications, park notes from
contemporary tribal events, historical and contemporary newspaper articles, and
a wide range of other archival materials. Also included are other materials, such as
archived notes from formal NPS consultation meetings with park-associated tribes
and meetings with these tribes specifically on matters of traditional plant community management. The database provides a wide range of searchable data, including
information on archeological, hydrologic, botanical, and other natural and cultural
resources with traditional cultural significance the Southern Sierra Miwuk, Mono
Lake Kootzaduka'a, Tuolumne Band of Me-Wuk Indians, Bishop Paiute Tribe,
Bridgeport Indian Colony, North Fork Rancheria of Mono Indians of California,
and the Picayune Rancheria of the Chukchansi Indians tribal communities with ties
to the park (Bloom and Deur, 2020b). This chapter avoids sharing sensitive
information and limits discussion to materials already available in the public realm.

Using this database, the authors identified all reported traditional plant habitat
management practices employed by tribes both historically and today, seeking to
identify recurring references in tribal oral accounts, written historical accounts, and
other documentation. We compared and cross-referenced these accounts with
ethnographic and oral history information shared with both of the authors of this
chapter by tribal representatives, especially members of the federally unrecognized
Southern Sierra Miwuk Nation. We then compared the contents of this record with
the published record of traditional resource use and management at Yosemite,
specifically noting the information on traditional resource use and management
that was available from these many unpublished sources but that was unrepresented
or underrepresented in published sources. Noting that wetland plants and environments appeared as a recurring underrepresented theme, we carried out visits to
traditionally managed riparian wetland environments with tribal members. We also
analyzed data relating to wetland plants and habitats that had been reported as
significant in the ethnographic interviews and in the available written record. As
part of this analysis, we undertook an assessment of plants and habitats identified as
"culturally significant" by tribal members in light of US federal wetlands criteria.
On the basis of this multidisciplinary investigation – including topical ethnographic
interviews with tribal members, field visits and analysis, and a nearly comprehensive review of the original archival and published record pertaining to traditional
Native American resource use and management – we have identified certain
consistent and recurring themes relating to the Indigenous management and uses
of wetland plants and habitats, as summarized in the pages that follow.

7.3 Results

Traditional Uses of Fire in Yosemite

Before elaborating specifically on the roles of wetland plant habitats within Native American land and resource management practices, we first summarize pertinent information relating to the use of fire. Again, the Native American use of fire for clearing vegetation has been widely reported for Yosemite Valley, even in the earliest written accounts. The Mariposa Battalion, for example, reported seeing "picket fires" ignited by Native people upon their arrival in 1851 – the first recorded entry of any non-Native people into Yosemite (Bunnell, 1880: 73). In 1861, H. Willis Baxley visited Yosemite Valley and reported the following: "A fire glow in the distance, and then the wavy line of burning grass, gave notice that the Indians were in the Valley clearing the ground, the more readily to obtain their winter supply of acorns and wild sweet potato root (huch-hau) [*Brodiaea* spp.]" (Baxley, 1865: 467). Other early accounts suggest that the goals of this burning, while principally focused on plant procurement, were diverse. Galen Clark reported on traditional management practices in the late-nineteenth century:

The Valley had then been exclusively under the care and management of the Indians, probably for many centuries. Their policy of management for their own protection and self-interests, as told by some of the survivors who were boys when the Valley was visited by Whites in 1851, was to annually start fires in the dry season of the year and let them spread over the whole Valley to kill young trees just sprouted and keep the forest groves open and clear of all underbrush, so as to have no obscure thickets for a hiding place, or an ambush for any invading hostile foes, and to have clear grounds for hunting and gathering acorns. When the forest did not thoroughly burn over the moist meadows, all the young willows and cottonwoods were pulled up by hand.

(Clark, 1894: 14)

This practice of fire management is widely reported to have helped prevent destructive wildfires in densely settled parts of the central Sierras, including Yosemite Valley, through the combustion of litter and the containment of conifer encroachment. A late nineteenth-century article in the *Daily Alta California* noted that:

[t]he Indians used fire and had none of the artificial means of confining it that we have, so that during their occupancy of the country the chance for fire, in use for cooking or signals, escaping and starting a disastrous conflagration was much greater than now, and it may seem singular that the forests were not all destroyed before the whites came. They were not, because the Indians used fire to make the forests fireproof . . . . He knew that if leaves and fallen wood were permitted to accumulate year after year they would finally form such a supply of fuel that when it was fired it would destroy everything inflammable. Therefore he carefully prevented such an accumulation by burning it every year.

(Daily Alta California, 1889: 4)

Likewise, in a report to the Secretary of the Interior on the park's fire suppression policies, Superintendent J. W. Zevely (1898: 1057) reported that, "prior to the inauguration of the present policy, fires occurred almost every year in all parts of the forest – in fact, they were frequently set by the Indians, but there was so little accumulation on the ground that they were in a great measure harmless, and did not in any sense retard the growth of the forest."

Similarly, the earliest accounts of Yosemite tribes by professional anthropologists depict the management of vegetation through fire and other methods as fundamental elements of traditional land use. Barrett and Gifford reported the extensive use of fire to clear vegetation, a process important in the enhancement and procurement of both plant and animal resources (Barrett and Gifford, 1933: 179, 182). Reviewing data compiled by Omer C. Stewart, Reynolds (1959: 139) listed the various reasons for which at least thirty-five California tribes utilized fire. The purposes of fire, in descending order of reported importance, included increasing the yield of desired seeds (including acorns), driving or creating habitat for game, enhancing the growth of vegetable foods and wild tobacco, improving access to food plants, improving visibility and open trails, and enhancing protection from dangerous species such as bears.

The work of M. K. Anderson (1988, 1990, 1993a, b) with the Southern Sierra Miwuk and that of Lewis (1973) addressing numerous California tribes confirmed these findings. Their work acknowledged that burning occurred at multiple scales. As Anderson (1993a: 18) observed, "The extent of burning varies from lighting individual plants on fire, to burning 'patches' of the plant, to burning whole hillsides." In addition, their work acknowledged other reasons for ground-clearing fires, as mentioned by tribal Knowledge Keeper including minimizing future fire potential and the reduction of pest insects, rodents, and oak mistletoe. Fire, they suggest, both clears away competing vegetation and temporarily imparts nutrients into the soil, fostering both the increased size and the quantity of culturally preferred plants. Such fires are also widely reported to improve hunting, creating productive and geographically predictable foraging places for such game as elk and deer and habitats for many other game and nongame animal species. Anderson also documented a number of plant management techniques used to enhance the output of culturally preferred species among Southern Sierra Miwuk Nation members. These methods principally involve pruning and coppicing, such as the regular pruning of willow (*Salix* spp.), redbud (*Cercis occidentalis*), and other materials used in basketry and other traditional crafts to enhance the output of long, straight shoots. Furthermore, Anderson documented the maintenance of plant communities through forms of tending and selective harvesting, such as selective digging of *Brodiaea* (Indian potato) bulbs (i.e., leaving smaller bulbs to grow for future harvests and possible transplanting) and the continuous revisiting of *Brodiaea* patches to turn the soil and remove competing vegetation. Taken together, these studies suggest that Native managers enhanced between 200 and 250 culturally important species of plants through the use of fire and other techniques.

Pollen studies conducted in Yosemite Valley in the 1990s highlighted the role of anthropogenic fire in the formation and maintenance of the valley's plant communities. Anderson and Carpenter (1991) identified a major change in the pollen assemblage, indicating that a significant shift in vegetation occurred approximately 700 years ago. Although climatic cooling and increased precipitation in that period should have favored an increase in conifers, the opposite happened. They noted "a decline in conifers and an increase in oak. Peaks in both charcoal, pollen, and sediment influx occur contemporaneously, indicating a period of erosion" (Anderson and Carpenter, 1991: 7). The authors attribute this change to a rapid increase in large-scale fires, probably of human origin. Large-scale fires would have had to occur regularly to maintain the oak woodlands indicated within the pollen record. However, between 1930 and 2003, no lightning-ignited fires occurred in Yosemite Valley, again providing strong support for the assertion that the relatively frequent fire intervals predating this period were the result of human intervention (NPS, 2002).

Plant communities resulting from traditional patterns of human management appear to have represented "mosaics" rather than the increasingly "monocultural" stands we see today (cf. Lewis and Ferguson, 1988). Meadow environments, riparian forests and wetlands, oak woodlands, and conifer forests of differing ages ensured a high level of biological diversity within a relatively small area on the Yosemite Valley floor. With this biological diversity came a diversity of resource procurement options for the valley's inhabitants. While Yosemite Valley and other nearby valleys were burned regularly, the surrounding highlands exhibited fire regimes more influenced by natural ignition sources, notably lightning, resulting in dense forests of conifers such as red fir (*Abies magnifica*) and lodgepole pine (*Pinus contorta*) (van Wagtendonk, 1986). Wildfire frequency on the valley floor arguably lagged behind natural ignition rates characteristic of these conifer-dominated portions of the Yosemite landscape. As such, these valleys with their mosaic of habitats became managed "islands" amidst surrounding conifer forests, which, while still utilized, provided resource-gathering opportunities that were much less enticing.

Most of the species identified as ethnobotanically significant in the current study are fire adapted in some way. Many of the tree and shrub species possess thick bark that is fire resistant in adult phases, such as species like incense cedar (*Calocedrus decurrens*) or gray pine (*Pinus sabiniana*). Repeat burning eliminates juveniles of these species while allowing the survival of adults (Show and Kotok, 1924). Fire suppression during the last century and a half has encouraged the proliferation of some tree and shrub species that would have been selected against, during their juvenile phases, if burning had been practiced. This is especially true in Yosemite Valley, where increasing conifer dominance has transformed the vegetation of the valley floor (Reynolds, 1959; Gibbens and Heady, 1964). Many of the plant species identified in the course of the interviews and literature review also exhibit rapid seed dispersal or germination following light ground fires. Some of these possess serotinous cones or seed pods that only open after fire scarification or have seeds that germinate well only on freshly exposed mineral soil. For example, manzanita (*Arctostaphylos* spp.) seeds germinate at much higher frequencies following the burning of the seed coating by light fire, and some manzanita species are largely dependent on fire for their reproduction, as they bear seeds that lie dormant in the soil until fire scarification occurs (Keeley, 1977). A number of the species identified in this study send out additional shoots or branches after being exposed to light to moderate fire. In some species, this process serves to enhance the abundance of culturally preferred plant parts. For example, some willow species exhibit rapid post-fire sprouting from the rootstock, resulting in the production of long shoots that are useful in basketry and other traditional crafts. In the case of black oak, many of these additional branches eventually bear acorns, often resulting in a denser concentration of acorn-bearing branches on specific trees, with acorns being a primary staple food for the Miwuk and other Native Californians (Lewis, 1973; Plumb and Gomez, 1983; Knowledge Keepers).

Evidence of Wetland Plant Use and Management

The uses and management of wetland plant habitats have received comparatively little attention in the available published record. Yet, from the beginnings of the written record, observers have commented on the biological richness of Yosemite's wetland and riparian habitats. In the first guidebook for Yosemite Valley, J. D. Whitney commented on the botanical variegation and diversity in these areas:

Along the banks of the river and over the adjacent rather swampy meadows, we find a somewhat varied vegetation, according to the locality . . . . Where the Valley widens out and the river banks become lower, so that the sloughs and swamps are formed, the Balm of Gilead poplar (*Populus balsamifera*) comes in . . . . The meadows are swampy, with deep peaty soil; their vegetation consists chiefly of carices or sedges and a few coarse grasses.

(Whitney, 1869: 73)

These observations reflect widespread, variegated wetlands on a hydrologically active valley floor. Historically, seasonal flooding caused localized inundation of the Merced River floodplain – an effect augmented by natural obstructions such as rock barriers, beaver dams, and logjams. Floods that inundated significant portions of the Yosemite Valley floor were not uncommon; indeed, tribal members have noted that floods coupled with inclement weather were among the reason for some families' descent from the valley to drier places like El Portal in the wintertime. Channel–floodplain connectivity and seasonal inundation of riparian and meadow environments restricted conifer growth near the river channel. Moreover, this fostered a diversity and abundance of riparian wetlands no longer seen in the study area. Riparian wetlands exhibited vertical biotic zonation associated with changes in frequency, depth, and velocity of inundation, but also considerable lateral differentiation, apparently influenced by varying flow velocities and sediment sorting, reflecting the gross stream morphology. Intermittent side channels, gravel bars, sand bars, and other riparian depositional features each possessed distinctive biotas that have been important in the traditional diet and pharmacopoeia.

Accordingly, in addition to fire-adapted plants, a large proportion of the gathered species identified in the literature review and a clear majority of the species reported by living Tribal Knowledge Keepers in the project interviews are wetland species (see Tables 7.2 and 7.3). To illuminate these findings, we assessed the full list of 101 plant species reported to us as "culturally significant in living memory" among tribal members (Deur, 2007; Bloom and Deur, 2022), referring to the wetland indicator status of these ethnobotanically significant species. US federal agencies identify specific "wetland indicator" plant species to guide the formal designation of wetlands, using a probabilistic assessment of the prevalence of particular plants within wetlands of a particular region. As defined by the US Fish and Wildlife Service (USFWS) and the US Army Corps of Engineers (USACE), all plants that may be diagnostic of the presence or absence of wetlands are assigned to five principal categories, as defined in Table 7.1: obligate wetland, facultative wetland, facultative, facultative upland, and upland (USFWS, 1988, 1997; USACE, 2021). The USFWS augments these indicator codes with a + or − symbol, the former indicating that a plant is typically found in wetter environments within its identified range and the latter indicating that a plant is typically found in the drier environments of the identified range. In addition, the USFWS identifies some species as "NI," meaning that their status as potential indicators has not been established.

Researchers assess the species compositions of plant communities on the basis of the criteria outlined in Table 7.1 as part of wetlands, including the delineation of "jurisdictional wetlands" holding special legal status. Jurisdictional wetlands are areas possessing the diagnostic conditions of wetlands and therefore fall under the jurisdiction of the USACE and other federal, state, and local agencies with mandates to regulate activities that might affect wetlands. Jurisdictional wetlands, by definition, must have a proportional dominance of species determined to be obligate wetland, facultative wetland, and/or facultative species, and must be cataloged accordingly in keyed species lists produced by the USFWS (USFWS, 1988, 1997; USACE, 2021).

A significant portion of the 101 vascular plants identified by tribal members in interviews with Deur (2007) are wetland species commonly occurring in riparian, freshwater pond, and wet meadow environments. Of the 62 species identified as potential indicators of wetland conditions in California by the USFWS's *National List of Vascular Plant Species that Occur in Wetlands*, a total of 44 (71 percent) are wetland species (i.e., identified as facultative, facultative wetland, or wetland obligate species). Table 7.2 reflects a selection of the wetland species identified by tribal members in interviews with Deur (2007), totaling one-quarter of the identified wetland species. (Note, the tables in this chapter merely present a few

Table 7.1 *USFWS classification of wetland indicator plant species*

Code	Habitat indicated	Characteristics
OBL	Obligate wetland	Almost always (estimated probability: 99 percent) under natural conditions in wetlands
FACW	Facultative wetland	Usually occurs in wetlands (estimated probability: 67–99 percent), but occasionally found in non-wetlands
FAC	Facultative	Equally likely to occur in wetlands or non-wetlands (estimated probability: 34–66 percent)
FACU	Facultative upland	Usually occurs in non-wetlands (estimated probability: 67–99 percent), but occasionally found on wetlands (estimated prob ability: 1–33 percent)
UPL	Upland	Occurs in wetlands in another region, but occurs almost always (estimated probability: 99 percent) under natural conditions in non-wetlands in the regions specified

Table 7.2 *Selected plants used in tribal Knowledge Keepers' living memory: USFWS wetland indicator plant species*

Plant species	USFWS wetland indicator status
Red willow (*Salix laevigata*)	FACW+
Wormwood (*Artemisia douglasiana*)	FAC+
Wild grape (*Vitis californica*)	FACW
Field mint (*Mentha arvensis*)	FACW
Rough sedge (*Carex senta*)	OBL
Showy milkweed (*Asclepias speciosa*)	FAC
Deer grass (*Muhlenbergia rigens*)	FACW
Flowering dogwood (*Cornus nuttallii*)	FACW
Cattail (*Typha latifolia*)	OBL
Hemp dogbane (*Apocynum cannabinum*)	FAC
Wild ginger (*Asarum lemmonii*)	OBL

Table 7.3 *Selected plants used in tribal Knowledge Keepers' living memory:*
USFWS upland indicator plant species

Plant species	USFWS wetland indicator status
White oak (*Quercus lobata*)	FACU
Hazelnut (*Corylus cornuta* var. *californica*)	FACU
Gumweed (*Grindelia nana*)	FACU
Harvest brodiaea (*Brodiaea elegans*)	FACU
Blue elderberry (*Sambucus mexicana*)	FACU
Sourberry (*Rhus trilobata*)	FACU

key examples, chosen from among the better-known species in each category; the bulk of each plant list is protected out of respect for the security and intellectual property of Native plant users.) Table 7.2 identifies the wetland indicator status of these culturally preferred species specifically from the California column of the USFWS's *National List of Vascular Plant Species that Occur in Wetlands: 1996 National Summary* (USFWS, 1997).

Of the forty-four wetland plant species identified in interviews, only eighteen of the species identified by living tribal Knowledge Keepers (or roughly 29 percent of the diagnostic species identified) are classified as being facultative upland or upland species in California by the USFWS. A third of those upland species are listed in Table 7.3 as a sample of the larger list. Still, it is important to note that some of the most important plants within tribal gathering traditions are upland species – such as oaks, hazel, elderberry, gumweed, harvest brodiaea, and sourberry. The species list in Table 7.3 identifies the wetland indicator status of culturally preferred species from the California column of the USFWS's *National List of Vascular Plant Species that Occur in Wetlands: 1996 National Summary* (USFWS, 1997).

Cumulatively, these findings suggest that a significant portion of plants gathered by tribal members of the last century are associated with wetland environments, particularly riparian wetlands and seasonally flooded wet meadows. This is reflected in a variety of statements by tribal members made in the course of this study and others. Riparian wetlands, in particular, were depicted as being "our supermarket, our drugstore . . . all those little flowers and grasses were our food and medicine. Everything we needed was there" (Deur, 2007: 51).

This is hinted at by past studies as well. Others have noted, "You can go down by the creeks, people don't realize it, they got their very own garden right there in the creeks" (Cramer, 1997: 3). Anderson also supports this characterization, describing the importance of these flood zones in the production of basketry materials:

Floods are regarded by Southern Sierra Miwuk basketmakers as an essential force in revitalizing the sandbar willow habitat and the sedge habitat. Plant populations are said to need periodic flooding from the river if they are to remain healthy. The best willow with the most flexible stems grow with their "feet" (roots) wet. Areas with sandbar willow which do not have active sand depositing yearly, such as abandoned flood plains where the river no longer travels, are undesirable gathering sites providing lower grade plant materials than the revitalized stands nearer the river.

(Anderson, 1988: 55)

Although underrepresented in the written literature, such descriptions of wetlands' significance – and of highly nuanced traditional ecological knowledge relating to the hydrologically dynamic wetlands of the valley floor – are consistent with the accounts of tribal members in the present study.

Intersection of Burning and Wetlands on the Valley Floor

Wetland habitats were not only significant as a source of plants within Yosemite tribes' ethnobotanical traditions, but also integral to their traditional plant management practices. In describing traditional burning practices, contemporary tribal members describe a system characterized by pyrodiversity – a complex and sophisticated burning process and schedule with fires occurring at different scales and in different wetland and wetland-margin environments over the course of a year. Burning patterns fluctuated, involving modifications to their timing, frequency, scale, and intensity in order to produce and maintain favored mosaics of habitats and species. Moreover, the accounts of living tribal members describe the historical use of riparian wetlands and saturated floodplains as natural firebreaks, which shifted in accordance with seasonal changes in valley floor hydrology, allowing for the containment of anthropogenic fires and the avoidance of large-scale wildfires.

To a limited extent, the complexity and sophistication of specialized burning in mosaic environments have been reflected in broader regional studies related to the Sierra Nevada, referencing practices of a range of ethnolinguistic groups (e.g., Anderson, 1988, 1993a; Anderson and Rosenthal, 2015). Particularly important to the current study, however, is the manner in which the specific characteristics of Yosemite Valley hydrology and the cultural preference for species reliant on both water and fire conditions were key in shaping fire regimes and the biodiversity of the valley.

Tribal members correctly note that traditional management of plant communities, including the use of fire, once played out on a far more complex and dynamic valley floor than one beholds at Yosemite today – the reasons for these changes, relating to park management over the last century, are addressed later in this chapter. Formerly, the valley floor was more intricately braided than it is today, with numerous riparian-influenced wetlands and ephemeral side-channels, and the water table was significantly higher. Then, and this happens to a more limited extent now, the water table of the valley would drop as the season dried, as did the water level of the Merced River – from peak runoff in late winter and early spring through the very dry season in late summer and early fall. This was reflected not only in a general drop in river levels and floodplain saturation, but also in the gradual downward movement of groundwater. This resulted in places with saturated soil or surface water drying out through the season, with wetlands and wetland margins exhibiting diminished soil saturation until precipitation rebounded in the winter. Following this pattern, all else being equal, vegetation tends to become dry at the high points first on alluvium and then lower in the alluvial terrain over the dry season until rains and snows resume. Most of the valley floor consists of alluvium, with areas of hydrologically integrated colluvium. Each year, as early-spring peak-stream flows dissipated and surface waters receded, the water table slowly dropped below the surface of the valley floor, even in wet meadows and ephemeral river channels. On the hummocky and irregular terrain shaped by this fluvial action, "islands" of relatively dry meadow emerged, surrounded by relatively saturated areas. As spring gave way to summer, these islands grew larger, connecting and ultimately encircling the low riparian areas, wet swales, and ponds that remained wet throughout the year (Heady and Zinke, 1978: 17; University of California Merced, n.d.; Yosemite National Park, n.d.) (Figure 7.1).

Inhabitants of the valley appreciated the burning opportunities fostered by these natural processes. Some Knowledge Keepers recalled oral traditions describing burning methods applied to this complex alluvial matrix of riparian and wet meadow habitats on the Yosemite Valley floor (Figure 7.2). Traditional burning, they suggest, began the moment that certain high places on the floodplain became sufficiently dry to ignite. This allowed fires to commence early in the season, but at small scales – the fires atop individual rises and hummocks, with drier conditions, being effectively contained by the saturated areas that encircled them. Anthropogenic fires then moved into lower elevation portions of the alluvium matrix through the rest of the season as conditions dried out (Lewis, 1973: 79). Accounts from living Elder align with certain written accounts too, such as that of Joaquin Miller, who reported in 1887 that:

[i]n the Spring . . . the old squaws began to look about for the little dry spots of headland and sunny valley, and as fast as dry spots appeared, they would be burned. In this way the fire was always the servant, never the master . . . . By this means, the Indians always kept their forests open, pure and fruitful, and conflagrations were unknown.

(Miller, quoted in Biswell, 1968: 46)

Figure 7.1 A detailed view of Yosemite Valley from Glacier Point by Carleton Watkins in 1866, a mere fifteen years after first Euro-American documentation of the valley's existence. The image hints at the extent of herbaceous plant communities and the complexity of the active Merced River floodplain prior to major hydrologic changes in the decades that followed – with riparian wetlands, ephemeral river channels, and freshly deposited sediments. By the time of this photo, early agricultural development was apparent on the valley floor. (Photograph by Carleton E. Watkins, from Gibbens (1962) [Figure 1A], courtesy of Yosemite National Park Archives.)

As the water table dropped into the early summer, traditional harvesters appear to have ignited a succession of fires in areas interdigitated between previously burned areas and the zone of enduring soil saturation. Here too, the fires remained contained; the spread of these fires into wet areas downslope was typically contained by poor ignition, while a lack of fuel in recently burned areas upslope restricted fire movement into those areas. In the late season, in summer or early fall, only lower elevation areas remained in the alluvium to burn – the drying plants of wetlands and intermittent river channels now being flammable enough by this time for these plant communities to be burned too, encouraging the growth particularly of herbaceous wetland species. Interviewee accounts suggest that Native

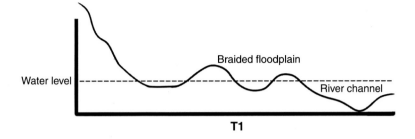

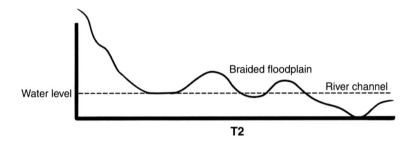

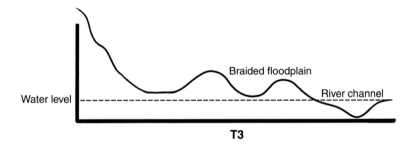

Figure 7.2 A highly idealized cross-section of the Yosemite Valley floor, with uneven alluvial deposits, swales, ephemeral channels, and other wetland environments. As described by tribal interviewees, traditionally burned areas were positioned to burn drying areas in response to the descending surface water and water table through the season, shown here diagrammatically on the Yosemite Valley floor, from time 1 (T1, spring) to time 2 (T3, summer) to time 3 (T3, late summer/fall).

burners repeated fires in this way until, cumulatively, much of the valley's alluvial zone had been burned. In places where soil moisture or other factors inhibited burning, woody species such as seedling pines (*Pinus* spp.), incense cedar (*Calocedrus decurrens*), Douglas fir (*Pseudotsuga menziesii*), or buckbrush

(*Ceanothus* spp.) were commonly pulled out by hand to maintain wet meadow environments (Clark, 1894: 14–15). The result was a mosaic of places burned at different times and a mosaic of plant communities of cultural significance – contributing to the apparent diversity of wetland species utilized by Yosemite tribes.

At the end of the year, after the gathering of staple acorn crops, as people prepared to descend from Yosemite Valley to communities at lower elevations east and west of the park, burning specialists ignited final large-scale clearing fires. While this probably included portions of the alluvial zone discussed here, much of this final burn was centered on the colluvial margins of the valley floor. By this point, moisture was no longer a factor in attempts to ignite other, often lower, parts of the valley floor. Interviewees attest that burning earlier in the season had sufficiently eliminated fuels at the surface, especially in the alluvial zone close to habitations and key plant-gathering areas, such that these late-season burns were easily controlled (Clark, 1894: 14; Reynolds, 1959: 134, 151–2). Additionally, the fires allowed a final clearing of unused plant materials from habitats such as the dampest wetland swales, which were finally dry enough to burn. Many of the oak prairies and isolated conifer stands of the valley margins were maintained by these fires too.

As such, the burning strategies for Yosemite Valley described by tribal Knowledge Keepers might be broadly categorized as "alluvial," centering on the riparian area and associated wetland and wet meadows on the valley floor, or "colluvial," centering on the areas lying on the valley edge outside the zone of principally riparian landforms. With a few exceptions (Lewis, 1973; Anderson, 1993a; Anderson and Rosenthal, 2015), the vast majority of the ethnographic literature tends to center almost exclusively on colluvial patterns, ignoring the complexity of the dynamic annual practices and their reliance upon hydrologic processes in wetlands, wetland margins, and active floodplains. Clearly, however, the alluvial patterns of traditional burning were no less significant in Yosemite Valley and suggest highly nuanced understandings and methods for traditional fire management among Native harvesters historically.

Together, as tribal members attest, the combined burning strategies enhanced the productivity and diversity of culturally preferred species in the following year. By practicing these management strategies every year, interviewees attest, Yosemite's precontact inhabitants maintained mosaics of culturally preferred species that allowed them to meet their needs for food, medicine, housing, implements, and basketry, among many other needs. This sophisticated and dynamic system allowed each season's fires to be contained by adjusting to moisture availability and incrementally burning down into wetter areas as conditions allowed. This is key to understanding many points: the variegated vegetation patterns described by nineteenth-century writers such as Whitney, the richness and wetland orientation of documented ethnobotanical practices, the apparent contradictions we see in many written accounts of burning scale and timing, and the many ways that the extirpation of fire *and* water affected Native use and management of the landscape. As one tribal member noted, "it is not just about using fire . . . it is about how we used fire and water." Through this manner of burning, harvesters could restrict each fire from engulfing villages or other inhabited places. Burning in such a constrained and densely populated valley was a delicate matter indeed. Anthropogenic burning of alluvial and wetland environments, within the dynamic Merced River floodplain, was one of the key ingredients that allowed Yosemite Valley tribal communities to inhabit the valley successfully for generations.

Fire Suppression and Hydrologic Modifications: Interconnected Factors in the Loss of Anthropogenic Plant Habitats

Today, a combination of state and federal fire suppression policies and changed hydrologic conditions have severely disrupted this complex system of traditional management. In turn, this has negatively impacted anthropogenic landscapes, plant communities, available plant and animal resources, and the endurance of many cultural practices. The suppression of fire, and especially the encroachment of dense conifer forests throughout much of Yosemite Valley, has resulted in reduced genetic diversity and a reduced range of resource-gathering opportunities. Fire suppression policies arrived with early park management – within only a few years of President Abraham Lincoln signing the Yosemite Valley Grant Act, making Yosemite a park on June 30, 1864 – and from that time forward most forms of traditional management were discouraged or prohibited, burning in particular. As early as the 1880s, park managers recognized that the meadow environments of Yosemite were disappearing under rapidly encroaching trees and brush – principally due to the suppression of both anthropogenic and natural fires. In a report to the Commissioners, dated May 20, 1882, William H. Hall reported:

The area of meadow is decreasing, while young thickets of forest or shrub growth are springing up instead. Members of your Board have observed this change; it is very marked, and it may be regarded as in a degree alarming, sufficiently so, at least, to prompt measures calculated to check it. The cause is alleged to be the abolition of the old practice of burning off the thickets, which practice formerly made new clearings almost every year for grass growth.
(Hall, 1882: 15–16)

The Secretary of the Commission, M. C. Briggs, reported in a letter dated December 18, 1882, that "[w]hile the Indians held possession, the annual fires kept the whole floor of the valley free from underbrush, leaving only the majestic oaks and pines to adorn the most beautiful of parks. In this one respect protection has worked destruction" (Briggs, 1882: 10–11). These observations resulted in modest changes in park vegetation management, although forest encroachment on the Yosemite Valley floor continued relatively unchecked. By the 1930s, the transformation of Yosemite Valley plant communities, especially the encroachment of dense conifer forests on meadows and oak groves, became the focus of modest press attention (Crowe, 1931; Taylor, 1931). The 1940s brought renewed attention to the issue within the NPS, particularly through the work of Emil Ernst (1943, 1949, 1961). In the years that followed, the causes and effects of fire suppression within Yosemite Valley gained attention from a broadening range of applied and academic researchers, fueled significantly by concerns regarding the aesthetic impacts of these changes (Reynolds, 1959; Gibbens and Heady, 1964; Heady and Zinke, 1978). Today, tribal Knowledge Keepers lament: "They [the plants] are all disappearing. Everything is overgrown, all those places we gathered plants are all covered with pines. They can't grow under brush and pine needles" (in Deur, 2007: 46; cf. Turner, 1991) (Figure 7.3).

The loss of fire in Yosemite Valley, however, is only half the story. The loss of culturally significant wetland environments in the last century has had interrelated and equally damaging effects, including a reduction in the quantity and quality of

Figure 7.3 An aerial view of Yosemite Valley from Glacier Point by Ralph Anderson in 1943. By the early twentieth century, conifer encroachment on the Merced River floodplain and associated wetlands was widespread, reflecting not only fire suppression policies of the NPS, but also significant engineered changes to valley hydrology. (Photograph by Ralph H. Anderson, from Gibbens (1962) [Figure 1B], courtesy of Yosemite National Park Archives.)

many culturally significant wetland species. Specific historical impacts included park demolition of two rock barriers that historically impeded surface runoff, contributing to wetland hydrology on the valley floor, namely the El Capitan moraine and the rock obstruction just below Mirror Lake on Tenaya Creek. The El Capitan moraine formed "a nearly straight dam across the valley just below El Capitan meadow" that impeded ancient Lake Yosemite across the valley floor during the early Holocene (Matthes, 1930). This moraine marked the downstream end of Yosemite Valley's 5.5-mile-long "central chamber," the valley's largest hydrologic unit, which includes the upper valley as far upstream as Tenaya Creek Pass. As the Pleistocene glaciers retreated, alluvial and lacustrine deposits accumulated behind this moraine, gradually producing the level valley floor in the millennia that followed. Galen Clark noted that, in the 1870s, the Merced River channel crossed the moraine but that "[t]he river channel at this place was filled with large boulders, which greatly obstructed the free outflow of the flood waters in the spring, causing extensive overflows of low meadow land above, greatly interfering with travel, especially to Yosemite Falls and Mirror Lake" (Clark, [1907] 1927: 15). The damming effect of this moraine appears to have not only kept river flows impeded and the water table "perched" through this central chamber of the valley, but also contributed materially to the complete flooding of the valley floor in high runoff years – a phenomenon noted in 1864, 1867, and 1871. Thus, the meadows of the valley floor visible in the photographic collections of C. E. Watkins of the 1860s appear to be differentiated by oxbows, sloughs, scour channels, natural levees, and fresh alluvial deposits from flooding and elevated water tables upstream from the moraine – geomorphic features that are much reduced and all but invisible to contemporary park visitors (Watkins, n.d.).

Not only were floods a threat to growing park infrastructure and proposed agricultural operations in the valley, but the perennially muddy conditions of the valley floor were a source of complaints among the rising tide of affluent park visitors. In 1879, in an effort to reduce flooding and lower the water table on the valley floor, Galen Clark demolished the El Capitan moraine at the point where the Merced River transected it. Almost instantly, this opened new land for agricultural use in the central valley and expanded the viable window of visitor access into the springtime, a season when floods and muddy conditions had previously been an obstacle. Using explosives, he eliminated the large boulders and leveled the remaining rock fragments, dropping the natural dam by an estimated four to five feet. The water table in the valley upstream from the moraine dropped proportionately (Clark, [1907] 1927; Milestone, 1978).

Additional park development soon compounded these impacts, such as the dredging and revetment of the Merced River under early NPS management (including the use of dynamite, in some cases, to produce visually appealing flat waters that reflect the mountain scenery), the dredging of sand from Mirror Lake, and the filling or development of wetland areas to foster park development. Beyond this, the park witnessed concentrated visitor activity along the riparian zone, gravel mining, and confinement of the river by bridge crossings. Removal of large woody debris to reduce the threat of floods to park facilities resulted in channel widening and disconnection of the river from its once-active floodplain. Flooding, which once occurred annually, today occurs only during significantly larger river flows, and much less frequently, as a result of these many changes (Booth *et al.*, 2020).

Combined with the impacts of Clark's blasting, these events produced an approximately 1- to 1.5-meter drop in the mean water table within many portions of Yosemite Valley, eliminating the hydrology necessary to maintain a number of historical wet meadows and swales. As tribal members note, as wetlands went dry in many parts of the valley, wetland plants disappeared, while declining soil saturation invited conifer encroachment into former wetland environments that was well beyond what would have occurred solely due to fire suppression (Ernst, 1943: 55). Only with the dewatering of certain wetlands were conifers able to expand their range across the valley floor. The combination of both anthropogenic fire and soil saturation had restricted conifer encroachment into wet meadow environments historically. Accordingly, most of the dense forests in the valley are said to date from approximately 120 years ago, suggesting a decade or two of succession in the meadows before the conifer forest became established. Most of those areas arguably could not have supported trees before the dewatering of the meadows. Even without factoring in the effects of global climate change, Reynolds explains that Yosemite's meadows "once remained wetter for longer periods than they do today ... water tables in meadowland fall more rapidly and farther today after the spring runoff period than they did during aboriginal times" (Reynolds, 1959: 57). Wetlands and alluvial complexity have been compromised, but the traditions of burning linked to these dynamic landscapes are undermined: The scale and timing of flooding no longer foster staged wetland burning on a historical scale. Visitor traffic also increased in these dewatered areas – diminishing remnant plant communities due to trampling, soil compaction, the inadvertent introduction of invasive plant species, and other impacts – as formerly wet lands became dry and navigable, and visitor numbers skyrocketed in the decades that followed.

While, over the course of the twentieth century, NPS policy exhibited an increasing reluctance to modify the valley's hydrology, a policy of reducing flooding and groundwater levels to protect visitor structures and visitor access continued into the 1970s. Only in the early 1970s did regular dredging at Mirror Lake and other scenic waterways cease; by this time, over 14,500 lineal feet of stream bank and riverbank had been lined with revetments, principally of riprap, in Yosemite Valley (Milestone,

1978). Although provisions of the Clean Water Act effectively ended direct wetland impacts, and although the park has taken an active role in wetland restoration in recent decades, a concern with flood control has persisted in attenuated form (Milestone, 1978; Madej *et al.*, 1991; Booth *et al.*, 2020).

Another cause for changing Yosemite Valley water tables was fire suppression itself. Reynolds noted that the increase of second-story vegetation under old-growth forest stands – which the cessation of anthropogenic fire precipitated – intercepted rainwater that would otherwise have percolated into the ground, causing it to be transpired or evaporated into the atmosphere. He describes this as "probably the most important single factor which has contributed to the desiccation of most central Sierra meadows during the past one hundred years" (Reynolds, 1959: 57). Another impact of fire suppression was the replacement of shallow-rooted herbaceous plants with deeper rooted species. This invasion of woody species with deeper roots decreased the quantity of groundwater storage in the meadows (Reynolds, 1959: 58). Therefore, in some parts of the valley, conifer encroachment and hydrologic change have been mutually reinforcing components in a positive feedback loop (Figure 7.4).

Recently, park researchers have established that meadows that dried out due to the historical changes in valley hydrology are also more prone to invasion by non-native

Figure 7.4 A view of Yosemite Valley from Glacier Point in the twenty-first century showing a nearly complete encroachment of conifer trees across undeveloped portions of the Yosemite Valley floodplain. (Image from Monument-Fruede [2013], courtesy of Wikimedia Commons.)

plant species. These places also experience more ground disturbance and predation on California black oak acorns and seedlings by such species as mule deer, pocket gophers, several birds including the acorn woodpecker, and insects such as filbert worms and filbert weevils than was apparently the case historically or when compared with the same variables in the valley's remnant wetlands (Yosemite National Park, 2011: 45). In turn, the presence of small mammals in these dewatered areas has been demonstrated to correlate with poor germination and survival of certain native, culturally important plant species, including California black oak, following predation (Yosemite National Park, 2011: 48, 68, 72; M. Downer, 2017, personal communication).

As a result of these many impacts on culturally significant plant habitats, tribal members suggest that many of the wetland species used historically are no longer used today (and are therefore not included in the tabulations presented here or in Table 7.2). In these areas, the plants are simply no longer available for use. In other contexts, they are so limited in their quantity and quality that they are no longer used. For example, Knowledge Keepers indicated that the best fern (especially *Pteridium aquilinum*) and sedge (*Carex* spp.) roots for basketry come from alluvial sand deposits near the upland–wetland margin. These species are adapted to rapidly colonizing fresh alluvial deposits with lateral rhizomes and do so annually in the riparian zone of the Merced River and its tributaries. Certain places, known for their abundant annual sand deposits, were historically important root-gathering areas for this purpose, with gathering places sometimes shifting from year to year in response to changing patterns of deposition. Today, traditional basket makers find very few active sand deposits and they describe the roots found in rocky soil matrices to be gnarled, short, difficult to extract, and difficult to use for traditional basket making. Some suggest that, as a result of that environmental change, the quality of modern baskets made from Yosemite Valley materials lags behind that of historical baskets. The cessation of the sophisti-cated annual cycle of burning practiced by tribal managers in these hydrologically dynamic places has adversely affected a number of traditionally gathered species such as sedges, rushes, ferns, cattails, and deer grass. For instance, tribal members observed that many desired wetland plants were producing desired leaves, shoots, flowers, or other parts only in spring, and that burning those areas in spring would actually have destroyed the plants and plant parts, making them unusable to Native harvesters. This, too, created strong incentives to burn incrementally, place by place, following the hydrology and creating managed mosaics across the Yosemite Valley floor.

7.4 Discussion and Conclusions

The Native peoples of Yosemite Valley and their caretaking practices have been the focus of generations of scholarly writing and firsthand accounts, and the tribes of

Yosemite increasingly tell their own story (Traditionally Associated Tribes of Yosemite, 2021). In spite of this, there are still aspects of Yosemite's heritage that have escaped the attention of the wider world. Among these is the centrality of wetland plants in the ethnobotanical traditions of Yosemite's Native peoples and the sophisticated methods employed by these peoples historically to modify wetland environments to enhance the output of culturally preferred wetland species. We contend that the traditional resource and environmental management of Yosemite tribes cannot be fully understood without considering this wetland specialization. We also anticipate that the practices described by Yosemite tribes, such as sequential burning of floodplain environments, may prove to be widespread upon more careful investigation of other Native traditions and landscapes across California and elsewhere in the Americas.

In Yosemite Valley, wetland species – sedges, rushes, milkweed, cattail, wormwood, and deer grass, to name but a few (see Table 7.2) – represented a large proportion of the plants utilized historically, and remain among the most valued and utilized species among modern tribes of the Yosemite region today. Of some of the species identified as being actively harvested within the living memory of tribal Knowledge Keepers in past ethnographic studies (Deur, 2007; Bloom and Deur, 2022), some 71 percent of the diagnosable sample qualified as wetland species, that is, were designated as wetland obligates, facultative wetland, and facultative species, using the US delineations of wetland areas and species (USFWS, 1988, 1997). Most are herbaceous species, well suited to growing in riparian wetlands and wet meadows situated in highly dynamic floodplains and alluvial soils, and also tolerant of regular burning. While certain upland species hold elevated significance in tribal traditions, they represented only some 29 percent of the species mentioned by living Knowledge Keepers. While these proportions are striking, they are also consistent with the historical condition of the Yosemite Valley floor, with formerly seasonally flooded riparian areas and other wetlands far more extensive than what can be observed in the valley today. In light of the prominence of wetlands and alluvial deposits throughout the valley prior to the major hydrologic changes engineered by the park in the late nineteenth and twentieth centuries, this nuanced focus on wetland habitats and environments is to be expected.

While anthropogenic fire clearly has been important in shaping the vegetation of Yosemite Valley, and this point has been celebrated in past studies (Anderson, 1988, 1990, 1993a, b), we find that many of Yosemite's culturally significant plants and plant habitats rely on a specific interplay of both fire and water. The strategies of traditional burning described by tribal members of the last generation provide a tantalizing glimpse into how harvesters navigated this damp and dynamic environment. Wetlands, rivers, and saturated floodplains served as important firebreaks within the larger fire-managed landscape. Early in the burning season, traditional fire managers focused on the drier, isolated high points on the valley floor where both surface water and groundwater had subsided, allowing for fires that were at once geographically focused and contained. As flooding ceased and the water table dropped through the summer season, these managers burned intermediate places that had dried in the interim – burning with a footprint that was significantly contained downslope by saturated areas and contained upslope by the absence of fuels in previously burned sites. Late in the dry season, with the water table dropping further and the higher places in the hummocky alluvial terrain cleared of fuel, these managers could burn remaining swales and wetlands. At each step, the fires remained contained and focused on a different range of species characteristic of the wetlands and wetland margins of Yosemite Valley. Tribal members attest that the abundance of culturally significant species was enhanced as a result: that the cattail and willow, wormwood and sedge, deer grass and milkweed, and countless other species that sustained their ancestors for food, medicine, materials, and more were made more predictably abundant in places known and accessible to Native peoples (Deur, 2007; Bloom and Deur, 2022). While these practices are not described in such intricate detail within the available archival record, there are tantalizing if fragmentary references in that archival record that cohere with these practices. In part, this may reflect the fact that these practices were clearly being suppressed and marginalized by park management prior to the arrival of professional anthropologists and the emergence of detailed written accounts of Yosemite tribes in the first half of the twentieth century.

In diverse and interconnected ways, colonization transformed not only the record of Yosemite's tribes, but also the landscapes, the plant communities, and the many cultural practices linked to the unique environments of the Yosemite Valley floor (Spence 1996). While fire suppression has been widely cited as a source of riveting cultural and environmental change, this is a necessary but insufficient explanation for the changes witnessed in the last century and a half. Instead, the expulsion of Native harvesters and managers had numerous measurable effects on the landscape. And yet, that is not the full story either; instead, tribal members point toward the combined effects of fire suppression and hydrologic changes occurring in tandem – the simultaneous removal of both fire and water across large swathes of the Yosemite landscape. With the demolition of the El Capitan moraine, the hardening of the Merced riverbank, and many other changes meant to improve visitor access to the new park, park managers effectively reduced river–floodplain connectivity and prompted changes in soil moisture and a drop in water table levels. This resulted in the rapid disappearance of wetland and riparian margin plant communities across the Yosemite Valley floor. While it is true that, as some authors have noted, fire suppression allowed conifers to encroach on historical meadows, we concur with the observation of living tribal Elders: that it was this rapid drop in the water table coupled with fire suppression that allowed conifers to occupy former floodplains, riparian wetlands, and many other habitats formerly utilized and managed by Native peoples. Together, these changes in land management eliminated opportunities for Native harvesting of a diverse range of wetland species and precluded the traditional management of habitats according to seasonal changes in the water table. In turn, these monumental changes in land management adversely affected a constellation of traditional cultural practices linked to the use of plant foods, medicines, and materials, as well as the many educational, cultural, and spiritual uses of plants traditionally harvested in the wetlands and wetland margins of Yosemite Valley.

While tribal members continue to gather plants at Yosemite today, the habitats that sustain these plants have been much eroded and degraded, undermining both the biological and the cultural integrity of Yosemite Valley while also raising the risk of wildfire on the valley floor and beyond. The implications of these findings for environmental restoration of culturally preferred habitats are compelling – at Yosemite and beyond. Prescribed burning programs are warranted in suitable settings. Yet, these programs require a consideration of seasonality, scale, and hydrologic conditions, informed by the recollections of tribal members and the teachings of ancestors who engaged and managed the landscape with fire across deep time – an approach that has been termed "ecocultural restoration" or "ethnoecological restoration" (Senos *et al.*, 2006).

These changes are underway at Yosemite, albeit slowly. These changes are also being made delicately in a valley that is now visible to worldwide audiences and contains buildings and infrastructure in places once flooded and burned annually. In recent years, Yosemite hydrologists have implemented restoration efforts including the removal of riprap, revegetating the riparian zone with native species, and increasing in-channel roughness, all to restore river structure and floodplain connectivity (Booth *et al.*, 2020: 10–11; Fong, 2020: 10) (Figure 7.5). In a few places, wetland restoration has also received separate and focused attention. Tribal members have suggested the water flow would have to be impeded again – dammed at El Capitan moraine or nearby – to allow a higher water table and the reestablishment of wetland environments throughout Yosemite Valley. In light of the potential effects on park infrastructure and visitors, the NPS is unlikely to embrace valley-wide restoration of floodplain connectivity and wetland hydrology. However, it continues to consider restoration activities interdigitated with park facilities, such as riparian wetland restoration involving limited grading to reconstruct swales and ephemeral river channels.

The NPS has also implemented prescribed burning in response to deteriorating vegetation communities and vistas and to the increased risk of destructive wildfire (Tait, 1999; Vasquez, 2019: 40–4) (Figure 7.6). These new burns often mimic conditions of traditional burning and are conducted with varying levels of tribal

Figure 7.5 Although dredging and the destruction of the El Capitan moraine have reduced flood heights and frequency throughout the valley, occasional floods still cover the Merced River floodplain, including lands traditionally managed by Native harvesters through sequential fires, including this major flood event in May 2018. (Photograph by R. Bloom.)

involvement. Native participants suggest that, in creating modern analog fires, the NPS has misinterpreted the timing and scale of their traditional burning practices, often with detrimental impacts on preferred species. They state that traditional burning was highly nuanced and dynamic, with careful attention to the interrelation between timing, scale, and habitat – including paying much attention to the effects of burning on the plants of wetlands and wetland margins. They have been critical of the NPS burning to date, describing these introduced methods as taking a "one-size-fits-all" approach, by burning most areas all at once, and avoiding some of the traditional burning times through the summer due to effects on views and visitors – often with unintentional negative consequences for culturally preferred species. Clearly, continued discussion and collaboration will be needed to find common ground and to develop ways to carry out traditional fire management in a landscape so transformed by the extirpation of anthropogenic fire, culturally important

Figure 7.6 NPS fire crews now actively burn the floodplain as part of "prescribed fire" programs that are meant to reduce the wildfire risk and impede conifer encroachment in the wake of a century and a half of park-mandated fire suppression and hydrologic change. Native burning traditions have served as inspiration for modern fire management, and the park sometimes consults with tribal experts in prescribed fire planning – yet, the complex interplay between floodplain hydrology, groundwater, and fire remains a significantly unexplored dimension in park land management. (Photograph by Kelly Martin [NPS], courtesy of Yosemite National Park Archives.)

wetlands, and the Native harvesters and managers who have long called this place home.

While these lessons are linked to the unique peoples, geographies, and habitats of Yosemite, they have broad implications for the wider world. Native harvesters clearly managed many montane valleys and other hydrologically active interior valleys across western North America (Lewis, 1973; Blackburn and Anderson, 1993; Boyd, 1999). While our understandings of traditional burning practices are thin in places, in some contexts Native knowledge holders have been able to recall such practices with precision – often revealing details that will be key if burning will continue to serve as a mechanism for landscape-scale plant habitat management into the future. Researchers are well advised to seek evidence of wetland and wetland margin burning, of nuanced methods for burning in wet and hydrologically active environments, and of the pyrodiversity and mosaic environments fostered by such practices when sustained by generations of traditional land managers. Such well-watered places are numerous and have idiosyncratic properties that require specialized knowledge to manage them successfully, as the traditional plant managers of Yosemite have known so well. With the guidance of Native harvesters and knowledge holders, we might yet recover aspects of these traditional practices and foster the restoration of biologically and culturally significant habitats. We might, as the NPS is compelled to do by the terms of its Organic Act (16 U.S.C. / 2–4), help preserve the natural and cultural legacies of these special places and – through the continuation of active traditional management – "leave them unimpaired for the enjoyment of future generations."

References

Anderson, M. K. (1988). Southern Sierra Miwok plant resource use and management of the Yosemite region: a study of the biological and cultural bases for plant gathering, field horticulture, and anthropogenic impacts on Sierra vegetation. Unpublished MA thesis, Department of Wildland Resource Science, University of California, Berkeley.

Anderson, M. K. (1990). Yosemite's native plants and the Southern Sierra Miwok. *California Indian Library Collections Yosemite*, **52**, 12–15.

Anderson, M. K. (1993a). *Indian Fire-Based Management in the Sequoia-Mixed Conifer Forests of the Central and Southern Sierra Nevada*. Report to the Yosemite Research Center, Yosemite National Park. Berkeley, CA: University of California.

Anderson, M. K. (1993b). The experimental approach to assessment of the potential ecological effects of horticultural practices by Indigenous peoples on California wildlands. Unpublished PhD dissertation, University of California, Berkeley.

Anderson, M. K. (2009). *The Ozette Prairies of Olympic National Park: Their Former Indigenous Uses and Management*. Port Angeles, WA: USDI National Park Service, Olympic National Park.

Anderson, R. S. and Carpenter, S. (1991). Vegetation change in the Yosemite Valley, Yosemite National Park, California, during the protohistoric period. *Madrono*, **38**, 1–13.

Anderson, M. K. and Rosenthal, J. (2015). An ethnobiological approach to reconstructing Indigenous fire regimes in the foothill chaparral of the western Sierra Nevada. *Journal of Ethnobiology*, **35**, 4–36.

Barrett, S. A. and Gifford, E. W. (1933). Miwok material culture. *Bulletin of the Public Museum of the City of Milwaukee*, **2**, 117–376.

Baxley, H. W. (1865). *What I Saw on the West Coast of South and North America and the Hawaiian Islands*. New York: D. Appleton Publishers.

Bibby, B. (1994). An ethnographic evaluation of Yosemite Valley: the Native American cultural landscape. Unpublished report, Yosemite National Park Research Library.

Biswell, H. H. (1968). *Prescribed Burning in California Wildlands Vegetation Management*. Berkeley, CA: University of California Press.

Blackburn, T. C. and Anderson, K. (1993). *Before the Wilderness: Environmental Management by Native Californians*. Ballena Press Anthropological Papers, No. 40. Banning, CA: Malki-Ballena Press.

Bloom, R. and Deur, D. (2020a). "Through a Forest Wilderness": Native American environmental management at Yosemite and contested conservation values in America's national parks. In K. Sullivan and J. McDonald, eds., *Approaching Place Ethnographically: Public Lands and Coasts in the West*. Lanham, MD: Lexington Books, Rowman & Littlefield, pp. 151–73.

Bloom, R. and Deur, D. (2020b). Reframing Native knowledge, co-managing Native landscapes: ethnographic data and tribal engagement at Yosemite National Park. *Land*, **9**(9), 335.

Bloom, R. and Deur, D. (2022). Yosemite Valley traditional use plant inventory and threat assessment: Yosemite Valley, Yosemite National Park. Unpublished report, USDI National Park Service.

Booth, D. B., Ross-Smith, K., Haddon, E. K., *et al.* (2020). Opportunities and challenges for restoration of the Merced River through Yosemite Valley, Yosemite National Park, USA. *River Research Applications*, **36**, 1–14.

Boyd, R. (1999). *Indians, Fire and the Land in the Pacific Northwest*. Corvallis, OR: Oregon State University Press.

Briggs, M. C. (1882). Report of the commissioners to manage the Yosemite Valley and the Mariposa Big Tree Grove. In *Biennial Report of the Commissioners to Manage the Yosemite Valley and the Mariposa Big Tree Grove*, so extended as to include All Transactions of the Commission from April 19, 1880, to December 18, 1882. Sacramento, CA: State Office, J.D. Young, Supt. State Printing.

Bunnell, L. H. (1880). *Discovery of the Yosemite and the Indian War of 1851 Which Lead to that Event*. Chicago, IL: Fleming H. Revell.

Clark, G. (1894). Condition of the floor of the valley. In *Biennial Report of the Commissioners to Manage Yosemite Valley and the Mariposa Big Tree Grove for the Years 1893–94*. El Portal, CA: State Office, On File, Resources Management and Science Library, pp. 14–15.

Clark, G. [1907] (1927). A Yosemite plea of 1907. *Yosemite Nature Notes*, **6**, 13–15.

Cramer, A. (1997). Interview with Brian Bibby. In A Native American Oral History of the Yosemite Region. Unpublished report, Yosemite National Park Research Library.

Crowe, G. C. (1931). Maria, Last of Chief Tenaya's Tribe, is Dead. May 2, 1931. (Newspaper clipping of unclear provenience, in files of Yosemite National Park Research Library).

Daily Alta California (1889). Fire to prevent fire. *Daily Alta California*, **80**(4), May 10, 1889. Available at California Digital Newspaper Collection: https://cdnc.ucr.edu/ (accessed September 3, 2020).

Deur, D. (2007). Yosemite National Park traditional use study: plant use in Yosemite Valley and El Portal. Unpublished report, USDI National Park Service.

Deur, D. (2009). "A caretaker responsibility": revisiting Klamath and Modoc traditions of plant community management. *Journal of Ethnobiology*, **29**(2), 296–322.

Deur, D. and Bloom, R. (2021). Fire, native ecological knowledge, and the enduring anthropogenic landscapes of Yosemite Valley. In T. F. Thornton and S. A. Bhagwat, eds., *Handbook of Indigenous Environmental Knowledge: Global Themes and Practice*. Oxford, UK: Routledge/Taylor & Francis, pp. 299–313.

Deur, D. and Bloom, R. (2022). *Yosemite Valley, Yosemite National Park: A Traditional Cultural Property Context Statement*. El Portal, CA: USDI National Park Service, Yosemite National Park.

Deur, D. and James, Jr., J. E. (2020). Cultivating the imagined wilderness: native peoples, plants, and United States National Parks. In N. Turner and P. Spaulding, eds., *Plants, People and Places: The Roles of Ethnobotany and Ethnoecology in Indigenous Peoples' Land Rights in Canada and Beyond*. Montreal, PQ and Kingston, ON: McGill-Queens University Press, pp. 220–37.

Deur, D. and the Knowledge-Holders of the Quinault Indian Nation (2021). *Gifted Earth: The Ethnobotany of the Quinault and Neighboring Tribes*. Corvallis, OR: Oregon State University Press.

Ernst, E. F. (1943). Preliminary report on the study of the meadows of Yosemite Valley. Unpublished manuscript, Yosemite National Park.

Ernst, E. F. (1949). Vanishing meadows in Yosemite Valley. *Yosemite Nature Notes*, **28**, 34–41.

Ernst, E. F. (1961). Forest encroachment on the meadows of Yosemite Valley. *Sierra Club Bulletin*, **October**, 21–33.

Fong, C. (2020). Restoring the Merced River: a commitment to good stewardship. *Yosemite Guide*, **45**, 10.

Fritzke, S. L. (1997). A California black oak restoration project in Yosemite Valley, Yosemite National Park, California. In N. H. Pillsbury, J. Verner, and W. D. Tietie, eds., *Proceedings of a Symposium on Oak Woodlands: Ecology, Management, and Urban Interface Issues*. Albany, CA: US Department of Agriculture, Forest Service, Pacific Southwest Research Station, pp. 281–8.

Gassaway, L. (2009). Native American fire patterns in Yosemite Valley: archaeology, dendrochronology, subsistence, and culture change in the Sierra Nevada. *Society for California Archaeology Proceedings*, **22**, 1–9.

Gibbens, R. P. (1962) *A Preliminary Survey of the Influence of White Man on the Vegetation of Yosemite Valley*. National Park Service Contract No. 14-10-0434-813. El Portal, CA: Yosemite National Park, National Park Service.

Gibbens, R. P. and Heady, H. F. (1964). *The Influence of Modern Man on the Vegetation of Yosemite Valley. University of California Division of Agricultural Sciences, Manual 36*. Berkeley, CA: University of California, Berkeley.

Hall, W. H. (1882). To Preserve from Defacement and Promote the Use of the Yosemite Valley. Annual Report of the State Engineer, 1882. In *Report of the Commissioners to Manage Yosemite Valley and the Mariposa Big Tree Grove, 1885–86*. Sacramento, CA: California State Printing Office.

Heady, H. F. and Zinke, P. J. (1978). *Vegetational Changes in Yosemite Valley*. Washington, DC: US Department of the Interior.

Jones, G. M. and Tingley, M. W. (2021). Pyrodiversity and biodiversity: a history, synthesis, and outlook. *Diversity and Distributions*, **28**(3), 386–403.

Keeley, J. E. (1977). Fire-dependent reproductive strategies in *Arctostaphylos* and *Ceanothus*. In H. A. Mooney and E. Conrad, eds., *Symposium on the Environmental Consequences of Fire and Fuel Management in Mediterranean Ecosystems: Proceedings*. General Technical Rept. WO-3. Washington, DC: US Department of Agriculture, Forest Service, pp. 391–6.

Lewis, H. T. (1973). *Patterns of Indian Burning in California: Ecology and Ethnohistory*. Ramona, CA: Ballena Press.

Lewis, H. T. (1979). *Fires of Spring*. Film. Edmonton, AB: University of Alberta/CRFN Television.

Lewis, H. T. (1982). *A Time for Burning*. Occasional Publication No. 17. Edmonton, AB: University of Alberta, Boreal Institute for Northern Studies.

Lewis, H. T. and Ferguson, T. A. (1988). Yards, corridors, and mosaics: how to burn a boreal forest. *Human Ecology*, **16**, 57–77.

Madej, M. A., Weaver, W., and Hagans, D. (1991). Analysis of bank erosion on the Merced River, Yosemite Valley, Yosemite National Park, California. Unpublished National Park Service report, Redwood National Park.

Matthes, F. E. (1930). *Geologic History of the Yosemite Valley.* US Geological Survey Professional Paper #160. Washington, DC: USDI Geological Survey, US Government Printing Office.

McGregor, S., Lawson, V., Christophersen, P., *et al.* (2010). Indigenous wetland burning: conserving natural and cultural resources in Australia's World Heritage-listed Kakadu National Park. *Human Ecology,* **38,** 721–9.

Milestone, J. F. (1978). The influence of modern man on the stream system of Yosemite Valley. Unpublished MA thesis, Department of Geography, San Francisco State University.

NPS (2002). *Draft Environmental Impact Statement for Yosemite Fire Management Plan.* Arcata, CA: USDI National Park Service, Yosemite National Park.

Plumb, T. R. and Gomez, A. P. (1983). *Five Southern California Oaks: Identification and Post-fire Management.* General Technical Report PSW-71. Berkeley, CA: US Department of Agriculture, Forest Service, Pacific Southwest Forest and Range Experiment Station.

Reynolds, R. D. (1959). Effect of natural fires and Aboriginal burning upon the forests of the Central Sierra Nevada. Unpublished MA thesis, Department of Geography, University of California, Berkeley.

Russell-Smith, J., Whitehead, P., and Cooke, P. (2009). *Culture, Ecology and Economy of Fire Management in North Australian Savannas: Rekindling the Wurrk Tradition.* Collingwood, Australia: CSIRO Publishing.

Senos, R., Lake, F. K., Turner, N., and Martinez, D. (2006). Traditional ecological knowledge and restoration practice. In D. Apostol M. and Sinclair, eds., *Restoring the Pacific Northwest: The Art and Science of Ecological Restoration in Cascadia.* Washington, DC: Island Press, pp. 393–426.

Show, S. B. and Kotok, E. I. (1924). *The Role of Fire in the California Pine Forests.* Bulletin No. 1294. Washington, DC: US Department of Agriculture.

Tait, D. J. I. (1999). Land managers struggle with tradeoffs: air quality concerns may hinder prescribed burn efforts. *California Agriculture,* **53,** 22–3.

Taylor, H. J. [Rose Schuster] (1931). The return of the last survivor. *University of California Chronicle,* **33,** 85–9.

Traditionally Associated Tribes of Yosemite (2021). *Voices of the People.* San Francisco, CA: Yosemite Conservancy.

Turner, N. J. (1991). "Burning mountainsides for better crops": aboriginal landscape burning in British Columbia. *Archaeology in Montana,* **32,** 57–73.

University of California Merced (n.d.). Hydrology. Yosemite Field Station. https://snrs .ucmerced.edu/natural-history/hydrology (accessed December 8, 2021).

USACE (2021). National Wetland Plant List (proposed revisions). *Federal Register,* **86,** 60449–52.

USFWS (1988). *National List of Plant Species that Occur in Wetlands: 1988 National Survey.* Biological Report 88. Washington, DC: US Department of the Interior USFWS.

USFWS (1997). *National List of Vascular Plant Species that Occur in Wetlands: 1996 National Summary.* Washington, DC: US Department of the Interior USFWS.

van Wagtendonk, J. W. (1986). The role of fire in the Yosemite wilderness. In *Proceedings of the National Wilderness Research Conference: Current Research, compiled by RC Lucas.* Ogden, UT: U.S. Forest Service Technical Reports, pp. 2–9.

Vasquez, I. A. (2019). Evaluation of restoration techniques and management practices of tule pertaining to eco-cultural use. Unpublished MA thesis, Humboldt State University.

Watkins, C. E. (n.d.). Photographs of Yosemite Valley from the 1860s. Photographs in the C. E. Watkins collection, Yosemite National Park Research Library.

Wharton, C. H., Kitchens, W. M., Pendleton, E. C., and Sipe, T. W. (1982). *The Ecology of Bottomland Hardwood Swamps of the Southeast: A Community Profile. National Coastal Ecosystems Team, Biological Services Program.* Washington, DC: US Fish and Wildlife Service.

Whitehead, P. J., Bowman, D., Preece, N., Fraser, F. and Cooke, P. (2003). Customary use of fire by Indigenous peoples in Northern Australia: its contemporary role in savanna management. *International Journal of Wildland Fire*, **12**, 415–25.

Whitney, J. D. (1869). *The Yosemite Guide-Book*. Sacramento, CA: California Geological Survey.

Yosemite National Park (2011). *2010 Assessment of Meadows in the Merced River Corridor, Yosemite National Park.* El Portal, CA: Resources Management and Science, Yosemite National Park, National Park Service, Department of the Interior.

Yosemite National Park (n.d.). Hydrology. Learn about the park. Yosemite National Park, National Park Service. www.nps.gov/yose/learn/nature/hydrology.htm (accessed December 8, 2021).

Zevely, J. W. (1898). Report of J. W. Zevely, Special Inspector and Acting Superintendent to Yosemite National Park to the Secretary of the Interior. Unpublished report, Yosemite National Park Research Library.

8

Indigenous Knowledge and the Kindergarten to Twelfth-Grade Science Classroom

SUSAN M. ARLIDGE[*]

8.1 Introduction

This chapter is a general introduction to why science teachers need to connect to Indigenous knowledge, teach science outdoors, use preceptory learning, and make teaching inclusive at the level of the learner to make science resonate with Indigenous students. As described in the other chapters, Indigenous knowledge has deep traditions of oral history, experience of inquiry, and knowledge of the land. It is natural to infuse this into contemporary science lessons informed by cultural wisdom to deepen student understanding of place (landscapes).

As a science outreach teacher, wilderness guide, and park naturalist, I have instructed students from kindergarten to twelfth grade, professional teachers, and undergraduates in multiple overlapping disciplines. I have observed thousands of teachers, university professors, and interpretive guides interacting with their classes or groups on field trips. Some of the field experiences lasted a day, while others were overnight programs, with some schools returning with students for more than two decades. Through these experiences, I have gained insight into teaching science on the land. My observations, backed by research, indicate that some of the best practices in teaching science overlap with Indigenous ways of knowing the land (Snively and Williams, 2016).

"The land has always been a teacher since humans have existed" (Actua, 2021: 6). Traditionally, Indigenous cultures used empirical knowledge to solve issues, ranging from acquiring fuel, food, and shelter to managing group hunts and moving through the landscape. Many of the skills required for science were traditionally part of life for Indigenous people, and they learned them

---

[*] I would like to thank all of the elders and traditional knowledge keepers who enrich everyone's understanding by sharing their teachings freely, in particular Margaret, Terry, and Travis Rider. From my heart, it is an honor to learn on the land together. I live in Exshaw, Alberta, a small town in the front ranges of the Canadian Rockies, just a few kilometers from the Îyârhe Nakoda. The places I live, work, and recreate are within the traditional territories of the Treaty 7 region of Southern Alberta. These people include the Blackfoot Confederacy (Siksika, Kainai, Piikani), the Tsuut'ina, the Îyârhe Nakoda Nations, and the Métis Nation (Region 3).

from the land. Global challenges need people trained in science with diverse ways of knowing to work on complex solutions. When interviewed, students state that they want to be part of future solutions (Hicks and Holden, 2007). The way that elders and knowledge keepers mentor young people and share critical lessons about the land is similar to the inquiry approach often used in science education.

In Canada, the topic has been examined by a growing body of scholars with many excellent new resources and perspectives (First Nations Education Steering Committee and First Nations Schools Association, 2016, 2019; Snively and Williams, 2016). Some describe the relationship between the nature of science and Indigenous ways of knowing, with its spiritual and cultural basis, as oppositional (Michell *et al.*, 2008; Aikenhead and Mitchell, 2011, Snively and Williams, 2016). Elder Albert Marshall chose the Mi'kmaw term *Etuaptmumk* (two-eyed seeing) as a way of describing the complementary nature of exploring science within the cultural context (Bartlett *et al.*, 2012). This chapter adopts the two-eyed seeing perspective, assuming that teaching students the two unique and insightful cultures of science and Indigenous ways of knowing deepens the learning experience for everyone (Marshall *et al.*, 2020). It also reflects the work done by knowledge keepers, researchers, and educators in expressing and creating this enriched relationship as "coming to know" (Cajete, 2000; Snively and Williams, 2016). As described by education researcher Gregory Cajete (2000: 66), "Coming to know reflects the idea that understanding is a 'journey, a process, a quest for knowledge and understanding' with all our relations."

Science is one of the world's most prevailing and dominant empirical knowledge generators because it works and provides humans with solutions for living on the land. Science, taught in tandem with traditional lessons from Indigenous elders and knowledge keepers, creates context for students to develop the knowledge and wisdom to experience and navigate the world they live in. Educators are aware of the education gap and the call to infuse Indigenous worldviews into the school culture, but have not been provided with the tools to act.

This chapter addresses some of the practical considerations of bridging the cultures of Indigenous knowledge, schools, science curricula, and teaching outdoors by addressing the following questions. What are the obstacles for Indigenous students in science education? What do science teachers need to learn about land-based learning from Indigenous advisors? What teaching approaches work? What do Indigenous advisors need to understand about science? What cultural protocols should be respected? How do we create mutually respectful collaborations that will be self-organizing and will be perpetuated? What are some examples of science and traditional knowledge intersecting in the school curricula?

8.2 The Gap and the Barriers

In Canada, the federal government controls Indigenous affairs, but the provinces are in charge of education. In the two northern territories, there is Indigenous co-governance, with Indigenous leaders creating curricula with the government. About 600 schools in Canada are on First Nations reserves and are run by band authorities. Those schools receive less than half of the funding of provincially funded schools. In 2018, only 44 percent of students in schools on First Nations reserves and 63 percent of Indigenous students in schools off reserve graduated from high school. Indigenous student graduation rates both on and off reserve fall far below the 88 percent graduation rate of other students (Indigenous Services Canada, 2018).

Worldwide, by all metrics of success, we have failed to educate Indigenous students, particularly females. The United Nations reports that "Indigenous students tend to have lower enrolment rates, higher dropout rates, higher absenteeism rates, higher repetition rates, lower literacy rates and poorer educational outcomes than their non-indigenous counterparts, with retention and completion being two important educational challenges" (United Nations, 2019). It follows that, compared with non-Indigenous students, few Indigenous students pursue or attain careers in science. In Canada today, there is a patchwork approach to implementing culturally informed science, technology, engineering, and mathematics (STEM) learning (Actua, 2021). The consequences of failing to fund and support Indigenous students is summarized as follows: "First you create a funding gap, and then you end up with a reading gap. This, in turn, gives the government, our country and Aboriginal people an achievement gap and then, of course, we end up with a terrible socio-economic gap" (Indigenous Awareness Canada, 2020).

In the United Nations Declaration on the Rights of Indigenous Peoples (United Nations, 2008), Article 14.1 states, "Indigenous peoples have the right to establish and control their educational systems and institutions providing education in their own languages, in a manner appropriate to their cultural methods of teaching and learning." The Canadian government has also promised action to close the gap, created by underrepresentation of First Nation students in careers and positions of leadership (Truth and Reconciliation Commission of Canada, 2015).

A discussion paper with input from 100 STEM outreach agencies and Indigenous advisors identified reasons for low retention and graduation rates of Indigenous students in Canada (Actua, 2021). This report found that Indigenous students receive unequal support and resources compared with non-Indigenous students; have teachers with low expectations of their students, inadequate training, and experience in working with Indigenous knowledge; and experience a non-inclusive structure and focus of provincial curricula. To change the success rate for

Indigenous students, there is a need to change schools and curricula, adding more cultural reference and co-mentoring students with Indigenous advisors. Schools should also be set up to have Indigenous partners co-teach curricula on the land. Despite all of the research on understanding how students learn, many of the learning challenges still come from within the education system and do not reflect student ability. Curriculum reform and learning to collaborate with Indigenous advisors are part of the solution. At the same time, teachers need training in using the land to teach curricula from science and math.

8.3 Land-Based Learning

One of the most important ideas for a science teacher to grasp is how the land is fundamental to Indigenous relationships. The land, spirit, and human are all connected and in a living relationship. Learning from keen observation and developing relationships with the natural processes are the essence of Indigenous life. "From an Indigenous perspective (as it was explained to me), the spiritual is inseparable from the physical: for example, the river is a living being with feelings and responses" (Haig-Brown, 2010). Because of the ubiquitous experience of place, there are multiple terms in education that are overlapping.

The Indigenous experience of land encompasses more than the physical space. It is the sum of experience that has passed on the land. Building on the research and conversations of many people, Michell *et al.* (2008) in *Learning Indigenous Science from Place* explain the breadth of the concept of *place* from an Indigenous worldview. This is summarized as part of the philosophy for the First Nations Education Steering Committee and the First Nations Schools Association teacher resource guides:

Place is multidimensional. More than the geographical space, it also holds cultural, emotional, and spiritual spaces which cannot be divided into parts. Place is a relationship. All life is interrelated. Place is experiential. Experiences a person has on the land give it meaning. Place is local. While there are commonalities, each First Nation has a unique, local understanding of place. Land is interconnected and essential to all aspects of culture. Making connections with place in science curricula is an integral part of bringing Indigenous Knowledge into the classroom. That means including experiential learning in local natural and cultural situations.

(First Nations Education Steering Committee and First Nations Schools Association,
2019: 16)

Preventing further loss of traditional culture and language is the domain of knowledge keepers who have been mentored by kin and elders while spending time on the land. While history reminds us of what culture has been lost, modern rediscovery initiatives of knowledge keepers are essential in contemporary tribal life. They have been taught to know the land in every season, every dawn and dusk, and during floods and droughts. They have been taught to track and hunt, to make traditional tools and medicines, and to observe keenly. They know the wind and clouds and how to navigate by landmarks and stars. This is where the interaction of the natural and cultural landscapes is united. In science, different people test and gather evidence about questions concerning the natural world. Science is also about *relationships* of things and natural processes (e.g., element cycling and trophic structure of ecosystems). In science, we describe and measure processes, predicting how biotic and abiotic factors interact and change over time and space.

Science and Indigenous knowledge collaborations often involve wider community participation in the lesson. Parents and siblings, community role models, elders, knowledge keepers, and mixed grades of students might take part in an experience together. In a hide skinning and tanning experience, for instance, older students may do the skinning, trying different scrapers (from traditional bone tools to modern metal scrapers) while an elder tells the story of a hunt and the second graders join for twenty minutes to watch, anticipating the time when they will be the ones scraping hide in a few years' time. The next week, the ninth graders could smoke the hide with different cultural teachers, and they could learn the science of tanning using brains. This activity could be the seeds of their science unit on enzymes and their role in body functions. Informal lessons like these align well with the Indigenous pedagogy of whole person learning, an approach used in many Indigenous interactions, expressed by one spiritual caregiver as follows: "emphasising cohort and peer education through an 'extended family' that results in whole person learning. Whole person learning requires moving beyond the textbook to include academic, professional, emotional, and spiritual growth" (Chartrand, 2020).

In British Columbia, the First Nations Education Steering Committee and the First Nations Schools Association recommend that curricula should be developed around specific *Indigenous principles of learning* (First Nations Education Steering Committee and First Nations Schools Association, 2019). These principles are as follows:

Learning supports the well-being of the self, the family, the community, the land, the spirits, and the ancestors. Learning is holistic, reflexive, reflective, experiential, and relational (focused on connectedness, on reciprocal relationships, and a sense of place). Learning involves recognizing the consequences of one's actions. Learning involves generational roles and responsibilities. Learning recognizes the role of Indigenous knowledge. Learning is embedded in memory, history, and story. Learning involves patience and time. Learning requires exploration of one's identity. Learning involves recognizing that some knowledge is sacred and only shared with permission and/or in certain situations.

(First Nations Education Steering Committee and First Nations
Schools Association, 2019: 5)

Lessons on the land taught by cultural leaders and science teachers provide critical context for students in developing their cultural identity and sense of belonging (Sobel, 2004). The more that students are attentive to, curious about, and interactive with the subject matter, the higher their motivation to learn and progress in their education (Lei *et al.*, 2018: 517). Infusing traditional knowledge into the experience is familiar and a foundation for other learning. "The most important single factor influencing learning is what the learner already knows; ascertain this and teach him accordingly" (Ausubel *et al.*, 1978). In examining multiple environmental education programs over two decades, researchers found positive learning impacts on multiple aspects of student learning. Improved academic performance as well as increased knowledge and skills in critical thinking, problem-solving, decision-making, and leadership are among the many measurable benefits (Ardoin *et al.*, 2018). Learning about the environment through hands-on inquiry provides opportunities that are more inclusive for every learner and, most importantly, develops skills for learners to have agency over their learning (Vander Ark *et al.*, 2020). There is simply no substitute for working outdoors with cultural mentors and science teachers. That is not to say that outdoor urban landscapes and schoolyards are to be excluded for land-based lessons. The distinction is that, when learning cultural lessons from Indigenous mentors, it is not the "playground" anymore, it is the land.

In Canada, schools off the reserve are multicultural. Experiencing Indigenous culture in a more immersive manner benefits everyone, creating deeper experience of community. In one study, after a visit to a museum, students reported having transformative experiences in their cultural empathy. Pre- and post-trip assessments indicate that, through active cultural learning, students changed their opinions on historic events in up to 6 percent of cases (Greene *et al.*, 2013). One of the major reasons that Indigenous students living off reserves report missing school is that they experience racism (Uppal, 2017). Learning about the land and science through the lens of an Indigenous guide deepens cross-cultural understanding and reduces Indigenous students' experience of isolation and racism.

School-reported benefits of Indigenous mentoring include positive behavior changes, both during the lessons on the land and later as carryover in the classroom, as well as higher student attendance on field days. Students struggling in the classroom show a more open attitude about learning from their community mentors, as they are held in esteem. Students who participate little in science class will speak with their cultural teachers in their own language. Hearing their own language and their ancestors' stories is relevant and engaging. Students don't learn the curricula if they are not at school: Getting them in the door, excited about learning science, is half the challenge.

The Canadian Education Association evaluated nearly fifty outdoor education programs to determine best practice. The program chosen as a case study saw an increase in students' social, academic, and intellectual engagement in all areas of student learning (Willms *et al.*, 2009). The report recommends three practices for successful Indigenous land-based learning programs in schools:

1. Foster relationships with the First Peoples of your school or school district's region, and involve them in your program's development and operations to ensure that Indigenous histories, cultures, ceremonies, languages, and protocols are taught and practiced in a manner that respects their cultural and spiritual integrity.

2. Develop programming with the premise that Indigenous knowledge and values are relevant to all students – both Indigenous and non-Indigenous . . .

3. Create a learning experience that is land-based, hands-on and experiential, as these are essential components of Indigenous pedagogy and are integral to gaining a genuine understanding of Indigenous Traditional Knowledge.

(Rebeiz and Cooke, 2017)

8.4 Traditional Methods of Learning from the Land

Traditionally, for many Indigenous groups, education was more incipient and informal. Lessons were everywhere. In *These Mountains Are Sacred Places*, Chief John Snow of the Stoney First Nation shares the following:

Parents, grandparents, and Elders told and retold stories and legends to the children by the campfires, in the teepees, on the hillsides, in the forest, and at special gatherings during the day and at night. It was an ongoing educational process about religion, life, hunting, and so on. Other topics were bravery, courage, kindness, sharing, survival, and foot tracks of animals, so it was a very extensive study of many things.

(Snow, 1977: 6)

Life lessons, spirituality, relations, kinship, and survival were all woven into the fabric of life.

"The best teaching occurs when the emphasis is less on imparting knowledge and more on joining the child on a journey of discovery" (Sobel, 1996). Discovery learning, whereby the student directs the experience, uses higher order thinking. It does not mean that there is no instruction; instead, the learner directs what knowledge or skills they require. Different discovery learning techniques, from guided to open inquiry and problem-based approaches can enhance how students learn science. Students enjoy the novelty of solving a riddle or setting off on a quest or adventure to discover a novel solution. Using more active teaching techniques matches traditional ways of teaching, including the use of storytelling or storylines as a way of expressing the processes and phenomena of science. An exemplary student reply to the question "What did you do in science today?" would be: "We figure out the science ideas. We figure out where we are going at each step. And we figure out how to put the ideas together over time" (Next Generation Science Storylines, n.d.).

Designing effective land-based student inquiry lessons takes practice. The easiest way to learn about how to implement inquiry teaching is to do it! In every grade, start small, building up student skills for outdoor investigations. Reinforce aspects of the nature of science to eventually have students asking and evaluating their own questions. For instance, have students design an experiment to learn something about the squirrels in the field and how frequently they visit placed items with human scent versus items with scent removed by wipes and placed with gloves. Alternatively, have them map where ants go or how long it takes for leaves with a smaller surface area to decompose compared with larger ones. Learn from challenging experiences. When you're learning from Indigenous mentors, watch how they interact with young people.

Model an investigation and work on solving it as a group. Students like to role play that they are specialists. As a teacher assigning students into working groups, make it fun and relatable. Tell the students that they are knowledge keepers, geologists, chemists, hydrologists, foresters, glaciologists, avalanche specialists, or oceanographers. Set up fake grant competitions in which students have to pitch their approach to solving made-up science problems, such as "How can we use what we know about the properties of snow in biotechnology?" Tell students about the real world and "cool" careers. Shake them up with a novel experience like digging into soil or exploring in a cold creek. Taking students out of their norm and keeping them safe and comfortable while exploring culture and science is a lot of work but it is incredibly rewarding. Being connected and learning with your friends and your Indigenous mentors is special. Learning science from place is more than curricula. Bring the hot chocolate thermoses and some dried meat, fish, or traditional food for a snack. Try to use campfires and circles as part of the experience. Be a bit radical; adding to the atmosphere of an event when students are working by playing traditional music can be very special.

8.5 Learners' Needs

Indigenous mentors need to spend some unstructured time on the land getting to know the students. Kinship and neighborly relations are part of the Indigenous learning landscape. When students learn from their cultural educators, it helps them develop a sense of place and community. Sharing lessons in a circle, at a campfire, or walking on the land with a knowledge keeper is familiar for Indigenous students. In Canada, many schools have a very multicultural population, and those students will have different learning needs. Co-designing Indigenous culture–science collaborations that meet multiple learning needs will take practice. However, at its most fundamental, education is about relationships. If you care about the students and their learning, the students will teach themselves.

Experiencing both Indigenous teachings and science on the land develops a sense of nature literacy: "the ability to learn from and respond to direct experience with nature" (Sobel, 1996). Two-eyed seeing experiences should consider students' learning development stages. Students in kindergarten to fifth grade are typically curious, active, and eager to problem solve and like to explore. In the earlier years (kindergarten to second grade), they learned through games, songs, and repetition. Looking for natural patterns, sorting, describing, observing, and imitating are naturally enticing to young students. By third to fifth grade, as well as all their prior skills, they begin using logic to solve problems, they learn by doing, and they can measure and quantify items, grasping basic mathematical functions. They can design simple experiments and test predictions. Students become more selective about what interests them between fifth and seventh grade, when many fundamental science skills start to get developed. Abstract problem-solving skills, understanding types of causation, and using math in problem-solving are some of the skills students gain at these ages. Developing inquiry skills through hands-on experiments, measuring data and processes, and examining relationships amongst objects, processes, and ideas are some of the skills to practice on the land. Challenging these students with group work and more complex multi-week field investigations with science tools and technology provides students with the chance to process other people's opinions and develop work ethics.

Teachers need to teach to students' level of scientific literacy. What do they know? Building in short skill sessions and lessons on the nature of science can develop students' skills quickly. To accomplish observations and experiments or inquiry, students will need to be taught how to work outdoors in a group; they will need to learn to use equipment and tools, collect data, read maps, take random samples, etc.

One technique I use early in the learning cycle is to lead a very open-ended investigation. For example, the following is a guided inquiry I have run with seventh graders, in which they can use some of the science ideas that they are building together. The curricula cover both the ecosystem unit and the ethnobotany, with specific reference to the water cycle and plant adaptations to seasonal change. It builds on sixth grade ideas from evidence, investigation, and photosynthesis. After a class conversation reviewing the processes within the water cycle and the unique properties of water, I ask students to consider where water in a plant goes in the winter. We head outside to a natural area and we explore how plants cope with what I describe as "the water crisis" in freezing temperatures. We debate whether plants are alive or dead in the winter. We hunt for evidence of life in plants. We find hairy leaves, evergreen leaves, Labrador tea leaves covered with orange hair, and "dead" grass leaves. I ask students why water matters for the plants. Students review photosynthesis and the role of water while instructors clarify naive ideas.

Students want to know if plants have to photosynthesize to be alive or why some plants are so fragrant, even in winter, and all kinds of other interesting questions. I use role playing and analogies when I teach, so the students know I will use metaphors when I teach. I talk about water like it is money for a plant – something the students are all interested in and understand. I have the students tell me the different ways that plants spend money in the winter. Students quickly jump on the idea of some plants spending all their money and making seeds fast (annuals) while the growing season is good and they do not have to deal with pesky water molecules puncturing their cell walls. Meanwhile, other students realize that some plants save their money and stay evergreen all winter. Most Canadian students know that maple syrup does not freeze, which quickly leads to conversations about sugar and salt as antifreeze. Some students ask if plants make antifreeze. Another student wants to know why some plants smell strong in the winter. We discuss the cool ways in which sugars move around, within, and between cells to lower the freezing temperatures. After further guided sensory investigations, student groups predict which plants appear to store the most water during the winter. This has all taken an hour.

Student groups are instructed to take samples of any living plant within the sample area and use them to test their mini hypothesis. Students use a scale to determine the wet mass of their samples; they then dry all of the samples in a drying oven and determine the dry mass (that is, by dividing it all by the wet mass). However, you should not give that instruction until someone starts to ask some pertinent questions about comparing sample size. Students usually follow the directions to this point; we have a quick lesson about how we compare samples gram/gram by dividing them by their original mass. Students use the typical style of the scientific process or a scientific poster by defining a question, a hypothesis, etc., but we call it quits before anyone can write a conclusion. This is often a good time for an active game or food break. The "experiment" is left intentionally raw and undefined.

We have now covered water cycle processes, photosynthesis, plant adaptations, the function of cells, osmosis, the plant life cycle, seasonal change, plant life forms, plant morphology, and a few other miscellaneous science processes and functions. The students really drive the conversation from here. For instance, I have had students ask: What defines life? Does sample type matter? Do evergreens or deciduous plants store more water? Is there more water in the stem or the leaf? Does the height off the ground affect how much water is in the plant? Is the snowpack affecting the plants? Is there a difference between moss and herbaceous plants or between plants and shrubs? Does the distance from ground water matter? Do aspect or slope matter?

The students' questioning has naturally led me into teaching about random sampling, setting up a student transect, or what kind of experiment we could set up to figure out if the water were in the moss or in the snow on the moss. A hands-on activity like this that explores the need to refine experimental design, get feedback from peers, reiterate the trials, etc., typically intrigues and engages most students. After your class is on the same page with the rules of science for their appropriate grade, be careful not to water down the science too much. If students are having trouble understanding what you are teaching, back up and give further background. Science is naturally engaging to students when it is brought out of the textbook and into their realm of understanding.

Local experiences are easier for students to relate to: "A richer curriculum results when educators connect with local Indigenous communities, as there is much knowledge that is locally held" (First Nations Education Steering Committee and First Nations Schools Association, 2016, 2019). In Canada there are fifty First Nations and more than fifty languages (Government of Canada, 2021). Each has its own culture and body of knowledge specific to its territory. The best practice includes locally developed content presented in collaboration with science curricula.

Many teachers and scientists need guidance on Indigenous ways of knowing and teaching. This is new ground for many of the instructors. Indigenous leaders should lead cultural lessons, stories, and songs. For instance, it is not up to the teacher to collect traditional plants with a group of students without a cultural guide. Having culture represented and expressed authentically is important to relationships, trust, and respect. Author Margery Fe (2000) notes, "Interpretation can be a kind of respectful listening, or it can be a kind of appropriation, and we always have to raise the issue that what we hope has been the first may in fact have been yet another example of power disguising itself as benevolence." Certain cultural items should not be part of class activities; for instance, creating totems is a spiritual expression that holds its value only within culture. Furthermore, non-Indigenous participants need to respect the domain of culture. Taking a weekend workshop on traditional medicines does not provide the permission to harvest plants.

Some Indigenous knowledge keepers and elders may also have little knowledge of science and the culture of science (e.g., directly questioning ideas may seem culturally disrespectful, but it is what we are expected to do in science). Teachers need to understand Indigenous culture and protocols to work with Indigenous mentors. In return, teachers may be teaching Indigenous advisors the nature and processes of the school science curricula.

8.6 Cultural Protocols and Indigenous Ceremony

The participants leading the Actua roundtable discussions stressed:

A fundamental guiding principle of Indigenous land-based education planning is that it is Indigenous-led. This means that Indigenous peoples are meaningfully and authentically engaged in all aspects of legislation, policy development, programming, curriculum development, and delivery. It means that educators need to have a strong awareness of Indigenous history, intergenerational issues, and barriers that disproportionately impact Indigenous youth.

(Actua, 2021)

Also often not mentioned is their need for an excellent background in the natural science of processes and not just natural history.

Science and Indigenous knowledge are different cultures. There are going to be difficulties overlapping the two. They are compatible, but they are constructed and conducted differently. Science is innately public, with a requirement to expose, test, and question. Culture and traditional knowledge are innately acquired and interpreted, and some of it is spiritual. Where we bring these experiences together is uncharted territory. Fields probing the human experience within the social sciences have documented traditional ways. The loss of many stories and cultural lessons, with some precious few recalled and recovered, makes them even more significant to culture. At the same time, social media, the internet, and globalism have broadcast cultural teachings, while some nations are just beginning to redefine their modern identity. This process is dynamic and centers on relationships.

The sense of urgency and the result-driven focus that is typical in schools is not part of Indigenous culture. Knowledge keeper Jared Qwustenuxun Williams offers this advice for communicating with Indigenous advisors:

I want to talk about how we speak. When we speak English, we try to speak as many words as fast as we can. Because we respect the audiences' time, we want to ensure we get as much information across as quickly as we can. Because the individual that has to listen to us has other things to do. In Hul'q'umi'num it is the opposite. We show respect to our listeners by slowing our words down to ensure that they can hear every sound. It is all there smooth and slow. And sometimes it gets misconstrued because in English we talk so fast that we often think those who speak slowly, are slow. But in Hul'q'umi'num, it is also the opposite. When we speak too fast, you are in a rush and there is nothing worse in our way than rushing the interaction you are having with whoever you are having it with. So, when you are working with the Elders, slow down your words, your mind, your questions, and your actions. Slow down.

(Williams, 2023)

Cultural cues indicate that there is a time for visiting, a time for ceremony and learning, and a time for business. I watched a scientist enthusiastically rush toward an elder after an education event and proceed to insist that the elder "must" come and speak at such-and-such event. Indigenous knowledge and stories are not commodities to be acquired. Lessons may be shared, in trust, once a relationship is established. The traditional knowledge holder shares lessons when the learner shows that they are ready to learn. Part of Indigenous knowledge is assessing when, with whom, and how to share.

If invited to a ceremony, ask about cultural protocols from an Indigenous advisor. Some Indigenous groups cover their bodies or heads or are positioned in a specific space in a ceremony. In preparing students to partake in Indigenous ceremonies, share expectations and protocols for local context. Many Indigenous lessons take place with the group in a circle and the elders open the conversation. Sometimes, there are offerings, blessings, songs, dances, or stories. Not all cultural lessons are meant for sharing with non-Indigenous learners. Some things will not be discussed aloud or at school. Some teachings are just for certain ages or sexes. Some stories are told at different times of year. Some are for different times in life. Unless requested, ceremonies and songs are not to be recorded. Before recording anything, ask everyone for permission.

In Canada, it is customary to offer elders tobacco before requesting any knowledge or lessons. Gifts – such as blankets, sweet grass, food, or colored fabric or pillows – are given to elders. Kinship relationships are deep, and non-Indigenous partners may not realize that Indigenous individuals in turn support their elders with gifts. It is customary in teaching situations to provide cash honoraria to Indigenous teachers and their helpers. It may be uncomfortable to discuss the honoraria, but it creates clear expectations if discussed with Indigenous organizers in advance. Easy financial transaction processes that provide cash should be part of the plan. School boards have policies and budgets for honoraria.

Part of the culture of science is to ask questions. This is not the typical style of conversing with most people and can be affronting with new acquaintances. When interacting with Indigenous knowledge keepers and elders, the best practice is to listen and ask questions sparingly. As a rule, educators should develop relationships before getting too nosy. Teachers and scientists will have a chance to learn important cultural lessons with the students. Traditional knowledge is not a commodity.

The culture of school and field trips may also be novel for the Indigenous instructors. Typically, the school staff are responsible for routines, group communication, class management and staying on schedule for transportation, and sessions with visiting instructors. School staff should clarify roles and routines with knowledge keepers before a trip. For instance, many schools do not allow staff to use their phones or smoke around students, even if on a break. At field sites, this may mean finding a spot out of view for adults to attend to their needs. In another example, an elder with a softer voice will be speaking to a class outdoors and it is a windy day. As the leader, you need to read the situation and set up for learning success. It may be best for the elder to greet the group in the controlled environment of the school bus or to have the students start with an active activity to use some of their energy. When done, remind them of listening routines, then bring the group to a sheltered spot and sit in a tight group before turning it over to the elder.

8.7 Developing a Shared Vision

To create meaningful relationships, teachers and scientists need to spend time and energy getting to know local Indigenous knowledge keepers. Some school districts have cultural liaisons acting as bridge makers between teachers, scientists, and schools, on the one hand, and educators and learning agencies or cultural leaders in the Indigenous communities, on the other. Similarly, each First Nation band authority has protocols for requesting cultural guidance and support. Contacting the Nation through the proper protocol is respectful, ensures that the request is appropriate, and prevents Indigenous advisors from getting overburdened with outside requests. These intercultural relationships are the foundation upon which all sides can learn to respect each other's autonomy while preserving the integrity of both Indigenous culture and the nature of science.

Building authentic relationships takes time and cannot be skipped. It might take years. Partially due to the history of reserves, many teachers and scientists have not had a lot of contact with knowledge keepers or Indigenous educators or advisors. Most teachers and non-Indigenous students have not been into schools or homes on a First Nation reserve, and vice versa. In Canada, teachers can self-educate and experience Indigenous culture in multiple ways beyond the internet. Indigenous arts events, powwows, rodeos, markets, museum exhibits, and school sports events are natural venues to meet outgoing and inspiring community members. Attending events at Indigenous institutions and supporting Indigenous-run businesses can lead to connections. Indigenous social media, radio, and television are all accessible. Creating a deeper understanding and widening the science teaching community is unfamiliar territory. It is an organic process and it may involve misunderstandings and mistakes. On all sides, the experience will be transformative.

The advisory committee should spend as much time together on the land as possible, including holding some of the planning session outside. A valuable use of school development time would see school administrators, educators, and neighboring Indigenous knowledge keepers and elders sharing experiences and setting their intent for co-teaching students on the land. One of the many problem-based approaches exposed in teacher education programs is the use of design thinking as an approach to curricula design. Design thinking, originally a tool developed for use by engineering design teams, is "a mindset and an approach for developing innovative and life-centered solutions to complex problems in collaborative teams" (Vander Ark *et al.*, 2020). It is a creative problem-solving approach that suits the cross-cultural and cross-disciplinary nature of designing two-eyed science experiences. Design thinking as a problem-solving process includes five stages: empathize, define, ideate, prototype, and test (Vander Ark *et al.*, 2020). Applying design thinking to science education could be a good model for creating the paradigm shift required to explore the natural world with multicultural perspectives. In the end, it is up to the advisory committee to define a learning strategy that will span twelve years of student learning, meeting students' growing cognitive needs.

8.8 Partnerships: Schools, Outreach Agencies, and Indigenous Advisors

Partnerships mean longer term commitment. Schools are not the only agencies that engage in collaborations. Libraries, science centers, outdoor camps, and science or engineering faculties all have outreach groups. It is possible that a partner has approached a school to offer outreach opportunities for students or a school is seeking funding partners. Providing services with outside agencies can create synergy and increase capacity. Partnerships can also provide resources and expertise. It is critical that each group's mandate and agenda is clear and focused on student success. Creating a memorandum of understanding between partners will help define roles, responsibilities, budgets, personnel, and terms of the collaboration. School boards have rules about working with outside agencies. While a teacher might get approached to collaborate on a grant-funded project, the project does require administrative approval. If partnerships last more than two years, revisiting the terms of collaboration will ensure continued success and highlight new opportunities.

If the project relies upon grant funds and the applicant is from an outside agency, such as the National Science Foundation in the USA and the Natural Science and Engineering Research Council in Canada, regular communication between partners is critical. Large grants often require up to a year lead time between creating the application and receiving funding. Two or more large institutions seeking joint funding need to have regular updates between project partners. Project parameters may change due to funds received. There are legal and reporting implications in obtaining grants. Such a dynamic situation needs to be well documented and communicated to all parties. Indigenous partners will have their own ways of staying connected and on target.

8.9 Program Advisory Committee

Establish an advisory committee with Indigenous knowledge keepers, science teachers, the school, and educational contacts from the band administration. Once levels of band administration are aware and they involve the appropriate agencies, it assumes informed consent between all parties. Without a project agreement in writing, the project may not have longevity. Creating a great learning plan takes time. From the outset, both cultures need to develop a trusting, mutually beneficial relationship. Each group needs to gain an understanding of the pedagogy and content of the other. Indigenous mentors create the cultural vision and methodology. Science educators identify the key curricula goals and objectives for each grade. The project vision will develop from discussion and collaboration throughout the program design phase. Learning goals for each collaboration should be clear and learning objectives should be measurable. Scaffold the learning experiences for each grade and create a school-wide curricula map. The map will include grades, dates, and specific activities, with the timeline clear and visible.

The advisory group should define the roles and responsibilities of different partners. For instance, Indigenous partners could be responsible for choosing learning sites, sourcing cultural resources such as animal hide or tipi poles, and creating cultural content important for sites and stages of youth development. The school administration could be responsible for scheduling and booking outdoor learning sites and scheduling and booking transportation. Cross-cultural teams could create guidelines for safety and for supporting cultural needs and student experience. Finally, science teachers are responsible for covering the curricula through land-based lessons. It is the teacher's job to take the lead from cultural advice and fit the science curricula into the framework.

The learning plan needs to be dynamic, adapting as the partnership deepens and Indigenous learning needs change. These are dynamic political times, both on and off reserves. Some schools have deep foundations of cross-collaboration through outdoor education and cultural camps but need to enrich the experience to add a new focus on the science curricula. Some highly multicultural schools have no Indigenous students but exist within the culture of Canada. What does inclusion of Indigenous culture with science look like in that setting? In Canada, a number of Indigenous land claims remain unsettled. The nuances of each community are different and need to be respected. This is why it is important that Indigenous advisors maintain agency over Nation members who should work in these collaborations. Bringing unknown teachers into an experience can create competing agendas. The best-case scenario is a wide body of trained science staff and Indigenous knowledge keepers and elders with land-based lessons and skills, stories, and songs important to culture.

A plan for program longevity or the realities of personnel turnover will disrupt relationships and impede progress. Succession planning conversations should take place frequently, with schools planning and allocating sustainable resources to keep initiatives running as part of the school culture. Creating a project binder to keep records of meetings, contacts, and brainstorming sessions is invaluable. Small lessons can eventually turn into annual projects, and that project binder can turn into a dedicated website with important cultural field sites, photos, land use history, elder interviews, language lessons, science curricula connections, and suggested STEM activities for each grade or each season.

Spend time planning before the first visit to a new site: walk about with cultural guidance and brainstorm the event together. Before activities, plan for transportation and childcare needs for co-presenters; arrange for payment, honoraria, and gifts; and discuss the knowledge keeper's requirements for their lesson on the land. Discuss logistics such as washrooms, food, water, clothes, shelter, on-site communication, and emergency transportation arrangements. Discuss the outdoor class arrangement; for example: Are the students sitting on the ground or at picnic tables? Consider the mobility needs of co-presenters and plan parking access for elders. Practically speaking, apart from the aesthetics, planning for a fire at learning sites is helpful for less mobile participants to teach. Portable screened fire pits are permitted in many sites and should be confirmed. When working with traditional teachers, flexibility in planning is essential. Community needs come first. Sometimes, community events come up that will change their availability.

8.10 Resource Assessment

Early in the planning process, completing a resource assessment will help to define the project scope. Identify what you have and what you will need: What staff, partner agencies, relationships, Indigenous partners, field sites, building access, funds, etc., does your organization have? What skills does your school team have besides class teaching? What knowledge and skills do key partners need for success? What is needed to accomplish project goals? Remember, these are long-term partnerships.

Funding for honoraria, food, travel, meetings, field trips, buses, and teacher training needs to be allocated. Indigenous partners who will cocreate and implement the program are often fulfilling so many other responsibilities. There are only so many knowledge keepers and elders, and those who are willing to share in STEM collaborations have many other responsibilities. The amount of time requested of partners in these new initiatives is substantial. Funds are also necessary for additional staff time and respectful compensation that considers the time that Indigenous partners spend in preparation.

A concrete way to assess resources is to plan one land-based learning theme. The following is one example of how it might work. Perhaps the working committee determines that the hunting theme best matches ninth-grade science. Knowledge keepers want to share songs, stories, and lessons about hunting and animal relations. They want to track animals, make animal calls out of wood, and make bows and arrows. The learning sequence starts with a winter field trip to walk on the land with a knowledge keeper, tracking wildlife and sharing stories. Next, a science lesson ties into a unit in the school curriculum or Nations goals on populations and wildlife management. Students learn to complete a forestry survey, gathering data and recording it in the field. Back in class, students use their data to determine species diversity, abundance, and plot density. Another venture outdoors includes an elder sharing a story of the land and animal relations. The unit finishes with three days of making bows and arrows: First, students walk on the land with a knowledge keeper, learning to choose suitable shrubs for the bows and the arrows. After harvesting the wood, they are taught how to carve the shape, first of the bow and then the arrow. During their experience of making bows, the students may be taught their language, learning vocabulary related to trees, feathers, wildlife, sinew, and hunting. Later, as integration into math class, students determine the force that can be applied on a pull to the bow before the wood breaks or loses its elasticity. Students can graph the slope of this relationship. Alternatively, perhaps students harvest willow tips to propagate in sand buckets and plant into the school yard or test which feather and which feather design flies the straightest while testing different methods of measuring distance.

Working backward, a needs assessment of the project will reveal the required resources. For instance, which knowledge keepers can support five afternoons of learning? Is there a site where willow bows can be harvested? Is the land suitable for students to visit and harvest? What permissions or permits, etc., are required? Are school bus bookings necessary? What tools are needed for a class to cut and whittle bows? What safety considerations are specific to this activity? Where can bow feathers and sinew be sourced? How will student learning be assessed? Should staff develop a process-based rubric or does it make more sense for students to share in creating the rubric, so they are aware of goals from the outset? Do cultural advisors have suggestions about the assessment? What are appropriate honoraria for the advisors? What is the project budget? How will finances be obtained and managed?

8.11 Teacher Development

Teachers undertaking Indigenous knowledge and science collaborations are developing a fusion style of lesson. Co-teaching, using the natural world as a springboard, requires shifting the learning outside. From the project outset, it is important to assess staff knowledge and skills and create a staff training program to fill the gaps. When key staff leave a school, their role and skill set need to be filled.

A skills assessment should be completed during the program planning process. Educators and knowledge keepers have life experience, contacts, and skills that will be really useful in creating collaborative programming. Skills such as hunting and trapping, cooking, gardening, videography, art, and more may be at your disposal.

A skills assessment also shows gaps and opportunities for training. Most teachers are not trained in designing and implementing successful outdoor science learning, especially with Indigenous collaborators. Professional teaching bodies, science outreach agencies such as interpretive or science centers, botanic gardens, libraries, museums, and zoos can provide teacher training. These agencies have experience and activities for developing students' curiosity, observation skills, outdoor learning skills, and, of course, science. Teaching skill workshops may include asking open-ended questions, using tools in the field, developing a disposition of inquiry, or creating process-based evaluations. Pedagogy experts may help by designing teacher training in active learning practices. Stories, games, simulations, and hands-on experiences – such as hide tanning or finding the height of a tree – engage students and match traditional teaching methods. Land-based collaborations require teachers and knowledge keepers with first aid and guiding certification. An inclusive professional learning plan for all collaborators should be developed.

Teachers need to establish class routines and expectations for successful learning outside the class. Provide the class with clear behavior and engagement expectations. Providing frequent land-based learning experiences creates anticipation of the routines for students. For instance, plan for every second Friday to be a land-based learning day for the whole school, with activities coordinated accordingly. If you scaffold the learning experiences to build on prior lessons, the experience will become a part of the fabric of learning. Dynamic lesson plans, with a combination of active and calm activities, will meet the needs of most students. Whether it is looking for the biggest dandelion on the way to a special spot by the river park to meet the knowledge keeper or listening for as many bird calls as they can hear, use students' developing relationship with nature to inspire the awe and wonder of science.

To summarize, there are a lot of moving pieces. Working on the land with knowledge keepers and students requires training in cultural protocols, first aid, off-site safety assessments, group management, curricula design, inquiry learning, pacing activities, scheduling, and program transportation and logistics. Appendix A provides a checklist for cultural collaborations in school science that covers the details.

8.12 Immersive Experiences and Frequent Collaborations

In schools, learning is compartmentalized into subject, time, and grade. Creating content around natural themes rather than specific curricula content builds in flexibility. Land-based themes such as water, forests, hunting, and navigating provide natural cultural and curricula overlap. For instance, the theme of hunting could cover the science found in the curricula for many subjects and grades, including animal behavior, ecosystem dynamics, population dynamics, wildlife biology, animal tracking, evidence and investigation, biodiversity, mapping, data analysis, making bows and arrows, the physics of flight, the flex and strength of wood, calculating biomass and carbon storage in a forest, tree and plant identification, propagating trees and shrubs in schoolyard gardens or greenhouses, anatomy, body systems, and cells and growth, to name a few of the possibilities. Math curricula is also found throughout the theme of hunting. Opportunities to collect, quantify, sort, map, graph, and manipulate data are found in each of these science strands.

Many Indigenous bands and family groups have developed culture camps. With the lessons to be learnt as unique as each family, experiences with these groups provide hands-on lessons in traditional life skills including hunting and tanning, knife skills, cooking, leatherwork and beading, childrearing skills, and more. Some schools working with tribal authorities have developed similar experiences, infusing science education into the mix. While the effort and resources required to offer such opportunities exceed what is typical in most schools, these overnight experiences are invaluable in creating foundations and developing teacher–student relationships. As immersive and deep learning experiences, I enthusiastically encourage teachers to collaborate with Indigenous advisors in developing school science and culture camps.

If a camp experience is not an option, teachers can identify outdoor learning sites that are unique for each grade. Weekly routines, such as each grade visiting its own special place, provides experiences in which students get to know the land, enriching their sense of place. For instance, if the kindergarten class always visits a special grove of trees while the sixth graders always go to the ravine, relationships between students and place can develop. Studying the same tree through the seasons, asking science questions about process, and observing change through time are ways of knowing through developing relationships.

Some school districts provide opportunities for Indigenous students to gain high school credits in courses outside the typical sciences. Teachers and Indigenous mentors can create programs with credits for students in forestry, engineering, and math.

8.13 Outdoor Learning Routines

A technique to initiate outdoor learning is to immediately destabilize the regular indoor routine. For instance, when I meet a group of students getting off a school bus, I do not take them into a building right away. I use nature or fun to bring them into a land-based learning frame of mind. I surprise them with a quick running or guessing game; I have them gather outside long enough to get too snowy, cold, or wet; I start right away with their feet in the water; or I have them stand with their eyes closed listening to birds. I find something to wake up their senses so that they know they are about to have a different type of learning experience.

Establish outdoor learning routines before land-based collaborations. Practice how you would like students to gather with their group at a designated spot with the appropriate field attire and equipment. Begin land-based inquiry outside the school's door. At the beginning of the year, outline acceptable behavior expectations and remain consistent with consequences. Keep behavior management issues to a minimum by planning engaging activities. Be very specific with expectations and directions, particularly with regard to safety. Have a clear outline of activities and keep the pace active for the first few experiences.

Even if you are limited to the school yard, you can regularly walk out the door and model thinking like a scientist by asking open-ended questions without providing any answers. This is a useful skill to hone when you are working with Indigenous partners. In *Coyote's Guide to Connecting with Nature*, Young *et al.* (2016) offer countless suggestions for awakening curiosity in the natural world. Using the metaphor of a coyote – playful, mischievous, and deeply connected to the land – the guide is full of techniques and activities for developing mentorship skills in exploring the land. As an example of how to use the coyote mentoring style, the guide suggests, "When people come in from the field, ask questions like where have you been and what did you see? Listen carefully for what gets them excited or stumped" (Young *et al.*, 2016: xxix). Beginning questions with phrases like "I wonder . . .," "Suppose . . .," or "What would happen if . . ." gets learners engaged in their own discovery.

Order the learning flow so that it begins with observation or immersion rather than teaching activities. Nature educator Joseph Cornell (1989, 1998) has written a number of activity books useful in outdoor learning. Beginning lessons with acclimatization activities for fifteen or twenty minutes helps settle the students into learning outdoors. Activities like scavenger hunts, interpretive walks, or observation games like Kit's Game are good openers for students of all ages (Cornell, 1989, 1998; Young *et al.*, 2016). Pacing activities and giving instruction keep students focused. Tasks should change every fifteen or twenty minutes, including active learning with instruction such as listening or storytelling. While some of the learning activities may be more unstructured than some of the ways we teach in school, they are still valuable. After a carefully designed "high–low" scavenger hunt in which students are sent off with a thermistor seeking the warmest or coldest spots, students are usually more comfortable learning outside and start to feel more comfortable in the natural environment. They can then move on to design an inquiry exploring where the moss is growing, for instance.

Having a few tough-to-answer questions for each level of the curriculum is another technique to get students thinking like scientists. For example, ask students to design an experiment to calculate the amount of snow landing on one tree in a year. As another example, tell students to imagine two large water tanks. Tell them one has lake water from the top of a mountain and the other has lake water from the valley. Remind the students that lots of microorganisms live in the water. Tell them that two tanks exist on the top of the mountain with water from high altitude and water from the valley. Ask students to predict which microorganisms will thrive over the years: those in the valley water that have been brought up to higher altitude or those in the mountain water brought down to lower altitude.

One of the challenging constructs we teach is that science is about processes, not things. Invisible processes are hard to understand. When working in the field, I might start a lesson with students listing all of the verbs they can find in the woods. I start by saying, "I see a log decomposing." Students are then tasked to look for processes and relationships. It is easier for them to understand that the essence of science is measuring the processes and how these change over time. Describing the nature of science as a story helps students express many of the invisible processes and phenomena at play in the natural sciences (www.nextgenstorylines .org/). The intersection with the storytelling tradition of most Indigenous cultures is natural. Stories help make sense of the world, making the invisible visible.

Student assessment should be designed into each project and linked back to the curricula. Indigenous advisors will know best what looks like success. Whatever style of student evaluation that is used, it should be culturally responsive and inclusive. Program partners need to talk and ask questions. Communication is key. What knowledge, skills, or processes will be evaluated? How will project outcomes be measured? Consider evaluating outcomes that are not just academic performance. For instance, has the project impacted student identity or self-efficacy? Have there been changes in student social skills, group behavior, participation, etc.? Program designers need to use a number of different means of assessment for these novel collaborations.

Assessments may include a five-minute one-to-one conversation with students, listening for key words as students describe the processes they are studying. Create a checklist for guided conversation to explore a student's idea development. Student journals, interviews, or videos express what students have to say about their learning. Have students design and then complete their own self-assessments. Design a poster depicting elders' lessons. Ask students to describe a science concept from a two-eyed seeing perspective. Pose problems in the past and have students discuss likely solutions based on technology. Ask students to reflect on specific cultural lessons and discuss how these teachings connect to the land.

8.14 Land-Based Learning Sites

Indigenous and scientist advisors will have ideas of where on the land they want to take students. When working with elders and knowledge keepers, using maps and satellite images and spending time driving to spots on the land may evoke ideas and stories. When choosing places to teach students, ensure that sites are safe and accessible and will be approved for a school visit. When setting up the program, listen to how the cultural advisors share stories, history, and protocols. Discuss what lessons can be shared in each space. Take time to note the land features that will be a springboard to science lessons. Consider habitat type, water, geology, soil, climate, plant communities, slope, and other obvious strands back to the science curricula. For instance, imagine the site is a slope that was used as a buffalo jump. While the cultural focus could be how a hunt was experienced, the science could look at comparing plant communities at the top and bottom of the ravine or measuring the rise and run of the slope or explore the soil composition and percolation rate along the slope. Learning about the land is not necessarily a wilderness nature study. Urban landscapes and cultural centers or museums might also be good teaching sites. This is how, at any site, students can experience culture and science together.

The project advisory group can identify and develop land-based sites within walking distance from the school. While these sites might look like parks, vacant lots, or even cultural landscapes, it is up to teachers, Indigenous mentors, visiting scientists, and outreach agencies to work together to develop and test the stories of time and land (in the past, the present, and the future).

8.15 Health and Safety

When questioned about why they do not take students out of the classroom more frequently, teachers cited the added safety concerns. Developing off-site routines to identify, communicate, and mitigate safety hazards makes teaching on the land less stressful. For instance, review bus safety and expectations before a trip. Gather pre-trip information regarding the health concerns of all participants, including Indigenous mentors. If necessary, share a participant's life-threatening conditions with all leaders. Leaders should be trained in the use of appropriate medical responses, which differ with each group. Discuss site-specific hazards pre-trip; for instance, if a student is allergic to bees, discuss the response protocol, etc. Other wildlife, environmental, or site-specific hazards should be identified and discussed.

On exiting the bus or arriving at the site, discuss specific hazards, boundaries, and meeting spots. If necessary, do a site tour with the whole group following. Identify certified first-aiders, the location of emergency supplies and first aid kits, and a specific group muster site. Practice the group gathering signal, such as an airhorn, so that participants are aware of its function. If hazards such as busy roads or steep banks must be navigated, have safety procedures and communicate them clearly. For instance, if it is necessary to cross a road, gather students a safe distance from the road and communicate that, when it is safe to cross, two teachers will go ahead and block traffic, only signaling for the students to walk across together when it is safe to do so. When walking as a class, have supervisors at the front, middle, and back of the group. One way to keep the slower participants engaged is to have them right behind the leader at the front of the group. Moving at the pace of the slowest will frustrate some students. Consider breaking into active and less active groups. This does not mean that the students will determine how far they go; instead, it provides opportunities for more flexible expectations.

During planning, teachers need to consider participant comfort. Keeping everyone, including teaching partners, warm, fed, hydrated, and near washrooms keeps learners' focus on learning. The most frequent source of injuries when learning outdoors is the students themselves. Set up routines for students to keep their hands to themselves and be clear with boundaries during free time. Model moving off trail in the forest by showing students to keep a body distance behind others to reduce branches in the face, etc. During first aid training, teachers should learn how to assess sprained ankles, as this is the primary injury resulting from trips and falls. The best prevention for such incidents is to challenge everyone to avoid tripping and falling. Good training for staff is to make up worst-case safety scenarios and talk them through with the team before off-site activities.

Some activities are more hazardous and may require specialized certificates and training. Teachers need to recognize terrain and situations to avoid. After near misses or safety incidents, leaders should debrief and make notes on future mitigations. Schools should maintain a site manual with safety concerns, emergency transportation, and communication details for every site, including the phone number and location of the nearest hospital and emergency services, maps, trails, and facilities for each off-site activity. Every time a teacher leaves the school site with students, they should have an emergency backpack with a first aid kit, a student list of parent/guardian contact numbers, duplicates of student medicine if required, a charged communication device, extra clothes, emergency devices like bear spray, and contact information for the school and for the transportation company.

8.16 Communication

Collaborating on the land means communication between busy schools, teachers, and Indigenous partners. Planning and contingency planning – including obtaining permissions, timelines, action plans, and communication plans – all take time. While a certain amount of organic growth is desirable, hosting focused and efficient meetings with Indigenous partners and agreeing on plans and contingency plans are important. Adults should determine how they will communicate in larger sites or when students split into groups.

Most schools have rules about cell phones in class, but many teachers do not make this clear on field excursions. In the field, I recommend that only leaders have cell phones and only for cameras or communicating with other leaders. If the class needs to have photos of the data, consider seeking a grant for outdoor science equipment and a class set of waterproof durable tablets, with science apps such as light meters, clinometers, or decibel meters. Reducing internet access during field time is more immersive and traditional. The idea is to develop a relationship on the land in the here and now.

During programs, each leader should have a means of contacting the others for purposes of staying on schedule and emergency contact. Depending on location, many places will not have cell coverage, so it is important to have access to radios or a device like a satellite phone. Before field trips, it is advisable to have a pre-trip site visit, checking for safety hazards, best on-site communication means, and access to the nearest emergency services.

8.17 Teaching Materials

As lessons develop, outdoor teaching backpacks can be created. Each pack has the lesson plan, laminated materials, photos of important land features, the communication and safety plans for each site, and more. Group equipment for student science investigations can be carried in a backpack or toolbox. Ensure field equipment is maintained and batteries work. Build in student accountability for ensuring tools and equipment are treated with respect. If Indigenous partners bring items to learn from, ensure permission is requested before touching or moving the item.

Of note is that some of the cultural artifacts that might be shared with students will have changed over time. For instance, on one occasion, a knowledge keeper brought hide-skinning tools used by his great grandparents and we compared them with modern skinning tools. The change in tools from bone to metal scrapers made with railway metal provided insight into how technology has changed over time for Indigenous peoples. This led to conversations about how climate change and melting sea ice affect modern northern Indigenous communities differentially. While some communities still hunt with dog teams and others mostly use skidoos, the changing sea ice means that the heavier skidoos are no longer safe on ice hunts.

A class set of rain gear, gloves, caps, backpacks, or anything else that will add to the repeated success of learning from the land is a good investment. Some experiences will be repeated year after year with each grade. Compiling tools, equipment, and class materials specific to the lesson helps to build in program continuity. For instance, a class set of carving knives, leather gloves, and safety glasses will be needed for bow carving for ninth graders every year. Keeping these items, along with a lesson plan linked to a video of a local elder talking about carving, provides depth and continuity. If new staff or knowledge keepers are asked to provide the lesson in the future, good records and resource kits are helpful.

8.18 Evaluate Land-Based Projects

Successful leaders know where they are going! Similarly, programs should be developed with evaluation and assessment goals in mind. School staff, cultural leaders, and learning specialists in assessment, pedagogy, and cultural competency should be consulted for input into creating a program evaluation plan. Program partners should discuss how program outcomes and impacts will be measured and set the evaluation structure into the program. Program pieces require ongoing assessment to ensure they are appropriate and match learning goals. For long-term collaborations, setting immediate, medium-term, and long-term objectives will keep the big picture in focus.

Instructors should schedule time to debrief with all program partners immediately after events. School staff should invite the knowledge keepers to share feedback and should implement their suggestions. Discuss all aspects of the experience, including the schedule, team teaching experience, challenges, future improvements, student feedback, etc. Keep notes and honor feedback and suggestions to enhance lessons. This will foster co-teaching experiences that are respectful and reciprocal. Teachers are used to taking charge, particularly regarding evaluating lessons and learning. Feedback sessions should be culturally appropriate, ensuring every participant is heard.

Some collaborations may produce student projects, recordings of stories, videos, public presentations, or other products. While the school may be very excited to showcase student work or Indigenous-led collaborations, it is important that this is under the advisement and consent of knowledge keepers. It may not be appropriate to share the lessons in a wider sphere. Some of the lessons may be appropriate to share in an in-school professional development workshop, but not on the internet or with the media. This is not to say that these collaborations should not be widely celebrated and lessons shared. Rather, this is another area of the collaboration that should be developed with Indigenous guidance.

8.19 Science Curricula Connections

A two-eyed seeing approach to infusing Indigenous knowledge of the land with science has many possible starting points. Elementary science themes that are found in curricula around the globe include weather and cloud watching; the needs of living things; ethnobotany and plant use; map making and representing ideas with models; measuring area, perimeter, circumference, and slope; observing ants, birds, or small mammal behavior; and the nature of science skills. By high school, science curricula become more challenging, with the disciplines of biology, physics, and chemistry separated into subjects examined in more discipline-focused and scientific ways. High school inquiry on the land covers environmental chemistry; tree height measurement using trigonometry; plant and forest community processes; population dynamics, species richness, and abundance; forest and stream ecosystems; food webs, energy flow and seasonal and climate science, to name a few of the obvious areas of intersection with traditional knowledge.

In creating lessons from the points of intersection, choose activities that best match the skills and intent of the Indigenous advisors. Themes could be matched to each grade, providing continuity in providing lessons, ensuring important Indigenous lessons are shared with the right ages. A larger project could be part of each grade's special lessons. In my previous example, for instance, the ninth graders all know they will make a bow and arrow as part of their lessons while the sixth graders know that they will run all aspects of the school garden with the knowledge keepers. They know they will cook and make dye from saskatoon berries.

Traditionally, young people learn skills from doing. By choosing appropriate projects, the same traditional experiences can be supported further through science.

The following longer activities are a few examples of how the multiple ways of knowing can be explored with Indigenous guides:

- Make bows and arrows (Box 8.1). What wood makes the strongest bows? How do we measure the maximum flex before the bow snaps? Why did arrows have feathers? What are the characteristics of feathers that make arrows fly better?
- Make flutes and animal calls from willows. Explore the physics of how sound travels. Set up an experiment to determine how the size of mouth opening changes sound.
- Have knowledge keepers teach hide scraping, and support the lesson in the science class by focusing on the chemistry of brain tanning or discussing the anatomy of deer or how the hide is made of multiple layers of skin.
- Make traditional drums and rattles. Have the students change the sound with a different design.
- Go on the land with hunters to set traps or track wildlife. Learn about herd behavior and wildlife management.
- Make shelters or tipis. Go with knowledge keepers to harvest and erect a tipi. Calculate tree volume or height. Calculate the circumference of the tipi.
- Go on the land to discuss plant use. Collect the plants to make dye/chemical compounds. Cook with traditional guidance and later explore the chemistry of cooking.
- Make bentwood boxes.
- Erect fences to prevent wild grazing on the field next to the school. What happens to the plant communities when you prevent horses and cows from grazing.
- Make traditional travel bags. Explore how to store food without it spoiling.
- What wood burns the best – a physics lesson from a campfire. What is the science of fire? What is the difference between a candle (turbulent flame) and a wildfire?
- Walk on the land with knowledge keepers and scientists to create and illustrate a local flower key, a guide on the geology of the area, an insect guide, a mammal guide, or a wildlife viewing guide for each grade's site.
- Create an interpretive trail and signs and add QR codes to keep building the story with additional information. Have students design interviews with elders or knowledge keepers to add their stories to the interpretive trail.
- Make maps and models of the land with natural materials and then represent the same place in two dimensions.

Box 8.1

Bows and Arrows Lesson Plan: Science and Indigenous Knowledge

Grades: 7,   Subjects: science,   Time: seven hours. Approximately one hour of instruction and
  8, 10, 11   math, physics   six hours of activity (most of the class is on the land, shaping
                                    and making bows, with knowledge keepers leading)

Curriculum connections

Grade 7 science: plants for food and fiber
Grade 7 science: structures and forces
Grade 8 science: mechanical systems
Science 10: energy flow in technological systems
Physics 20: kinematics
Physics 20: dynamics

Learning objectives

Students will:
1. gain an understanding about forces and mechanical systems
2. understand how plant-derived materials vary in their properties, which effects their ability to be manipulated by forces

Knowledge keeper and teacher background

1. A local Indigenous knowledge keeper who will share the process of finding wood, carving wood, shaping the bow, language, etc.
2. Knowledge of bow and arrow use and history
3. An understanding of the forces involved in bows and arrows
4. Identification of local tree and shrub species – willows, saskatoon, etc.
5. Physics 20: Young's modulus/elastic modulus

Student prior knowledge

1. A familiarity with graphs and interpreting graphs
2. Data-recording and experimental design skills
3. An introduction to measuring force and distance
4. An introduction to potential and kinetic energy

Resources

Bow and arrow physics (details about this lesson are given after this table)
Guide to the Common Native Trees and Shrubs of Alberta (https://open.alberta.ca/publications/1711129)
YouTube video: What is elastic modulus? (www.youtube.com/watch?v=pahoQeY4WC0)
YouTube video: Plotting stress vs strain in Excel to find elastic modulus (www.youtube.com/watch?v=H_08pyYJIDg)

Materials needed

1. Samples of fresh bow wood
2. 1.5-meter lengths of local wood for bows
3. Handsaws
4. Carving knives
5. Gloves
6. 2-meter 2×4 boards
7. Rope or twine
8. Spring scale

Three-Lesson Sequence: thirty-minute pre-trip class before the day on the land; half-day field trip led by knowledge keepers to harvest wood, carve wood, and tie bows over wood to create the warp in the bow; forty-five-minute post-trip class.

Box 8.1 (cont.)

1. Pre-trip class

Tell students that bows (Figure 8.1) and arrows have been around at least 4,000 years and have been invented by multiple groups in different places. Ask them to imagine they are on horseback or in dense bush and need to hunt deer. Ask: What do you need to know to make a bow and arrow?

Have a guided discussion with the class, pausing for student feedback. Ask: What materials can you use for the bow? What are all of the local trees or shrubs we could possibly use? (Students list trees and shrubs.) What kind of wood would be best? (Have samples that students can touch.) Does each tree have different kinds of wood? (Show pictures of a tree with dead and living wood, or old and new growth, etc. – review parts of the tree from sixth grade). Should we make the bow from the dead seasoned wood in the middle of trees (hardwood) or should it have some amount of the sapwood that is part of the tree on the outside which is still wet and transporting water? What about the number of rings? Would it be better if it were made from a tree which had very narrow rings or wide rings?

Think, pair, and share: Students get into pairs to brainstorm. Ask: If we needed to design an experiment for which bow was made from the best wood, what variables could we consider? (Species and amount of each type of wood [sapwood/hardwood], length and diameter of branches, narrow rings versus wider rings, etc.). Directing them to these types of comparison questions helps guide their thinking.

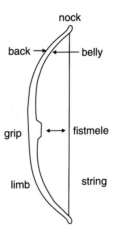

Figure 8.1 Parts of a simple bow.

Box 8.1 (cont.)

Tell students that a bow is a means by which you bend a piece of wood and then transfer the energy of that bent wood through the bowstring to the arrow. Ask: What are the relationships we need to explore to understand a bow and arrow? How do we know how long the arrows should be? (Give a hint if they need it.) With the bowstring on my bow, how far back can I draw the string? Why does this matter? (This will be the length of the arrow.)

Demonstrate with a bow and arrow. Ask: Does the tip of the bow bend differently at different draw lengths? Is a different amount of weight required to pull the string back at different draw lengths? What happens if you pull too hard? This is the relationship we are digging into because it is what we need to know to make bows that do not break! And, of course, every engineer needs to know this kind of stuff before a big beam or pole is erected.

If we think about these two measurements, we are trying to decide the relationship between what we will call the draw force and the displacement of the bow tip (Figure 8.2). This will give us some characteristics of the material of the bow: how elastic it is and how much energy will transfer to the bow, which will allow us to compare different bows. We are exploring how the elastic potential energy in the bow is transferred to kinetic energy as the arrow is released.

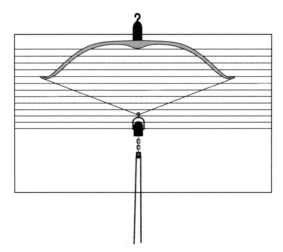

Figure 8.2 The relationship between strain (the displacement of the bow tip) and stress (the draw force). The straight parts of the curve have different slopes and indicate the difference in wood elasticity. The curved parts of the line indicate the points at which the bow is past the point of springing back.

Box 8.1 (cont.)

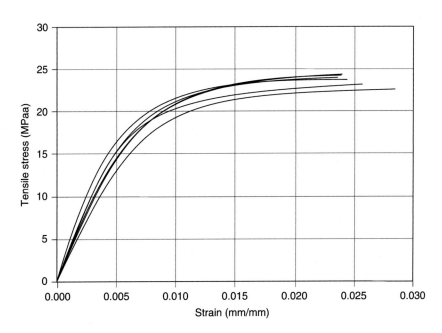

Figure 8.3 The arrangement for plotting stress versus strain for different woods.

Notice that the relationship of the draw force and the displacement of the bow tip form a straight line from zero displacement and draw force to a certain displacement (Figure 8.3). This is straight with a certain slope. If the line stays straight, the bow is retaining its elasticity, which is to say that it will spring back to its original shape. The graph also tells us the amount of energy the bow is storing at different amounts of displacement of the bow tip. However, if you kept pulling on the bowstring, at some point when you let go the bow would not spring back and would stay permanently bent. That point of displacement and draw force tells us the limit of the elasticity of the bow material. Of course, if you keep pulling after this point you will eventually get to the point where the bow will break. Sometimes, bows will break closer to the point where the bow has lost its elasticity.

So, what have we learned using a simple graph? We could decide to compare different numbers of rings, the amount of sapwood, and different kinds of wood in different bows. This is an opportunity to have the students decide which variable they are most interested in comparing to help decide which materials make the best bows. It is also a chance to get them excited about going out on the land to get traditional lessons on making bows and arrows.

2. Field trip: Land-based experience making bows

Spend a lesson gathering material and beginning to create the bows with an elder/ knowledge keeper. Indigenous leaders meet students on the land to share traditional knowledge of how to find materials and build traditional bows and arrows. As the

Box 8.1 (cont.)

knowledge keeper shows safe knife use, and how to strip the bark, the class can chat about the parts of a bow (see Figure 8.1) and arrow, the types and sizes of traditional bows and their associated uses, local tree species, the best season to harvest, etc. After watching how to carve and shape their bows, students spend a few hours carving bows. Before the end of the field trip, students will secure their bows over blocks on plywood to flex the ends.

3. Post-trip class

Students evaluate and create their own graph to compare different bow materials. Figure 8.2 shows the bow arranged on the board and how to make these two measurements.

With the bow on the top of the board, pull the string down with your scale and mark on the board the weight it requires to pull the string to that notch on the board. Also keep track of how far the tip has bent from the point where it sits when the bow is unstrung. Using these relationships between the amount of draw and the tip deflection, one can determine by the slope of the line (see Figure 8.3) and the elasticity of the wood that you are using for any variation in its characteristics, for example the number of rings for the number of sapwood areas.

For each bow made from varied materials, pull the string back to pre-determined draw lengths using a spring scale (see Figure 8.2). Example: Use draw lengths at intervals of 5 cm: 0 cm, 5 cm, 10 cm, 15 cm, 20 cm, 25 cm, and 30 cm. Record the data at each draw length. Record the amount of force required to pull back the string (stress) and the amount of displacement of the bow tips (strain). If the bow is made well, the tips should displace equally, but if there is variation in the displacement on the left and right, it can be averaged.

Graph the plot points for each different bow. Calculate the slope of each line to determine the elastic modulus, or Young's modulus, of each material type. A lower slope equals a more flexible material.

Note that if the material is bent past the point where it returns to its original position, it has now undergone plastic deformation, as opposed to elastic deformation. Elastic modulus can be calculated only on a line that has a linear slope, so if the line curves you will need to determine the area of the line that is elastic deformation and perform the slope calculation only on that part of the line. See the following resource video for more information: www.youtube.com/watch?v=H_08pyYJIDg.

Evaluation

There are many variations for the evaluation of student understanding of the concepts covered in this lesson, ranging from individual lab reports on the experiment to simplified worksheets that guide students to calculate line slope or interpret from a graph which material is more flexible. Once bows and arrows are cured, it is fun to have a contest on a sports field to shoot the arrows into targets.

Taking It Further

Further concepts in math, science, and physics can be explored using bows and arrows, including looking at arrow trajectory, examining how seasonal changes affect wood elasticity, or exploring social studies connections to bows and arrows.

8.20 Conclusion: Coming to Know

In Canada, teaching science with two-eyed seeing seems like the obvious solution to retaining Indigenous student engagement in science. Using traditional knowledge and science to explore natural phenomena are complementary experiences that enrich student understanding of the world. Students thrive when engaged in active learning experiences, which match well within traditional methods of teaching. Part of the Indigenous worldview is the sense of responsibility toward place. The world is complex and scary: Students need tools for them to have the agency they want over their future. For Indigenous students to understand the world, their worldview needs to be presented, explored, and understood. Non-Indigenous participants learn Indigenous lessons and culture at the same time as Indigenous students, which helps create a shared understanding and deepens community. Schools and educational institutions are experts in current pedagogy. Indigenous knowledge keepers, mentors, and educators are experts at understanding their young people. All students, including Indigenous young people, have higher academic success in curricula-based learning experiences on the land. If this were easy or obvious, it would already be happening across the country.

Schools, like most institutions, have multiple layers of administrative oversight. Add to that Indigenous governing bodies and their structures, federal funding of Indigenous education, and provincial curricula and things get complicated. However, we cannot allow these obstacles to fail Indigenous students again. In many ways, coming to know the nuances of understanding the world is uncharted territory, but it is the natural path. We need to work at building bridges with Indigenous knowledge keepers to provide the land-based, Indigenous-led science curricula that will prepare Indigenous students with the understanding of the world that is needed to create a fulfilling future.

APPENDIX A

Checklist for Cultural Collaborations in School Science

Submit trip outline to administration for approval.
Research and follow all school policies.
Determine project resources including supervisors, knowledge keepers or elders, finances, equipment, etc.
Secure funds and resources.
Outline cultural and curricular objectives.

Create a rubric outlining what success looks like from the students' perspective.

Work with Indigenous advisors to create a learning sequence and outline of
  activities.

Discuss pre-trip preparation with parents, students, and co-teaching partners.

Discuss behavior and learning expectations with students.

Co-develop and share the assessment rubric with students and adult leaders.

Choose a field site.

Complete site and activity safety and communication assessments.

Provide a site map for students and leaders with emergency meeting spots,
  washrooms, etc.

Determine dates (plan alternative dates).

Inform other school staff of plans to be off site and ensure it works.

Provide families with trip details and student food, clothing, and behavior
  requirements.

Retrieve permission, health, dormitory, or waiver forms before departure.

Book transportation and facilities if required. If the site is remote, keep the bus on
  site as emergency shelter.

Provide clear directions with maps and meeting times for all drivers.

Determine transportation and supervisor plans for non-life-threatening student
  evacuation in case of an emergency.

Have a communication plan for drivers; arrange to drive to the field site in
  convoy with the bus in front and a teacher at the rear. Have a communication
  plan for changes in plans.

Plan teacher/supervisor roles for field event (i.e., coordinator, first aider,
  volunteer liaison, timekeeper, special student or elder support, etc.).

Create student groups; ensure they have enough materials or equipment/group.

Build field equipment/tool accountability into group expectations.

Confirm transportation the day before departure.

Bring all supplies and equipment, including extra water, spare parts, and teaching
  supports.

Ensure students have the clothing, equipment, and supplies required before
  departure.

Review emergency procedures with all supervisors; ensure you have safety gear,
  first aid kits, and charged communication devices and that supervision ratios
  and expectations of supervisors are understood.

Check for emergency weather, road closures, etc.

Evaluate field trip and record improvements and feedback for future.

Take photos of site-specific hazards, features, and trails and, if appropriate, record
  cultural lessons and ideas and create a field resource guide as a teaching aid.

References

Actua (2021). Indigenous land-based STEM education: discussion paper. https://actua.ca/wp-content/uploads/2022/03/Indigenous-Land-Based-STEM-Education-Discussion-Paper-Actua-Canada.pdf.

Aikenhead, G. S. and Mitchell, H. (2011). *Bridging Cultures, Indigenous and Scientific Ways of Knowing Nature*. Toronto, ON: Pearson Canada.

Ardoin, N. M., Bowers, A. W., Roth, N. W., and Holthuis, N. (2018). Environmental education and K-12 student outcomes: a review and analysis of research. *The Journal of Environmental Education*, **49**(1), 1–17.

Ausubel, D. P., Novak, J. D., and Hanesian, H. (1978). *Educational Psychology: A Cognitive View*, 2nd ed. New York: Holt, Rinehart and Winston.

Bartlett C., Marshall M., and Albert M. (2012). Two-eyed seeing and other lessons learned within a co-learning journey of bringing together Indigenous and mainstream knowledges and ways of knowing. *Journal of Environmental Studies and Science*, **2**, 331–40.

Cajete, G. (2000). *Native Science: Natural Laws of Interdependence*. Santa Fe, NM: Clear Light Publishers.

Chartrand, C. (2020). Neecheewam Inc. – whole person learning. The National Center for Collaboration in Indigenous Education, First Nations University. www.nccie.ca/story/neecheewam-inc-whole-person-learning/.

Cornell, J. B. (1989). *Sharing the Joy of Nature: Nature Activities for All Ages*. Nevada City, CA: Dawn Publications.

Cornell, J. B. (1998). *Sharing Nature with Children of All Ages*, 2nd ed. Nevada City, CA: Dawn Publications. https://asknature.org/resource/sharing-nature-nature-awareness-activities-for-all-ages/.

Fe, M. (2000). Reading aboriginal lives. *Canadian Literature*, **167**(4), 5–7.

First Nations Education Steering Committee and First Nations School Association (2016). Science First Peoples teacher resource guide, grades 5–9. www.fnesc.ca/science-first-peoples/.

First Nations Education Steering Committee and First Nations Schools Association (2019). Science First Peoples secondary science teacher resource guide. www.fnesc.ca/learningfirstpeoples/.

Government of Canada (2021). First Nations. www.rcaanc-cirnac.gc.ca/eng/1100100013791/1535470872302.

Greene, J. P., Kisida, B., and Bowen, D. H. (2013). The educational value of field trips. Taking students to an art museum improves critical thinking skills, and more. *Education Next*, **14**(1), 78–87.

Haig-Brown, C. (2010). Indigenous thought, appropriation, and non-aboriginal people. *Canadian Journal of Education/Société canadienne pour l'étude de l'éducation*, **33**(4), 925–50.

Hicks, D. and Holden, C. (2007). Remembering the future: what do children think? *Environmental Education Research*, **13**(4), 501–12.

Indigenous Awareness Canada (2020). Indigenous Awareness Canada: the world leader in indigenous awareness training. https://indigenousawarenesscanada.com/.

Indigenous Services Canada (2018). Quality education, background. www.canada.ca/en/indigenous-services-canada.html.

Lei, H., Cui, Y., and Zhou, W. (2018). Relationships between student engagement and academic achievement: a meta-analysis. *Social Behavior and Personality: An International Journal*, **46**(3), 517–28.

Marshall, A., Popp, P., Reid, A., and McGregor, D. (2020). Etuaptmumk/two-eyed seeing and beyond. https://rwok.ca/dialogue-4.

Michell, H., Vizina, Y., Augustus, C., and Sawyer, J. (2008). *Learning Indigenous Science from Place. Research Study Examining Indigenous-Based Science Perspectives in Saskatchewan First Nations and Métis Community Contexts*. Saskatoon, Canada: Aboriginal Education Research Centre, University of Saskatchewan. https://aerc.usask.ca/downloads/Learning-Indigenous-Science-From-Place.pdf

Next Generation Science Storylines (n.d.). Supporting the framework for K-12 science. Northwestern School of Education and Public Policy. www.nextgenstorylines.org/.

Rebeiz, A. and Cooke, M. (2017). *Land-Based Learning: A Case Study Report for Educators Tasked with Integrating Indigenous Worldviews into Classrooms*. Toronto, ON: Canadian Education Association. http://cea-ace.s3.amazonaws.com/media/CEA-2016-IITS-REPORT-LAND-BASED-LEARNING.pdf.

Snively, G. and Williams, W. L. (2016). *Knowing Home: Braiding Indigenous Science with Western Science*, Book 1. Victoria, BC: University of Victoria. https://pressbooks.bccampus.ca/knowinghome/.

Snow, J. (1977). *These Mountains Are Our Sacred Places: The Story of the Stoney Indians*. Calgary, AB: Fifth House.

Sobel, D. (1996). *Beyond Ecophobia: Reclaiming the Heart of Nature Education*. Great Barrington, MA: Orion Society.

Sobel, D. (2004). *Place-Based Education: Connecting Classrooms and Communities*. Great Barrington, MA: Orion Society.

Truth and Reconciliation Commission of Canada (2015). *Truth and Reconciliation Commission of Canada: Calls to Action*. Winnipeg, MB: Truth and Reconciliation Commission of Canada. https://publications.gc.ca/collections/collection_2015/trc/IR4-8-2015-eng.pdf.

United Nations (2008). Declaration on the Rights of Indigenous Peoples. Article 14.1. www.un.org/development/desa/indigenouspeoples/wp-content/uploads/sites/19/2018/11/UNDRIP_E_web.pdf.

United Nations (2019). *State of the World's Indigenous Peoples: Education*, 3rd vol. New York: United Nations. www.un.org/development/desa/indigenouspeoples/publications/state-of-the-worlds-indigenous-peoples.html.

Uppal, S. (2017). Young men and women without a high school diploma. Statistics Canada. www150.statcan.gc.ca/n1/pub/75-006-x/2017001/article/14824-eng.htm.

Vander Ark, T., Liebtag, E., and McClennen, N. (2020). *The Power of Place: Authentic Learning Through Place-Based Education*. Alexandria, VA: Association for Supervision and Curriculum Development.

Williams, J. (2023). TikTok. www.tiktok.com/@qwustenuxun/.

Willms, J. D., Friesen, S., and Milton, P. (2009). *What Did You Do in School Today? Transforming Classrooms Through Social, Academic, and Intellectual Engagement*. First National Report. Toronto, ON: Canadian Education Association.

Young, J., Haas, E., and McGown, E. (2016). *Coyote's Guide to Connecting with Nature*. Santa Cruz, CA: OWL-Link Media Corporation.

Index

Printed in the United States
by Baker & Taylor Publisher Services